Intermediate Algebra

Intermediate Algebra

Second Edition

Ray Steege, M.A.
Former Professor of Mathematics
Laramie County Community College

Kerry Bailey, M.A.
Professor of Mathematics
Laramie County Community College

Schaum's Outline Series

New York Chicago San Francisco Lisbon London Madrid
Mexico City Milan New Delhi San Juan Seoul
Singapore Sydney Toronto

RAY STEEGE received his B.A. in mathematics from the University of Wyoming and his M.A. in mathematics from the University of Northern Colorado. The first 10 years of his teaching career were at East High School in Cheyenne, Wyoming. He continued his professional career at Laramie County Community College in Cheyenne, Wyoming, for an additional 25 years prior to his retirement in 1994. Among his many achievements and honors are: past president of the Wyoming Mathematics Association of Two-Year Colleges, Wyoming Mathematics Coalition Steering Committee member, newsletter editor, and recipient of the Outstanding Faculty Member of the Physical Science Division award at the college.

KERRY BAILEY received his B.A. in mathematics from San Diego State University and his M.A. in mathematics from the University of Colorado. He has been teaching at Laramie County Community College in Cheyenne, Wyoming, for the past 25 years. Prior to this position, he taught for 10 years at Pikes Peak Community College in Colorado Springs, Colorado. Among his achievements and honors are: Wyoming Mathematics Coalition Steering Committee member, newsletter editor, and recipient of the Outstanding Faculty Member of the Physical Science Division award, and corecipient of the Outstanding Faculty Member award for the entire college at Laramie County Community College.

Schaum's Outline of
INTERMEDIATE ALGEBRA

6 7 8 9 0 CUS/CUS 1 5

ISBN: 978-0-07-162998-0
MHID: 0-07-162998-X

Sponsoring Editor: Anya Kozorez
Production Supervisor: Tama Harris McPhatter
Project Supervision: Paradigm Data Services (P) Ltd.

Library of Congress Cataloging-in-Publication Data

Steege, Ray.
Intermediate algebra / Ray Steege, Kerry Bailey. – 2nd ed.
p. cm.
Includes index.
ISBN-13: 978-0-07-162998-0
ISBN-10: 0-07-162998-X
1. Algebra—Problems, exercises, etc. 2. Algebra—Outlines, syllabi, etc.
I. Bailey, Kerry. II. Title. III. Title: Theory and problems of intermediate algebra.
IV. Title: Intermediate algebra.

QA157.S728 2010
512.9—dc22 2010004016

Preface

The primary purpose of the book is to provide an effective tool for students that will help them understand and master the basic concepts and techniques of Intermediate Algebra. The book can be useful to the reader in several ways. It may be used as a valuable supplementary text for the second algebra course to assist in clarifying and simplifying algebraic concepts and procedures. The book may also be used for self-study or as the text for a course in Intermediate Algebra. Additionally, it is an excellent book to review in order to clarify concepts and improve skills prior to enrollment in College Algebra. We have attempted to provide a treatment of Intermediate Algebra that is more easily understood, and therefore more useful than most available texts at this level.

The book possesses numerous benefits or strengths. Concepts are introduced and explained at the student's level in a thorough, brief manner. The processes employed make the algebra involved as simple and concrete as possible. Each concept is illustrated completely by one or more solved problems which clarify and illuminate the relevant ideas. Definitions, properties, and so on, are expressed in clear, concise, understandable words as well as in mathematical form. Calculator procedures using RPN and Algebraic Entry calculators are illustrated and employed where it is appropriate. The graphing calculator is used to great advantage in many instances. A large number of graphs are included to help the student visualize abstract concepts. Step-by-step procedures accompanied by clarifying statements are employed in many instances. Word problems are often very difficult for students. We have provided the most detailed step-by-step treatment of word problems available. Students are led through the process in phases, thereby resulting in manageable steps. The book includes recommendations to the student for correct, efficient use of mathematics. A concise summary of effective study skills and suggestions on proven techniques is included. There are 886 solved problems with explanations and step-by-step solutions included. Solved problems are referenced to similar supplementary problems. There are 1,100 supplementary problems with answers included for student practice. Problems are arranged in an easy to more difficult order.

The book covers the concepts typically found in the second algebra course including: fundamental concepts; polynomials; rational expressions; first- and second-degree equations and inequalities; exponents, roots, and radicals; systems of equations and inequalities; relations and functions; exponential and logarithmic functions; sequences, series, and the binomial theorem. The terms and notation employed are those commonly used by other authors.

We thank Mr. Stephen Koch for planting the seed that grew into the final product. Our sincere appreciation goes to those people at McGraw-Hill who played a part in the development of the work.

Finally, and most importantly, for their encouragement, tolerance, and never-ending support, we thank our families: Ray's wife Marge; and Kerry's wife Jan and children Matt, Sara, and Abby.

RAY STEEGE
KERRY BAILEY

Contents

Intermediate Algebra

Study Skills

Notes to the Student: Effective Suggestions for Success

Book Bits

The book is the basis for course study. The book contains the essential course information; complements class lectures and notes; supplies visual aids; includes relative examples and problems; and provides material needed for review and exam preparation. It can also be a valuable reference in the future.

Your book is probably the most valuable tool available to assist you through a course. It is a reliable study companion that you can take anywhere. Read it carefully; underline or highlight important ideas; make notes in the margins; and employ any aids which it provides when applicable.

Survey the entire book. Become familiar with its various parts. Read the preface. Scan the table of contents. Look at the final pages of each chapter. Authors sometimes summarize the main points of the chapter there. Look through the entire book from beginning to end. Become familiar with all of the aids it provides.

Identify the main ideas and concepts in each section and/or chapter. Concentrate on the material you are reading so that you understand each chapter, section, paragraph, sentence, word, and symbol. Think about what you have read and summarize it. Study each example, table, graph, and illustration. Try to interpret and understand each of them.

Question yourself as you read. What does the chapter title or section heading mean? What are the meanings of the subheadings? Why did the author develop ideas in a particular order? How does the instructor's presentation correspond to the author's?

Time Requirements

Learning algebra requires more time on task than most students realize. It takes considerable time to acquire and perfect algebraic skills. High school students normally enroll in the course for one academic year. One to two hours doing homework outside of class is usually sufficient. If you are a college student, chances are that the time you spent doing homework in high school will be entirely inadequate to be successful in college. Typically college courses cover more material in half as much time as is allotted in high school. Therefore, the student is expected to do much more outside of class in college than is normally done while in high school. One hour per day spent reading the text and doing exercises is entirely inadequate for almost all college students. We recommend that most students work two to four hours daily studying algebra in addition to the time spent in class. Of course there are some who may not find it necessary to spend that much time daily; we are all different in ability. Most will profit greatly by following the recommended guideline.

You should take a short break periodically when working for an extended length of time. Our concentration wanes after a certain length of time has elapsed; be aware of the time when your ability to concentrate is declining.

Preview Material

It is highly desirable to preview the material prior to attending class. Do a quick reading in order to identify the theme of the section(s) to be explained in class. You will find that the instructor is more easily followed as the material is explained. The ideas presented are more easily followed and they will be more meaningful to you as well. Previewing also allows you to identify difficult concepts; you can question the instructor about them if you do not understand what was presented. You will conserve time in the long run.

What's Next?

After class you should reread the section(s) slowly and carefully. (Speed reading is not appropriate for mathematics books.) Read with paper and pencil in hand so that you may work solved problems as they are encountered. Reflect on what the instructor said and what you have read. Relate what you read to previous material. Think about what the problems illustrate; try to understand what is being done and why it was done in a particular manner. The material will be more meaningful to you and you will remember it better as a result.

Begin learning definitions, terms, properties, theorems, formulas, and the meaning of symbols as they are introduced. The material will be meaningless to you if you do not know and understand these important items. Statements in the book will not be understood if you do not know what the terms or symbols employed mean. Communication is impossible if statements are not understood; it's as though we were using a foreign language.

What Else?

Attend class daily unless you are ill. Some students feel that missing just one day per week is acceptable. Consider this. If you miss one day of a class which meets three days per week, you have missed one-third or over 33 percent of the material. If the class meets four days per week and you are absent one day, you have missed one-fourth or 25 percent of the material presented. A significant amount of explanation and discussion has been lost in either instance.

One of the worst situations is that you fall behind in a mathematics class. It is extremely difficult to catch up in a mathematics class once you fall behind. Exert as much energy as you have to avoid missing material. It is very difficult to understand new material being presented if you are unfamiliar with concepts presented previously. Attend class religiously.

Of course you must do more than show up for class. You must be an active participant in the class presentations and discussions. It is ineffective and boring to just sit there. Listen to the instructor carefully. Try to anticipate the next step or concept. Take notes to help you understand and remember the explanations. Taking notes will also keep your mind on the task. Don't try to write down everything the instructor says or does; that isn't practical. Be selective in your note taking; summarize the important ideas. Ask questions when you don't understand something. Most instructors want their students to participate actively in class.

The learning of mathematics is not a "spectator sport." It is an activity in which you must participate actively. One must do mathematics in order to learn it. It is impossible to learn how to ride a bike, drive a car, read, or write by watching someone else do it. Similarly, it is not possible to learn algebra by watching your instructor, classmate, friend, or tutor do it. Algebra appears easy when a knowledgeable person is doing it. You may encounter stumbling blocks which you did not anticipate when you attempt a problem. There is a point at which we must practice doing algebra in order to become proficient at the skill. It's a cliché, but "we learn by doing."

Doing Problems

Read the directions. Problems sometimes appear alike, but instructions may differ. Determine what is being asked and what must be done to answer the question(s). Be certain you have copied the problem correctly. It's very frustrating and time consuming to find that you miscopied the problem. Take your time initially—develop your speed after you understand the concepts better.

Think about what you are doing and what your rationale is for doing it. Avoid guessing about what you should do. There should be a reason for what you do. Observe how new concepts are applied and how they may relate to skills you learned previously.

Arrange your work systematically, even if you are working on scratch paper. If your work appears haphazard, your thinking is most likely haphazard and illogical also. Include sufficient detail as you write in order to enable you and others to follow your train of thought. Your statements should make sense. You need not write all of your thoughts or steps down; that takes too much time. Some operations should be performed mentally. On the other hand, don't try to do too many operations simultaneously; errors will occur.

Learn how to check your answers when possible. Most books do not give answers to all problems and some answers may be wrong. Furthermore, answers are not available to you when you take an exam. The ability to check solutions helps eliminate errors and builds confidence.

After you have arrived at a solution, answer all questions that were asked. Sometimes the solution to an equation which you have solved may not be the answer to the question that was asked. It depends upon how the question was stated and how the problem was set up. Ask yourself if your answer is reasonable; does it make sense? It is usually best to respond to all questions with a complete sentence.

Are You Stuck?

It's a frustrating situation, but don't give up too quickly. Keep your cool! We must be persistent in order to be successful. Trying a little of "this" and some of "that" probably won't work. Do you know how to begin? To work a problem, you must begin it. Don't just stare at the book or your paper. Write what is known down and try to think logically about how the known quantities are related to the unknown. Don't try to solve the problem all at once. Take a small step initially, even if it may seem trivial. Sometimes starting with the easiest task is best. Analyze the principles involved and try to apply them to your problem. Reread the text and your notes. Look for pertinent examples that relate to the circumstance at hand.

If you are still unsuccessful after working a "reasonable" length of time, proceed to another problem or take a short break. (A few minutes are probably not a "reasonable" length of time. A complex problem may require twenty or thirty minutes to analyze and work.) Try it again later. The mind can keep analyzing the problem even when you are engaged in other tasks. Perhaps a good night's rest is needed. We don't think well when we are tired. You may even awake in the middle of the night with a solution. Try again the next day.

Don't keep trying the same process if what you are doing isn't working. You must do something different in order to obtain a different outcome. Try talking through your approach out loud. It's amazing how simply verbalizing it can pinpoint gaps and errors in your logic.

Classmates are another important resource that is available to you. Talk to other students about the ideas they might have regarding a particular problem. It may be useful to form a study group or find one or two other students to work with. Others may have a different approach in a given situation. Keep the conversation on the task at hand, however. Time is too valuable to waste on idle talk, unless the group needs a break.

If you are still stuck, ask your instructor for assistance. Be as specific as possible when asking questions, rather than simply asking the instructor to work the problem. Determine why you are having difficulty. The same difficulty will very likely arise again unless you take corrective action.

Don't give up until you have tried the suggestions made above. Frustrations will be reduced and time will be saved in the long run.

Review Regularly

Regular review is neglected by most students; however, it is one of the most useful activities a student can do. Review includes rereading or writing important definitions, properties, theorems, etc. It also includes working problems from sections and chapters previously concluded. Most new algebraic concepts or skills depend on concepts or skills developed previously. If you are not proficient in certain skills, your ability to be successful will be seriously affected.

Regular review helps us retain concepts and/or techniques which we previously learned. Review also helps us relate topics which were introduced previously to those we are currently learning. Less time is needed to prepare for exams when we review daily or every other day.

Tuning up for Tests

There will be limited time available while taking a test. The time restriction will most likely result in your being more anxious and stressful than you are ordinarily. You will therefore be more likely to make errors that you normally would not make when doing homework.

The book and your notes are not available to refer to when you are uncertain about what you should do. This contributes to your stress level also. Additionally, the test will involve many concepts which you need to distinguish from one another rather than a few which you might apply in a particular section. This makes it more difficult to recognize where to begin, and which steps should be taken.

When, Where, and What?

Don't try to do all of your studying in one day. Avoid studying into the early morning hours prior to the test. Do some studying each day for several days. Studying late results in your being tired and reduces your ability to concentrate and think clearly during the test. Begin studying several days prior to the day the test is scheduled. This allows you time to ask your instructor questions about concepts which you may not fully understand.

Study in a quiet area free of distractions. Avoid areas where people might be coming and going or talking. Don't sit in front of the television or play music loudly while you are trying to concentrate. Try to set aside a special area where you can have all pertinent materials readily available for your use during study periods.

You should have done all assigned problems previously. Now is the time to review the book and your notes. Concentrate on the definitions, properties, and theorems of the relevant chapters. You must know and understand the meaning of these important items. Reflect on the material to be covered and think about how it relates to the processes encountered. Review exams and quizzes previously taken. They may indicate the types of questions to expect. Drill with a classmate. Make up potential test questions to ask one another. Try choosing problems randomly from all problem sets covered. Write them in random order on a sheet of paper. Put it away for a day so you won't remember where the problems were in the chapter(s). When you then try to work those problems, you won't be influenced by the content of nearby problems. Do the end-of-chapter tests if provided in your book and time yourself.

You should be well rested before taking the test; you can think more clearly when rested. Don't study right up to the last minute. You may try to look at everything to be included on the test and end up confusing yourself. You may also encounter concepts in which you are not completely competent. That may result in your becoming anxious and nervous. Take a break of at least thirty minutes duration prior to the time the test is to begin.

Arrive with plenty of time to spare. This allows time for you to settle down and get your thoughts and test-taking materials organized. If you arrive late, you will be hurried and it will be more difficult for you to concentrate and relax. Additionally, you will deprive yourself of valuable time which you very likely will need for doing the test.

Taking a Test

Listen closely to any oral instructions or corrections given by the instructor. Follow written directions carefully. Write down any pertinent formulas which you may need initially. You can then concentrate fully on the test questions. Count the number of questions and divide that number into the time allotted. This allows you to determine the average time you can spend on each question. Of course some questions will require very little time, while others may require a larger block of time to complete.

Begin by working on problems or answering questions which are easiest. This may require looking over the entire test in order to find the easy problems, but it is time well spent. This allows you to "get into the groove" and helps build your confidence.

Show your work legibly and arrange it in a logical, orderly manner. Your instructor will appreciate it and can probably award partial credit in a fair way. Your thinking will be more orderly also. Avoid working everything on scratch paper and transferring the work to the exam. This procedure introduces copying errors unnecessarily, takes additional time, and shows lack of confidence. Errors in sign can be reduced by slowing your thinking down to the speed at which you write. This is accomplished through concentration. Think about what you are writing as you are writing it.

Next proceed to the more difficult questions. Don't be too alarmed if there is one which you cannot answer immediately. Take a deep breath, relax, maintain control. It is only one of many questions. Think about the concepts presented in the chapter(s) and how they might relate. Be systematic in your approach. Reject techniques which aren't working. Do something differently if the process you are applying leads nowhere. Return to problems you have completed; do they provide a clue? If not, put the problem aside and return to it later if time permits. Try to avoid omitting high-point valued questions.

Be aware of the time. Don't look at the clock every five minutes, but do not neglect to check the time occasionally. Don't allow time to expire before you have done all that you are able. Allow time to check your answers. Don't leave early without checking your results. It's surprising how many careless errors we tend to make while taking a test. If there is time once finished, turn your test over and think about something else before you turn it in. This allows your mind a period of relaxation and may allow it to overcome a difficulty you had on the test. One of the most frustrating test experiences is to turn in the test, walk out into the hall, and then know immediately how to solve a problem that was causing you great difficulty. Allowing your mind to relax, to "give up" before turning the test in, frees it from a great deal of stress. Finally, are your answers reasonable? Did you follow directions?

After the Test

Look your test over carefully when it is returned to you. Determine why points were deducted. Identify topics and concepts in which you are strong and those in which you are weak. Take steps to clarify concepts which you do not understand fully; they will be encountered again. Reread any applicable sections of the text and your notes. Ask your instructor to clarify the concept if you are still having difficulty. If you don't understand why your answer or process was incorrect, be sure to have your instructor work through the problem completely. Don't assume you'll learn "why" later. Try the problem again on a different sheet of paper when you think you understand what you did incorrectly.

Were your errors careless arithmetic or sign errors? If so, try to concentrate more intensely on what you are writing while you are writing it. Avoid letting your thoughts outrun your pencil.

Finally, if allowed, retain your exam for future studying. It can be a valuable tool while preparing for comprehensive exams like those encountered during finals.

Work diligently and you will be successful. Remember not to give up too easily; you will be rewarded by being persistent.

Fundamental Concepts

1.1 Definitions

A *set* is a collection of objects. The collection should be well-defined. That is, it must be clear that an object is either in the set or is not in the set. The objects in the set are called *elements* or *members* of the set. The members of a set may be listed or a description of the members may be given.

We list the members or describe the members within braces { }. Capital letters such as A, B, C, S, T, and U are employed to name sets. For example,

$$A = \{2, 4, 6, 8, 12\} \qquad B = \{3, 6, 9, 12\}$$
$$U = \{\text{people enrolled in algebra this semester}\}.$$

The symbol used to represent the phrase "is an element of" or "is a member of" is "$\in$." Thus we write $4 \in A$ to state that 4 is a member of set A. The symbol used to represent the phrase "is not an element of" is "$\notin$." Hence $4 \notin B$ is written to indicate that 4 is not an element of set B.

Sets are said to be equal if they contain the same elements. Hence $\{1, 5, 9\} = \{5, 9, 1\}$.

Note: Order is disregarded when the members of a set are listed.

Sometimes a set contains infinitely many elements. In that case it is impossible to list all of the elements. We simply list a sufficient number of elements to establish a pattern followed by a series of dots "...". For example, the set of numbers employed in counting is called the set of *natural numbers* or the set of *counting numbers*. We write $N = \{1, 2, 3, 4, \ldots\}$ to represent that infinite set. If zero is included with the set of natural numbers, the set of whole numbers is obtained. In this case the symbol used is $W = \{0, 1, 2, \ldots\}$.

Set B above could be described as the set of multiples of three *between* 0 and 15. Note that the term "between" does not include the numbers 0 and 15. Set-builder notation is sometimes used to define sets. We write, for example,

$$B = \{x \mid x \text{ is a multiple of three between 0 and 15}\}$$

There are occasions when a set contains no elements. This set is called the *empty* or *null* set. The symbol used to represent the empty set is "Ø" or "{ }." Note that no braces are used when we represent the empty set by Ø. {negative natural numbers} is an example of an empty set.

Definition 1. Set A is a *subset* of set B if all elements of A are elements of B. We write $A \subseteq B$.

Hence, if $A = \{2, 4, 6\}$ and $B = \{1, 2, 3, 4, 5, 6, 7\}$, $A \subset B$. A is called a *proper subset* of B. If $C = \{4, 2, 6\}$, $A \subseteq C$ since the sets are the same set. A is called an *improper subset* of C.

New sets may be formed by performing operations on existing sets. The operations used are union and intersection.

Definition 2. The *union* of two sets A and B, written $A \cup B$, is the set containing all of the elements in set A or B, or in both A and B.

If set-builder notation is used, we write $A \cup B = \{x \mid x \in A \text{ or } x \in B\}$. If set $A = \{1, 5, x, z\}$ and $B = \{3, 5, 7, z\}$, then $A \cup B = \{1, 5, x, z, 3, 7\}$. Recall that the members of a set may be listed without regard to order.

Definition 3. The *intersection* of two sets A and B, written $A \cap B$, is the set containing the elements common to both sets.

If set-builder notation is used, we write $A \cap B = \{x \mid x \in A \text{ and } x \in B\}$. Hence, if set $A = \{1, 5, x, z\}$ and $B = \{3, 5, 7, z\}$, then $A \cap B = \{5, z\}$.

Venn diagrams are sometimes used to illustrate relationships between sets. Figure 1.1 (*a*) and Figure 1.1 (*b*) shown below illustrate the concepts discussed above. The shaded regions represent the specified set

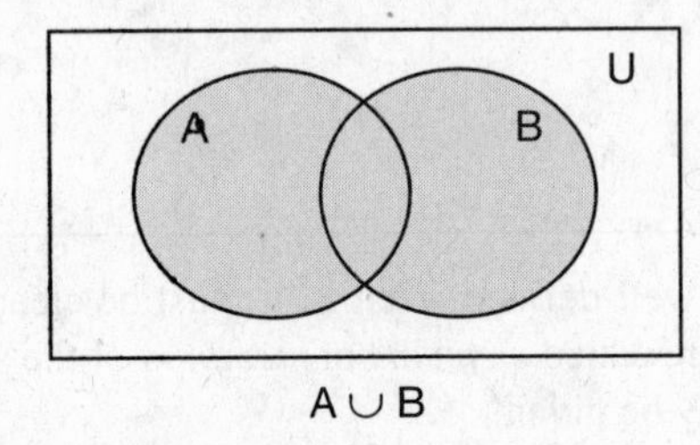

Figure 1.1 (*a*) Set union.

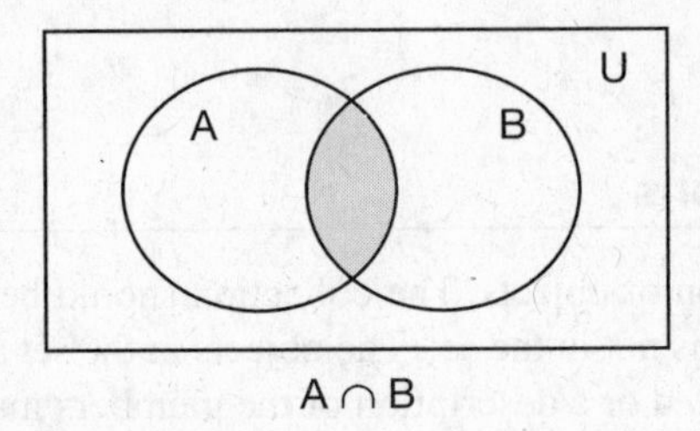

Figure 1.1 (*b*) Set intersection.

U represents the *universal* set. It is a set which contains all of the elements under discussion in a given situation. The universal set is typically represented by a rectangular region.

Figure 1.2 shows a Venn diagram which illustrates $A \subset B$.

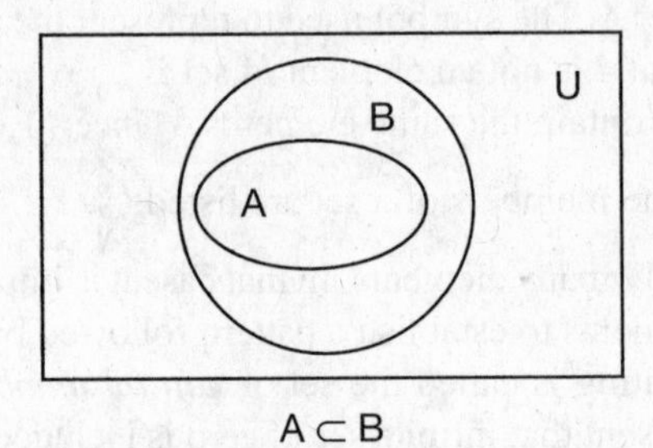

Figure 1.2 Subset.

Note that all of set A is completely contained in set B.

If set S is not a subset of set T, we write $S \not\subset T$. This occurs when S contains at least one element which is not in T.

There are several additional sets of numbers which will be referred to often. Their definitions follow.

The set of *integers, J,* is the set of natural numbers combined with their negatives and zero. Symbolically $J = \{\ldots, -2, -1, 0, 1, 2, \ldots\}$. Some examples are -4, -1, 0, 13 and 22. Some authors sometimes use the letter Z to represent the set of integers. We shall employ Z to represent the set of complex numbers in a subsequent section.

The set of *rational numbers, Q,* is the set of elements which can be expressed in the form $\frac{a}{b}$, where a and b are integers and $b \neq 0$. Rational numbers can also be expressed as terminating or repeating decimals. Some examples are $-2, 0, 5, \frac{2}{3}, \frac{9}{8}, 0.34$, and -3.588. Note that $\frac{2}{3}$ is equivalent to the repeating decimal 0.66666… .

The set of *irrational numbers* is the set of elements whose decimal representations are either non-terminating or non-repeating. There is no single letter which is commonly used to name this set, although various authors identify this set with some capital letter. Some examples are $-1.010010001\ldots$, $\sqrt{5}$, π, and $\sqrt[3]{7}$.

The set of *real numbers, R,* is the set of rational numbers combined with the set of irrational numbers. That is, R is the union of the sets of rational and irrational numbers.

A *constant* is a symbol that represents only one number. Letters near the beginning of the alphabet, like a, b, and c, are typically used to represent constants. A symbol which represents a constant has only one replacement.

A *variable* is a symbol that represents any value in a specified set. Letters near the end of the alphabet, like x, y, and z, are typically used to represent variables. The set of possible replacements of the variable is called the *domain* of the variable. The domain of a variable is assumed to be the set of real numbers R if no other domain is specified.

EXAMPLE 1. Suppose $A = \{-3, 0, 4, 6, 9\}$, $B = \{2, 4, 6, 10\}$, $C = \{-3, -1, 0, 1, 3, \ldots\}$, and $D = \{2, 4\}$; then it follows that:

(*a*) $D \subset B$ (*b*) $A \cup B = \{-3, 0, 4, 6, 9, 2, 10\} = \{-3, 0, 2, 4, 6, 9, 10\}$ (*c*) $D \not\subset A$

(*d*) $A \cap B = \{4, 6\}$ (*e*) $A \cup C = \{-3, -1, 0, 1, 3, 4, 5, 6, 7, 9, 11, 13, \ldots\}$ (*f*) $N \subset Q$

(*g*) $B \cap C = \varnothing$ (*h*) $A \cap C = \{-3, 0, 9\}$ (*i*) $4 \in B$

(*j*) $4 \notin C$ (*k*) The subsets of D are $\{2, 4\}$, $\{2\}$, $\{4\}$, and $\varnothing$.

See solved problems 1.1–1.3.

1.2 Axioms of Equality and Order

In mathematics we make formal assumptions about real numbers and their properties. These formal assumptions are called *axioms* or *postulates*. The terms *property, principle,* or *law* are sometimes used to refer to those assumptions although these terms may also refer to some consequences of those relationships we assume to be true. The assumptions can be made without restriction although their consequences should be useful.

An *equality* is a statement in which symbols, or groups of symbols, represent the same quantity. The symbol used is "=", read "is equal to" or "equals."

We shall assume that the "is equal to" relationship satisfies the following properties. Assume that a, b, and c are real-valued quantities.

Reflexive property: $a = a$. (A quantity is equal to itself.)

EXAMPLE 2. $x - 6 = x - 6$.

Symmetric property: If $a = b$, then $b = a$. (If the first quantity is equal to a second, then the second is equal to the first; an equation may be reversed.)

EXAMPLE 3. If $5 = y + 7$, then $y + 7 = 5$.

Transitive property: If $a = b$ and $b = c$, then $a = c$. (If the first quantity is equal to a second and the second quantity is equal to a third, then the first and third quantities are equal.)

EXAMPLE 4. If $x = y - 3$ and $y - 3 = z$, then $x = z$.

Substitution property: If $a = b$, then b may be replaced by a and vice versa. (If two quantities are equal, one may be substituted for the other.)

EXAMPLE 5. If $x - 4 = y$ and $x = z$, then $z - 4 = y$. Think of x being replaced by z in the first relationship.

See solved problem 1.4.

We associate each real number with one and only one point on a line. The line is called a *number line*. Each number is paired with only one point and each point is paired with only one number. The number line is helpful in visualizing relationships between numbers. The numbers are called the *coordinates* of the points and the points are called the *graphs* of the numbers. A number line is shown in Figure 1.3 below. Each mark represents one-half unit. This is known as the *scale* in the drawing. You must always indicate the scale that is being employed in a drawing.

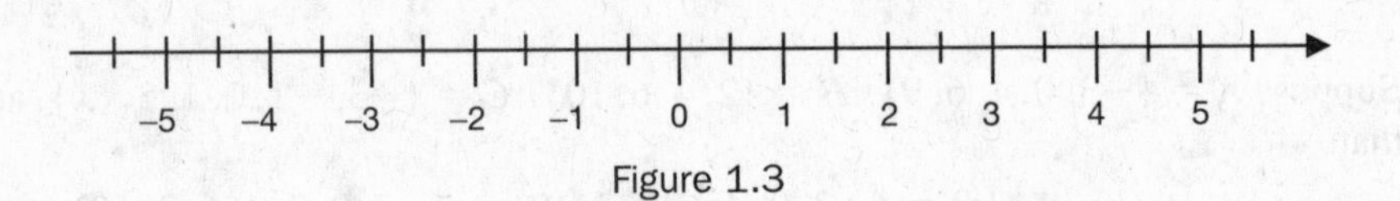

Figure 1.3

The point associated with zero is called the *origin*. Numbers associated with points to the right of the origin are *positive* while numbers associated with points to left of the origin are *negative*. The number 0 is neither positive nor negative.

Definition 4. The number b "is less than" the number a if $a - b$ is positive.

Equivalently, the point associated with b on the number line lies to the left of the point associated with a. The symbol used to represent the "is less than" relation is "$<$." We write $2 < 4$ and $-1 < 3$ for example.

It is also true that if b is less than a, then a "is greater than" b. The symbol used to represent the "is greater than" relation is "$>$." We could express the above relationships equivalently as $4 > 2$ and $3 > -1$.

We sometimes combine the relations "is less than" and "is equal to" into one statement. In this case the symbol used is "$\leq$." It is read "is less than or equal to." Similarly the relations of "is greater than" and "is equal to" may also be combined. The symbol used to represent this relation is "$\geq$." It is read "is greater than or equal to." For example, $3 \geq 0$ and $x \leq 2$ are common types of statements.

Furthermore, two conditions such as $a < b$ and $b \leq c$ are sometimes combined into the compound or double inequality $a < b \leq c$. It is read "a is less than b and b is less than or equal to c." If it is true that $a < b < c$, then b is *between* a and c. Compound or double inequalities are used to indicate "betweenness."

It is not appropriate to combine expressions such as $a < b$ and $b > c$ into the compound inequality $a < b > c$. Inequality symbols which are used in compound inequality statements must always be in the same sense. That is, they must all be "less than" relationships or "greater than" relationships.

Some "same sense" inequalities may not be combined into a compound inequality statement. Consider $-2 < t < -6$. This statement does not make sense, since there is no number t such that t is greater than -2 and simultaneously is less than -6. Remember that compound or double inequalities imply "betweenness."

A number that is zero or positive is said to be *nonnegative*. If $a \geq 0$, then a is nonnegative. Also a number that is zero or negative is said to be *nonpositive*. Hence if $a \leq 0$, then a is nonpositive. For example, 6 and 0 are nonnegative while -3 and 0 are nonpositive.

The following property of order of real numbers is assumed to be valid.

Transitive property: If $a < b$ and $b < c$, then $a < c$. (If the first quantity is less than a second and the second is less than the third, then the first is less than the third.)

EXAMPLE 6. If $x < y$ and $y < 7$, then $x < 7$.

See solved problems 1.5–1.7.

Subsets of real numbers are sometimes represented using interval notation. For example, $\{x \mid a < x \leq b\}$ is written as $(a,b]$. The parenthesis is used to indicate that a is not included in the interval. The bracket indicates that b is included in the interval.

Other intervals are represented as indicated below.

(a,b) represents the real numbers between a and b. It is an *open interval*. Neither endpoint is included.
$[a,b]$ represents the real numbers between a and b inclusive. The endpoints are included. It is a *closed interval*.
$[a,b)$ represents the real numbers between a and b, including a but not b. It is a *half-open interval*. The endpoint "a" is included, while the endpoint "b" is not included. The interval $(a,b]$ is also half-open.

The graphs of some subsets of the real numbers extend forever in one or both directions. The intervals used to represent those subsets are represented using the *infinity* "∞" or the *negative infinity* "$-\infty$" symbols. The infinity symbol does not represent a real number; it represents a concept. For that reason, it is never included in an interval. Some representative intervals are illustrated below.

(a, ∞) represents all real numbers greater than a.
$(-\infty, b)$ represents all real numbers less than b.
$(-\infty, b]$ represents all real numbers less than or equal to b.
$(-\infty, \infty)$ represents all real numbers.

See solved problems 1.8–1.10.

1.3 Properties of Real Numbers

The following eleven properties of real numbers are assumed to be true.

Closure property of addition $a + b$ is a unique real number.
Closure property of multiplication ab is a unique real number.

(The sum and product of two real numbers is a unique real number.)

EXAMPLE 7. $2 + 4 = 6$ $\quad -3 + 8 = 5$
$3(4) = 12$ $\quad -4(6) = -24$

Associative property of addition $(a + b) + c = a + (b + c)$.
Associative property of multiplication $(ab)c = a(bc)$.

(Real numbers may be grouped or associated differently when adding or multiplying without affecting the result.)

EXAMPLE 8. $(3 + 4) + 7 = 3 + (4 + 7)$ $\quad (x + 5) + 1 = x + (5 + 1)$
$(4x)y = 4(xy)$ $\quad (8s)(tu) = 8(stu)$

Commutative property of addition $a + b = b + a$.
Commutative property of multiplication $ab = ba$.

(When adding or multiplying real numbers the result is the same if the order of the addends or factors is changed.)

EXAMPLE 9. $3 + 6 = 6 + 3$ $\quad 4 + x = x + 4$
$2(5) = 5(2)$ $\quad r(3) = 3(r)$

Identity property of addition $a + 0 = 0 + a = a$.
Identity property of multiplication $a(1) = 1(a) = a$.

(Zero added to a number or a number added to zero is the number; a number times one or one times a number is the number. Zero is the additive identity. One is the multiplicative identity.)

EXAMPLE 10. $2 + 0 = 0 + 2 = 2$ $\quad$ $0 + (x - 5) = (x - 5) + 0 = x - 5$

$3(1) = 1(3) = 3$ $\quad$ $(x - 9)1 = 1(x - 9) = x - 9$

Inverse property of addition $\quad a + (-a) = (-a) + a = 0.$

Inverse property of multiplication $\quad a(1/a) = (1/a)a = 1, a \neq 0.$

[There is a unique real number $(-a)$ such that when added to a results in the additive identity zero; if $a \neq 0$, there is a unique real number $(1/a)$ such that when $(1/a)$ is multiplied by a, the result is the multiplicative identity one. $(-a)$ is the opposite of a. $(1/a)$ is the reciprocal of a.]

EXAMPLE 11. $2 + (-2) = 0$ $\quad$ $(-t) + t = 0$

$3(1/3) = 1$ $\quad$ $(1/s)s = 1, s \neq 0$

Distributive properties $\quad a(b + c) = ab + ac$ and $(b + c)a = ba + ca.$

(A factor may be distributed over a sum from the left or the right. Alternatively, a common factor may be factored out.)

EXAMPLE 12. $3(x + 4) = 3x + 3(4)$ $\quad$ $(s - t)5 = 5s - 5t$ $\quad$ $6y + 8 = 2(3y + 4)$

See solved problems 1.11–1.12.

1.4 Operations with Real Numbers

The additive inverse property was discussed in Section 1.3. Recall that the additive inverse of a number is sometimes called the opposite or negative of the number. The opposite or negative of a number is indicated by placing a minus sign, "−," in front of the number. Hence, the following are additive inverses of one another.

$$2 \text{ and } -2;\ -6 \text{ and } 6;\ x \text{ and } -x;\ x + 3 \text{ and } -(x + 3) \text{ or } -x - 3.$$

The *absolute value* of a real number is the distance between that number and zero on the number line. We write "$|q|$" to represent the absolute value of a quantity q. The expression is read as "the absolute value of q." The definition of absolute value of q follows.

Definition 5.
$$|q| = \begin{cases} q & \text{if } q \geq 0 \\ -q & \text{if } q < 0 \end{cases}$$

In other words, the absolute value of a positive quantity is that quantity, while the absolute value of a negative quantity is the opposite of that quantity. The absolute value of zero is zero.

Note that the opposite of a positive number is negative and the opposite of a negative number is positive. Hence, the opposite of 5 is -5 which is less than zero and the opposite of -5 is $-(-5) = 5$ which is greater than zero. Some examples of the absolute value operation are shown in the example below.

EXAMPLE 13.
$$|3| = 3; |-4| = -(-4) = 4; \text{ and } |x - 4| = \begin{cases} x - 4 \text{ if } x - 4 \geq 0, \\ -(x - 4) = 4 - x \text{ if } x - 4 < 0 \end{cases}$$

It is also true that $(-1)x = -x$, the additive inverse of x. This follows since $x + (-1)x = (1)x + (-1)x = [1 + (-1)]\,x = 0x = 0$. Hence, (-1) times a quantity is the additive inverse of the quantity. Additionally, $-1(-x) = -(-x) = x$ since $-(-x)$ is the additive inverse of $-x$ and x is also the additive inverse of $-x$.

See solved problems 1.13–1.14.

Addition of two quantities is the operation which associates those quantities with a third quantity called the sum. We employ the plus sign, "+," to represent the operation of addition. Thus we write $s + t$ to represent the sum of s and t.

Addition of real numbers may be thought of as movement along a number line. Adding a positive number is associated with movement to the right while adding a negative number is associated with movement to the left. We must always start at zero.

Hence, $(-4) + 3 = -1$ can be shown on the number line in Figure 1.4 below as follows:

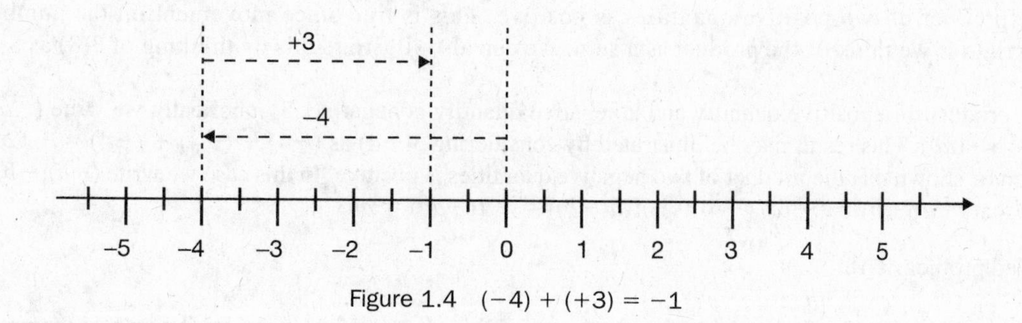

Figure 1.4 $(-4) + (+3) = -1$

See solved problem 1.15.

It is not necessary to employ a number line when adding signed numbers. The results can be obtained from the properties stated previously. The next example illustrates how to add -13 and -25.

EXAMPLE 14.

$(-13) + (-25)$	$= (-1)13 + (-1)25$	$-q = (-1)q$
	$= -1(13 + 25)$	Distributive law
	$= -1(38)$	Number fact
	$= -38$	$-q = (-1)q$.

Example 14 illustrates the fact that the sum of two negative numbers is negative. It is also apparent that the sum of two positive numbers is positive, since movement on the number line will be to the right. The sum of a positive number and a negative number may be positive or negative. The sum is positive if the absolute value of the positive number is larger than the absolute value of the negative number. On the other hand, the sum is negative if the absolute value of the negative number is larger than the absolute value of the positive number.

See solved problem 1.16.

We define the *difference* of two quantities a and b as $a + (-b)$. In other words, the difference of a and b is the sum of a and the additive inverse (opposite) of b. Differences may be obtained by referring to the number line given above also. The process of finding the difference of two quantities is called *subtraction*. We normally write $a - b$ to represent the difference of a and b. Note that the minus sign is used for both a negative number and subtraction. Be sure you know how it is being used each time you see it. Also understand that $a - b = a + (-b)$ whereas $b - a = b + (-a)$ and the results are not the same (unless $a = b$). In other words, there is no commutative property of subtraction.

EXAMPLE 15. (*a*) $5 - 6 = 5 + (-6) = -1$; (*b*) $-8 - (-18) = -8 + [-(-18)] = -8 + 18 = 10$.

See solved problem 1.17.

Multiplication is an operation that associates with each pair of quantities a third quantity called the *product*. We write st or $s(t)$ or $(s)(t)$ to represent the product of s and t. Each of the quantities we multiply is called a *factor*.

Multiplication may be thought of as repeated addition. For example, 3(4) may be written as $4 + 4 + 4 = 12$. That is, 3 times 4 or equivalently the product of 3 and 4, is the same as the sum of three four's.

Zero factor property: The product of a quantity and zero is zero. Symbolically we write, $q(0) = 0(q) = 0$. (The product of two quantities is zero if at least one of the factors is zero.)

If we think of 3(0) as three zeroes or $0 + 0 + 0$ for example, the result is 0 by the additive identity property.

The product of two positive quantities is positive. This is true since movement on the number line is to the right if we think of the product as a sum. We can also illustrate this by thinking of 3(8) as $8 + 8 + 8 = 24$.

The product of a positive quantity and a negative quantity is negative. Symbolically we write $(-a)(b) = (a)(-b) = -(ab)$. This result may be illustrated by considering $3(-4)$ as $(-4) + (-4) + (-4) = -12$.

It can be shown that the product of two negative quantities is positive. In this case we write $(-a)(-b) = ab$. Observe that $(-a)(-b) = (-1a)(-b) = -1[a(-b)] = -1(-ab) = ab$.

See solved problem 1.18.

If we are determining the product of more than two signed quantities, it can be shown that the product is positive if the number of negative factors is even, but the product is negative if the number of negative factors is odd. The product is positive if all of the factors are positive.

For example, $(8)(-7)(-4)(-2)(3)$ is negative since there is an odd number of negative factors. Also, $(-8)(-6)(7)(-5)(-8)$ is positive since there is an even number of negative factors.

We define the *quotient* of a and b or a/b as the product of a and $1/b$, $b \neq 0$. In other words, dividing a by b is equivalent to multiplying a by the multiplicative inverse (reciprocal) of b.

Since a quotient is expressed in terms of multiplication, the sign of the result can be determined from the results we obtained when multiplying. For example, $12/4 = 12(1/4) = 3$; $12/(-3) = 12(-1/3) = -4$; $\frac{-12}{-3} = -12\left(\frac{-1}{3}\right) = 4$. These examples illustrate the fact that the quotient of quantities with like signs is positive while the quotient of quantities with unlike signs is negative.

Note that division by zero is undefined since zero has no multiplicative inverse. On the other hand, $0/q = 0$ since $0(q) = 0$ (i.e., answer times divisor equals the dividend). (Technically, $\frac{0}{0}$ is indeterminate, not undefined, but that is left for a study in Calculus. In this book, division by zero will always be referred to as undefined.)

See solved problem 1.19.

It was stated previously that multiplication by a positive integer is another way to accomplish repeated addition. That is, $4 + 4 + 4 = 3(4)$. In a similar manner, *exponential notation* is another way to express repeated multiplication. For example, $4^3 = 4 \cdot 4 \cdot 4 = 64$. In general we define exponential notation for a positive integer n as follows. $b^n = b \cdot b \cdot b \cdots b$. In this expression, b occurs as a factor n times. b is called the base and n is called the *exponent*.

Note: The exponent only applies to the symbol immediately preceding it, unless that symbol is a right hand grouping symbol where it would apply to the entire grouping. So, $-2^2 = -2 \cdot 2 = -4$ but $(-2)^2 = (-2)(-2) = 4$.

If no exponent is written, it is understood to be one. That is, b means b^1.

See solved problem 1.20.

1.5 Order of Operations

If multiple operations are to be performed in an expression, we must agree which operations are to be performed first in order to avoid ambiguity. The following order must be adhered to.

1. Operations within symbols of inclusion such as parentheses, brackets, fraction bars, etc., must be performed first. Begin within the innermost group.
2. Begin by evaluating powers in any order.
3. Next, multiplications and divisions are performed in order from left to right.
4. Last, perform additions and subtractions as encountered in order from left to right.

See solved problem 1.21.

1.6 Evaluation by Calculator

Evaluation of some arithmetic expressions can become very tedious and time consuming, if the expressions are complex at all. Expressions that are terribly complex, or involve more advanced operations, are just not reasonable to evaluate with pencil and paper. (Think of an incredibly difficult expression involving many fractions and powers. Now consider evaluating it within the next hour.) Fortunately, calculators and computers can do much of the work for us. Calculators can do most of the problems at the level of Intermediate Algebra; computers may be required for much more complicated problems (maybe like the one you thought about). An increasingly popular type of calculator employs a Computer Algebra System, or CAS. This type evaluates expressions that used to require a computer. Since CAS calculators perform the algebra and not just the arithmetic, we do not recommend their use for learning algebra (though they can be useful for checking your algebra skills and results). This section is dedicated to evaluating expressions by using a calculator. If you don't have a calculator to use, this section may not be very useful to you until you have one.

Types of Calculators

There are many different types of calculators. The type you use will determine how you evaluate expressions with it. This section will give a general overview of the different types and should help you to use yours. The brand name of a calculator may identify its type but not all calculator manufacturers use the same type for all of their calculators. A different way of identifying types is needed.

The two main calculator types now being used are set apart by their method of entering expressions: RPN (for Reverse Polish Notation) and Algebraic Entry. The RPN calculators are arguably the more efficient, but they are more difficult to learn too. The Algebraic Entry calculators are easier to learn since the expressions are entered similar to the way they are written. Let's consider the Algebraic Entry type first.

Algebraic Entry Calculators

Algebraic Entry calculators are normally set apart by including an "=" key, [×], on the bottom right side of the keypad. (Some Algebraic Entry calculators have replaced that key with [ENTER] or [EXE] for more advanced operations – such as graphing – but the keys serve the same purpose for evaluating expressions. The [×] will be used for this section.) Not all calculators with the "=" key are Algebraic Entry, however. Some calculators are best identified as Last Entry calculators.

A simple calculation using the [×] key detects whether an Algebraic Entry or Last Entry calculator is being used. As we saw in the previous section, to evaluate expressions correctly, the order of the operations must be followed. (Calculators use "*" to display multiplication as expressions will in this section; most still use "×" for the multiplication key on the keypad, so [×] will be used for a multiplication keystroke.) So, to evaluate $5 + 2 * 3$, the following keystrokes can be used (a keystroke is simply a push of the indicated "key" on the calculator keypad): [2] [×] [3] [+] [5] [=]. This will give the correct answer, 11, because the correct order of operations was keyed in. Now, to detect if the calculator is an Algebraic Entry calculator, use keystrokes in the order the expression is written: [5] [+] [2] [×] [3] [=]. If this also gives the answer as 11, then the calculator is Algebraic Entry, not Last Entry. (A Last Entry calculator will give an answer of 21 for that ordering since it will evaluate 5 + 2 as 7 and <u>then</u> multiply by 3, which violates the order of operations.) Last Entry calculators will always evaluate current operations immediately using the "last entry"; Algebraic Entry calculators will evaluate expressions by using the correct order of operations automatically.

Algebraic Entry calculators can easily be divided into two kinds: *single-line* and *multiline* displays. The single-line display will display only the last number entered even though others may be pending evaluation due to the order of operations being controlled by the calculator. Multiline display calculators will show the entire expression as entered (up to the point of the [×] key press). Although their displays differ, both evaluate the same way.

RPN Calculators

The other general type of calculator is identified by its evaluation method. That method is called Reverse Polish Notation, or RPN for short. (This method is identified as postfix in computer jargon.) Calculators identified as RPN rarely have an "=" on the keypad — those that do generally only include it for programming purposes. RPN calculators have a different way of entering expressions for evaluation and a few simple examples may help you understand. Notice in these examples that the arithmetic operation to be used between two numbers, or intermediate results, is entered after both are entered. *Either* [↑] *or* [ENTER] *is used to let the calculator "know" a number was entered; this section will use* [ENTER].

EXAMPLE 16.

Expression	RPN Evaluation	Answer
8 + 7	[8] [ENTER] [7] [+]	15
9 * 5 + 4	[9] [ENTER] [5] [×] [4] [+]	49
26 − 3 * 8	[2] [6] [ENTER] [3] [ENTER] [8] [×] [−]	2

Note the last expression in the previous example required the 3 and 8 to be multiplied before that answer was subtracted from 26. The process can be visualized as a stacking process where the top two numbers are always used for an operation and the answer is put back on the stack. The evaluation steps for $26 - 3 * 8$ could be visualized by the following:

KEYSTROKES	STACK	
[2] [6] [ENTER]	26	⟸ level 1
	26	⟸ level 2
[3] [ENTER]	3	⟸ level 1
	26	⟸ level 3
	3	⟸ level 2
[8]	8	⟸ level 1
	26	⟸ level 2
[×]	24	⟸ level 1
[−]	2	⟸ level 1

All values on the stack would be displayed for this problem if the calculator has a multiline display. Only the values at **level 1** would be displayed at each step for single-line displays.

Special Keys

There are several special keys that will be needed for more involved calculations. Unfortunately for this discussion, different calculators use different keys for the same needs. Some of the different keys will be discussed, but not all. You may need to find the specific keys for your calculator from your calculator's reference manual.

Changing Signs

A most important special key allows us to change the sign on a number. Since the calculator only shows positive digits on the keypad, this ability is very important if any negative numbers are to be used in a calculation (or, at least, easily used). Probably the most frequently used keys on different calculators for sign changing are [+/−], [CHS], and [(−)]. (The [CHS] key stands for CHange Sign; the others are more obvious.) The [(−)] key is normally on the keypad of calculators that have graphic displays and would be pressed before the number is entered to show the negative sign. The other keys are generally used on calculators where the key is pressed after the number is entered to change the sign. For example, to display −5 on the calculator, you would use [5] [+/−], [5] [CHS], or [(−)] [5] depending on the calculator used. The [−] key will normally only be used for subtraction. The key for sign changing cannot be used for subtraction and vice versa.

Exponentiating

Another important key for evaluating expressions on the calculator gives us the capability to exponentiate, that is to find powers. Once again, there are three widely used keys on different calculators that perform this function: [x^y], [y^x], or [^]. The first two are very similar (the x and y are simply switched) and they operate the same way. The last key type is the operator for exponentiating generally used by computers; it operates the same as the other two.

EXAMPLE 17. To calculate 9^6 on the calculator, press [9] [x^y] [6] [=], [9] [y^x] [6] [=], or [9] [^] [6] [=], again depending on the calculator used. The answer is [531441].

Very Large and Very Small Numbers

When very large or very small positive numbers, or their negatives, are used a method is needed to enter and interpret their display. Two common keys for performing the input of these numbers on different calculators are [EE] and [EXP]. (Don't confuse the second with the exponential function on computers.) To display 372×10^{53} you would enter either [3] [7] [2] [EE] [5] [3] or [3] [7] [2] [EXP] [5] [3]. For very small numbers, the exponent on the 10 needs to be negative; 72×10^{-13} would be entered as [7] [2] [EE] [1] [3] [+/−] (Note: [EXP] might be used for [EE], and [CHS] or [(−)] might need to be used for [+/−]). The display of numbers like these varies greatly from calculator to calculator and the value is displayed in scientific notation (which will be explained in Section 5.2). Many calculator displays simply leave one or more spaces between the number and the exponent of 10 to show the correct value; thus 2.91×10^{35} would be displayed as [2.91   35]. Other displays would show the same number with the "EE" or possibly simply an "E" designation: [2.91 E35]. Some calculators even display the entire expression just as it is written: 2.91×10^{35}. To see your display of a large number, calculate 123456789 * 987654321. The answer is approximately [1.219326×10^{17}] (though the exact answer is 121932631112635269, your calculator won't show that many digits).

Memory Storage and Retrieval

The last special keys to consider in this section are for simplifying your work in extended calculations and circumstances when intermediate results are required. These keys have to do with the storage capabilities of the calculator. Most calculators will allow values to be stored for later retrieval so that you don't have to reenter all of the digits to use the number again. To store the number displayed on the calculator, various keys may be available, such as STO, X→M, M+ and others. (M+ adds the displayed value to whatever is currently stored in the calculator's memory.) Some more advanced calculators can store more than one value at a time and a combination of keys is required to store any value. (Normally the combination would be a "store" type key followed by a key designating a letter of the alphabet.) To retrieve a value stored in the memory, the available keys may include RCL, M→X, RM or others.

Note: For the following problems, STO, RCL, +/−, and y^x are used for convenience. You need to use the appropriate keys for your calculator. The example keystrokes for each calculator type should be read left to right, top to bottom.

See solved problems 1.22–1.28.

1.7 Translating Phrases and Statements into Algebraic Form

The application of mathematics normally requires a translation process before any actual mathematics can be performed. That translation converts word phrases into mathematical expressions. The mathematical expressions are "models" of the real problem much like airplane kits are models of real airplanes; both promote the understanding of their real counterpart through analysis of the model. With mathematics, like airplanes, more complete and detailed models offer better understanding. With experience, those models can be assembled. At this level, however, the mathematical detail is not nearly complete, though it still gives a good model with which to work.

Once a problem is converted to a mathematical form, the well-defined rules of mathematics generally allow a simplification of the problem. The results of that simplification are interpreted in light of the original problem (and its restrictions), thus frequently solving it. This provides a highly successful, structured process for solving problems, which is our ultimate goal. In this section, we concentrate only on the translation process.

Careful reading (and probably rereading) of the words in a problem is essential for the translation process to be successful. Even at that, trying to make the translation in one step is more difficult than need be. A frequently used aid for breaking this process into parts or steps is to identify key words and key phrases. Many of these words have been in your vocabulary for a long time: add, multiply, take away, etc. When these are identified, the mathematical symbols can be used in their place.

EXAMPLE 18. The phrase

"multiply the first number by five and add the second number"

would be converted perhaps initially to

"5(first number) + (second number)."

A bit of reordering of the elements of the phrase was used for clarity, but was not required.

EXAMPLE 19. The phrase

"take 7 away from twice a number"

translates into

"(twice a number) − 7."

Notice the reordering was required here since twice the number is not being "taken away from" the seven but just the reverse. This example illustrates the use of another key word, "twice" which means "two times" or "multiply by 2." The original phrase could then be more fully translated into

"2(number) − 7."

Besides key words being replaced by their mathematical symbols, unknown quantities may be replaced by variables that would be used to identify them. When doing so, be very specific about the identity of the variable. *If you can't specifically write the identity of the variable, you don't understand its purpose well enough to continue with the problem.*

Table 1 illustrates translations of numerous commonly encountered word phrases into algebraic form (algebraic expressions).

TABLE 1 Key phrase examples

TYPE	WORD EXPRESSION	ALGEBRAIC TRANSLATION
ADDITION	5 more than x	$x + 5$
	n increased by 17	$n + 17$
	t added to 9	$9 + t$
	the sum of r and 2	$r + 2$
	four plus z	$4 + z$
SUBTRACTION	17 less than y	$y - 17$
	6 subtracted from x	$x - 6$
	8 minus z	$8 - z$
	subtract 3 from a	$a - 3$
	difference of p and q	$p - q$
	t decreased by 9	$t - 9$
MULTIPLICATION	12 times x	$12x$
	twice q	$2q$
	one-half of n	$\frac{1}{2}n$
	t multiplied by 9	$9t$ (or $t(9)$ or $t \times 9$)
	product of 4 and p	$4p$
	three factors of y	yyy or y^3
DIVISION	78 divided by p	$78/p \left(\text{or } 78 \div p \text{ or } \frac{78}{p} \right)$
	the quotient of x and 4	$x \div 4$
	ratio of 2 to 3	$2 \div 3$ or $\frac{2}{3}$
EQUALITY	x is 7	$x = 7$
	y equals 4	$y = 4$
	14 is the same as t	$14 = t$
INEQUALITY	x is not less than y	$x \not< y$ or $x \geq y$
	p is at least 19	$p \geq 19$
	r is not as great as v	$r < v$
	17 is more than z	$17 > z$
	8 is not equal to t	$8 \neq t$
	w is no more than 10	$w \leq 10$

EXAMPLE 20. In Example 18 above,

"multiply the first number by five and add the second number," the letter "x" could be used to designate the "first number" and "y" could be used for the "second number." Then the phrase could be completely translated into

$$"5x + y."$$

For most of this text, mathematical expressions that result from translating word phrases (or statements) will be limited to using a single variable. (To do so in Example 20 would require an additional statement providing the mathematical relationship between the first and second numbers.)

EXAMPLE 21. For Example 19, "n" could represent the "number"; the mathematical expression then becomes

$$"2n - 7."$$

See solved problem 1.29.

When the word phrase is actually a statement, the translation becomes either an equation or an inequality. The key word "*is*" (or an equivalent such as "*equals*") becomes a guide as to whether the words form a phrase or a statement. When the word "*is*" (or an equivalent) is included, a statement is normally indicated; when not included, a phrase is normally indicated. (If it is a statement, the translation is referred to specifically as either an equation or as an inequality. The word "expression" is normally reserved for those translations that aren't statements.) With "*is*" included, the words in the statement preceding "*is*" can form the expression on the left side of the equation (or inequality if designated). Those following "*is*" (or the words associated with it for designating an equation an inequality) can form the right side.

EXAMPLE 22. "Five more than twice a number is 37"

becomes "(5 more than twice a number) = 37"

Then using "n" as the number, the left side expression can be symbolized as before to give

$$"2n + 5 = 37."$$

See solved problem 1.30.

For help in translating word statements, try to identify the general type of problem stated. The problems above are categorized as number problems. This section will deal with a few types of problems (specifically: number, coin, and geometry problems) with more types introduced in later sections. The advantage in identifying the type of problem stated is that you need not "reinvent the wheel"; use your knowledge about solving the *type* of problem to solve the *specific* problem. Even professional "problem solvers" use this technique when possible.

Each type of problem will be introduced by a simple statement and then translated. The problems may seem useless to the student, but the purpose here is simply to develop skill in translating various phrases and statements into algebraic form. That will be extremely useful.

See solved problems 1.31–1.33.

SOLVED PROBLEMS

1.1 Specify each set by listing its elements within braces.

(*a*) {natural numbers greater than 8}
$\{9, 10, 11, \ldots\}$

(*b*) {integers between -4 and 4}
$\{-3, -2, -1, 0, 1, 2, 3\}$

(*c*) {first four whole numbers}
$\{0, 1, 2, 3\}$

(*d*) {odd rational numbers between -4 and 4}
$\{-3, -1, 1, 3\}$

1.2 Let $A = \left\{-4, -\sqrt{8}, \frac{-3}{7}, 0, \frac{1}{4}, \frac{7}{5}, 3.4, 5.222\ldots\right\}$. Specify each set.

(*a*) All members of *A* that are whole numbers.

$\{0\}$

(*b*) The members of *A* that are irrational.

$\{-\sqrt{8}\}$

(*c*) $\{x \mid x \in Q\}$

$\left\{-4, \frac{-3}{7}, 0, \frac{1}{4}, \frac{7}{5}, 3.4, 5.222\ldots\right\}$

(*d*) $\{x \mid x \in J$ and x is negative$\}$

$\{-4\}$

Refer to supplementary problems 1.1 and 1.2 for similar questions.

1.3 Suppose $A = \{0, 2, 5\}$, $B = \{1, 3, 5, 8\}$, and $C = \{3, 5\}$. Find the following:

(*a*) $A \cap B$ (*b*) $A \cup C$ (*c*) $B \cap C$

(*d*) $B \cup C$ (*e*) $A \cap N$ (*f*) $B \cup W$

(*a*) $\{5\}$ (*b*) $\{0, 2, 3, 5\}$ (*c*) $\{3, 5\}$

(*d*) $\{1, 3, 5, 8\}$ (*e*) $\{2, 5\}$ (*f*) $\{0, 1, 2, \ldots\}$

Refer to supplementary problem 1.3.

1.4 Make each statement an application of the given property of equality by replacing each question mark with the appropriate quantity.

(*a*) Symmetric property — If $s = t - 6$, then $? = s$.

(*b*) Transitive property — If $4 = w + 3$ and $w + 3 = y$, then $4 = ?$

(*c*) Substitution property — If $4 = w + 3$ and $w = y$, then $4 = ?$

(*d*) Reflexive property — $r - s = ?$

(*a*) $t - 6$ (*b*) y (*c*) $y + 3$ (*d*) $r - s$

See supplementary problem 1.4.

1.5 Express each relation using appropriate symbols.

(*a*) 7 is greater than 1. (*b*) -4 is less than -3.

(*c*) $x - 2$ is positive. (*d*) t is between 0 and 5.

(*e*) s is greater than or equal to -4 and less than 3.

(*a*) $7 > 1$ (*b*) $-4 < -3$ (*c*) $x - 2 > 0$ (*d*) $0 < t < 5$ (*e*) $-4 \leq s < 3$

1.6 Replace each question mark with the appropriate order symbol to form a true statement.

(*a*) 2 ? 8 (*b*) 2 ? − 3 (*c*) −3 ? −1 (*d*) −1.4 ? −1.5

(*a*) $<$ or $\leq$ (*b*) $>$ or $\geq$ (*c*) $<$ or $\leq$ (*d*) $>$ or $\geq$

Refer to supplementary problems 1.5 and 1.6 for similar questions.

1.7 Graph $\{-2, -3/2, 2.5, 4.0\}$ on the number line in Figure 1.5 below.

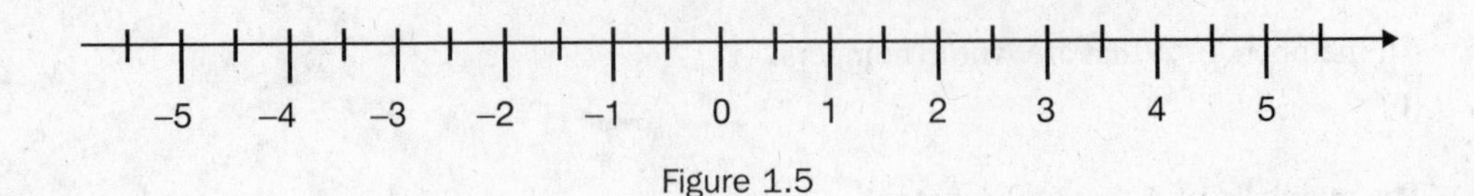

Figure 1.5

See Figure 1.6.

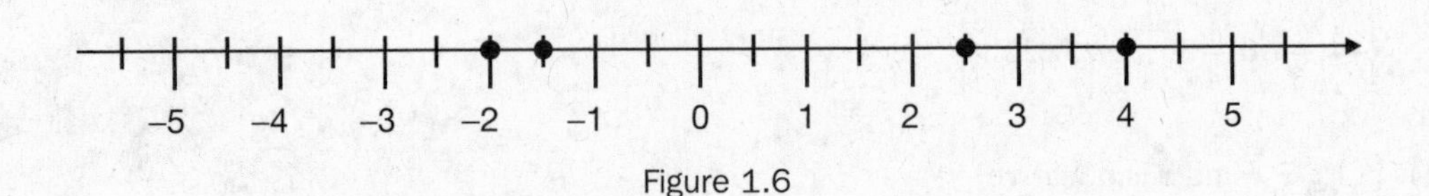

Figure 1.6

1.8 Graph the indicated intervals on a number line.

(*a*) $(1, 3]$ (*b*) $[-2, 2]$ (*c*) $[0, \infty)$ (*d*) $(-\infty, -1)$

(*a*) See Figure 1.7. (*b*) See Figure 1.8. (*c*) See Figure 1.9. (*d*) See Figure 1.10.

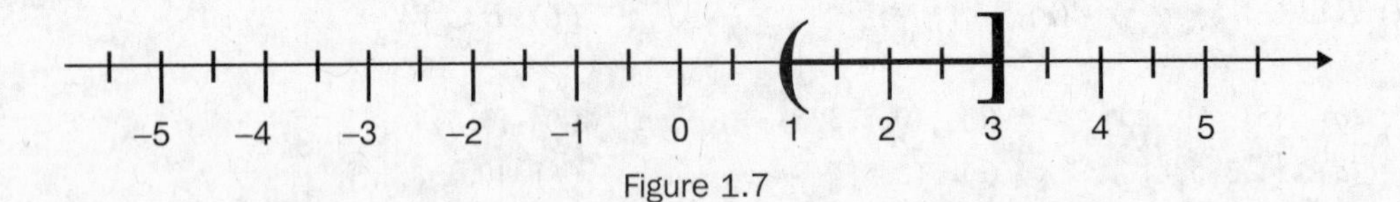

Figure 1.7

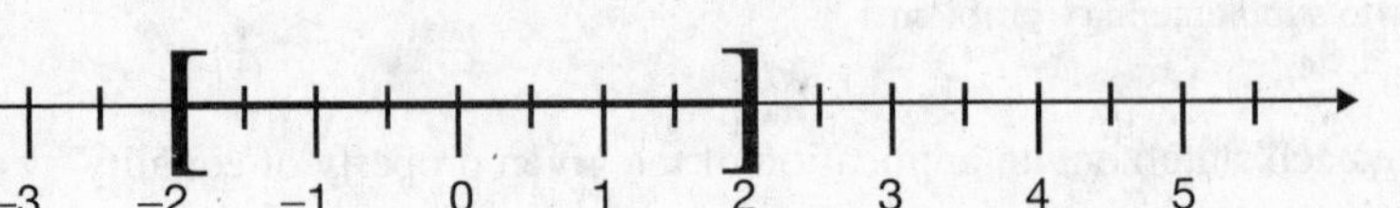

Figure 1.8

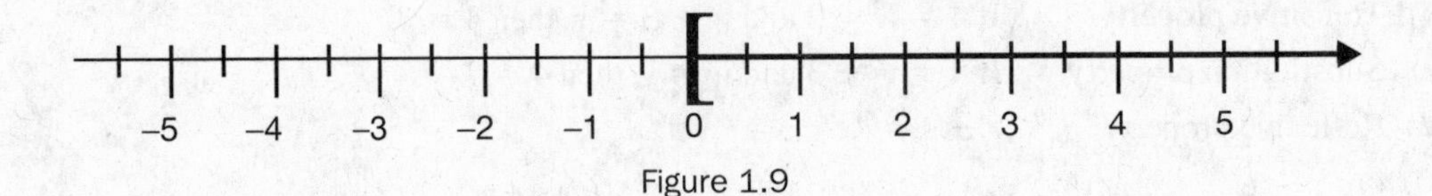

Figure 1.9

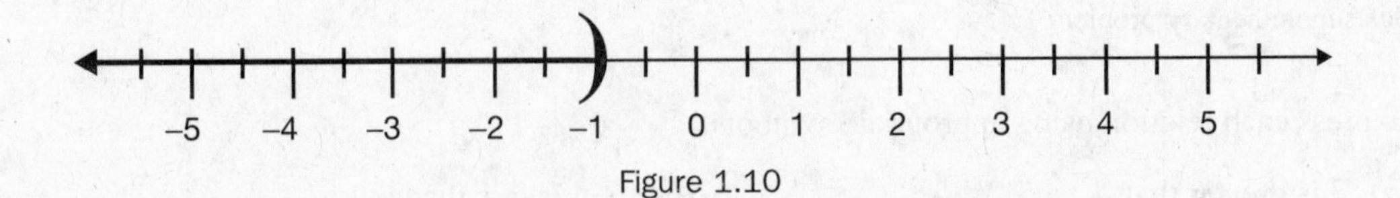

Figure 1.10

1.9 Describe the intervals in Solved Problem 1.8 above using set-builder notation.

(*a*) $\{x \mid 1 < x \le 3\}$ (*b*) $\{x \mid -2 \le x \le 2\}$ (*c*) $\{x \mid x \ge 0\}$ (*d*) $\{x \mid x < -1\}$

1.10 Graph each of the indicated sets of real numbers on a separate number line.

(*a*) $\{x \mid -8 \le x < 4\}$ (*b*) $\{t \mid t < 6\}$ (*c*) $\{s \mid s \ge -3\}$ (*d*) $\{y \mid -10 < y \le -4\}$

(*a*) See Figure 1.11. (*b*) See Figure 1.12. (*c*) See Figure 1.13. (*d*) See Figure 1.14.

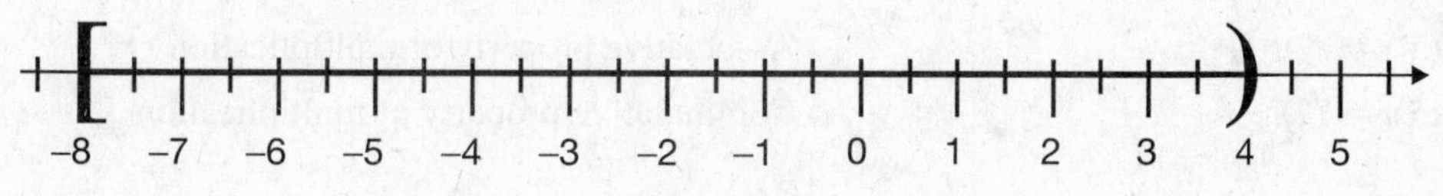

Figure 1.11

Figure 1.12

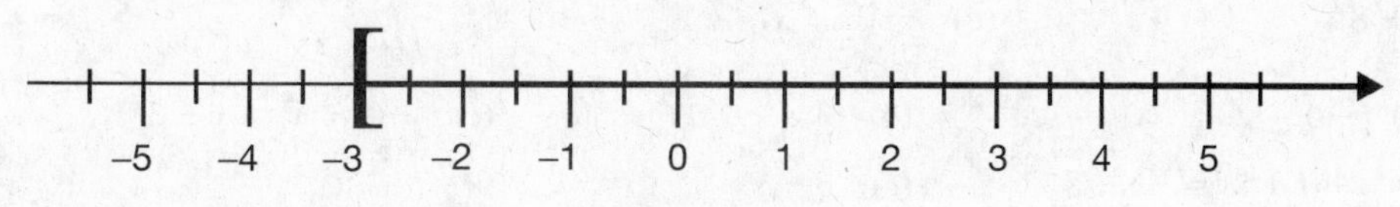

Figure 1.13

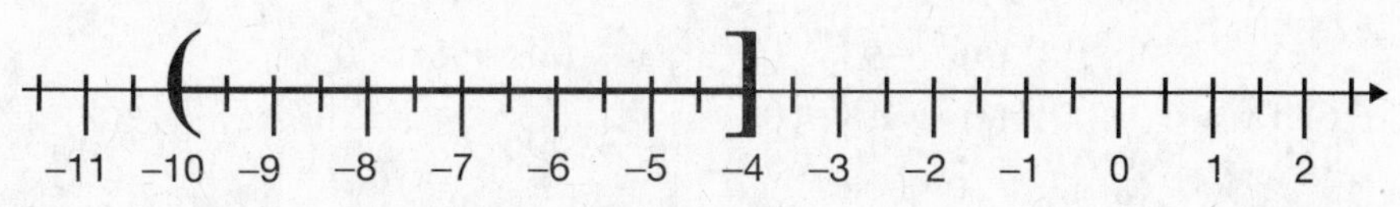

Figure 1.14

See supplementary problems 1.7, 1.8, and 1.9.

1.11 State the name of the property illustrated in each of the following.

(*a*) $4 + 2x = 2x + 4$
(*b*) $(x - 3)8 = 8(x - 3)$
(*c*) $3t + 6 = 3(t + 2)$
(*d*) $6 + (-6) = 0$
(*e*) $3(t + 2) = 3(2 + t)$
(*f*) $1(5) = 5$
(*g*) $3(4)$ is a real number
(*h*) $1 = [1/(s - 2)]\,(s - 2)$
(*i*) $r + (7 + t) = (r + 7) + t$
(*j*) $r + (7 + t) = r + (t + 7)$

(*a*) Commutative property of addition
(*b*) Commutative property of multiplication
(*c*) Distributive property
(*d*) Additive inverse property
(*e*) Commutative property of addition
(*f*) Identity property of multiplication
(*g*) Closure property of multiplication
(*h*) Inverse property of multiplication
(*i*) Associative property of addition
(*j*) Commutative property of addition

1.12 Replace each question mark so that the given statement is an application of the given property.

(*a*) $4 + ? = 4$ — Identity property of addition
(*b*) $s + (-s) = ?$ — Additive inverse property
(*c*) $x + (3x + 9) = (x + 3x) + ?$ — Associative property of addition
(*d*) $r + 8 + ? = r + 0$ — Additive inverse property
(*e*) $3(?)t = t$ — Multiplicative inverse property
(*f*) $3(x - 5) = ? - 15$ — Distributive property

(*g*) $s(5t) = (?)t$ Associative property of multiplication

(*h*) $(s5)t = (?)t$ Commutative property of multiplication

(*a*) 0 (*b*) 0 (*c*) 9 (*d*) −8

(*e*) 1/3 (*f*) $3x$ (*g*) $s5$ (*h*) $5s$

See supplementary problems 1.10 and 1.11 for additional drill.

1.13 Find the opposite of each of the following expressions.

(*a*) 6 (*b*) −3 (*c*) $3 - x$

(*d*) $t + 8$ (*e*) $-(6 - s)$

(*a*) −6 (*b*) 3 (*c*) $-(3 - x) = x - 3$

(*d*) $-(t + 8) = -t - 8$ (*e*) $6 - s$

1.14 Find the absolute value of each of the following expressions.

(*a*) 7 (*b*) −9 (*c*) t if $t > 0$

(*d*) $-(x - 6)$ if $x > 6$ (*e*) s if $s < 0$

(*a*) 7 (*b*) 9 (*c*) t

(*d*) $x - 6$ (*e*) $-s$

1.15 Refer to the number line in Figure 1.15 below to find the following sums.

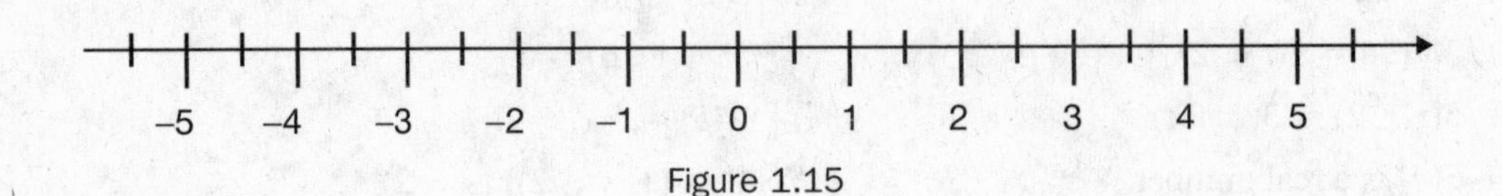

Figure 1.15

(*a*) $-3 + 4$ (*b*) $2 + 3$ (*c*) $6 + (-4)$

(*d*) $(-1) + (-3)$ (*e*) $0 + (-2)$

(*a*) 1 (*b*) 5 (*c*) 2

(*d*) −4 (*e*) −2

1.16 Find the indicated sums.

(*a*) $6 + 3$ (*b*) $-5 + 3$ (*c*) $4 + (-6)$

(*d*) $-3 + (-7)$ (*e*) $-10 + 17$

(*a*) 9 (*b*) −2 (*c*) −2

(*d*) −10 (*e*) 7

See supplementary problem 1.12.

1.17 Find the following differences.

(*a*) $5 - 9$ (*b*) $-3 - (-7)$ (*c*) $22 - 14$ (*d*) $-7 - 14$

(*a*) $5 - 9 = 5 + (-9) = -4$ (*b*) $-3 - (-7) = -3 + 7 = 4$

(*c*) $22 - 14 = 22 + (-14) = 8$ (*d*) $-7 - 14 = -7 + (-14) = -21$

See supplementary problems 1.13 and 1.14.

1.18 Find the following products.

(*a*) $5(7)$ (*b*) $4(-7)$ (*c*) $-4(4)$

(*d*) $(-5)(-6)$ (*e*) $11(0)$ (*f*) $0(-8)$

(*a*) 35 (*b*) -28 (*c*) -16

(*d*) 30 (*e*) 0 (*f*) 0

See supplementary problem 1.15.

1.19 Find the indicated quotients.

(*a*) $28/7$ (*b*) $-20/5$ (*c*) $\frac{0}{-6}$

(*d*) $\frac{12}{-3}$ (*e*) $-36/(-4)$ (*f*) $7/0$

(*a*) 4 (*b*) -4 (*c*) 0

(*d*) -4 (*e*) 9 (*f*) undefined

See supplementary problems 1.16 and 1.17.

1.20 Write the following without exponents and evaluate.

(*a*) 2^4 (*b*) 3^3 (*c*) $(-4)^3$

(*d*) $(-1)^5$ (*e*) 0^4

(*a*) $2 \cdot 2 \cdot 2 \cdot 2 = 16$ (*b*) $3 \cdot 3 \cdot 3 = 27$ (*c*) $(-4)(-4)(-4) = -64$

(*d*) $(-1)(-1)(-1)(-1)(-1) = -1$ (*e*) $0 \cdot 0 \cdot 0 \cdot 0 = 0$

See supplementary problem 1.18.

1.21 Perform the indicated operations.

(*a*) $2 + 3 \cdot 6$

$2 + 3 \cdot 6 = 2 + 18 = 20$

(*b*) $33 - 4 \cdot 5 + 8$

$33 - 4 \cdot 5 + 8 = 33 - 20 + 8 = 21$

(*c*) $4[8 - (-3)]$

$4[8 - (-3)] = 4[8 + 3] = 4[11] = 44$

(*d*) $(6 - 9)(8 - 12)$

$(6 - 9)(8 - 12) = (-3)(-4) = 12$

(*e*) $12 - (4 \cdot 3 + 8)$

$12 - (4 \cdot 3 + 8) = 12 - (12 + 8) = 12 - (20) = -8$

(*f*) $\frac{-20 + 4}{-2 - 6}$

$$\frac{-20 + 4}{-2 - 6} = \frac{-16}{-8} = 2$$

(*g*) $\dfrac{2(-4)(-5)}{-2(4)+4}$

$$\frac{2(-4)(-5)}{-2(4)+4}=\frac{-8(-5)}{-8+4}=\frac{40}{-4}=-10$$

(*h*) $\dfrac{15}{-3}-\left(\dfrac{8}{-4}\right)$

$$\frac{15}{-3}-\left(\frac{8}{-4}\right)=-5-(-2)=-5+2=-3$$

(*i*) -7^2

$$-7^2=-(49)=-49$$

(*j*) $3+2-4^2$

$$3+2-4^2=3+2-16=5-16=-11$$

(*k*) $-(3+2-4^2)^2$

$$-(3+2-4^2)^2=-(3+2-16)^2=-(5-16)^2=-(-11)^2=-(121)=-121$$

(*l*) 6^2+3^3

$$6^2+3^3=36+27=63$$

(*m*) $(3-3^2)^2$

$$(3-3^2)^2=(3-9)^2=(-6)^2=36$$

(*n*) $3[7-4(5-2)]$

$$3[7-4(5-2)]=3[7-4(3)]=3[7-12]=3[-5]=-15$$

(*o*) $2(-1)(-5)^2$

$$2(-1)(-5)^2=2(-1)(25)=-2(25)=-50$$

(*p*) $\dfrac{3\cdot 7+4^2-1}{2+(7-3)}$

$$\frac{3\cdot 7+4^2-1}{2+(7-3)}=\frac{3\cdot 7+16-1}{2+4}=\frac{21+16-1}{2+4}=\frac{37-1}{6}=\frac{36}{6}=6$$

(*q*) $\dfrac{3[4-3(-2)^2]}{2^2-4^2}$

$$\frac{3[4-3(-2)^2]}{2^2-4^2}=\frac{3[4-3(4)]}{4-16}=\frac{3[4-12]}{-12}=\frac{3[-8]}{-12}=\frac{-24}{-12}=2$$

(*r*) $\dfrac{4^2-2\left(\dfrac{3+5}{4-8}\right)+4}{-2+3\left(\dfrac{6-3}{1-4}\right)+9}$

$$\frac{4^2-2\left(\frac{3+5}{4-8}\right)+4}{-2+3\left(\frac{6-3}{1-4}\right)+9}=\frac{16-2\left(\frac{8}{-4}\right)+4}{-2+3\left(\frac{3}{-3}\right)+9}=\frac{16-2(-2)+4}{-2+3(-1)+9}=\frac{16+4+4}{-2-3+9}=\frac{24}{4}=6$$

(*s*) $10^2-\dfrac{10+2\left(\dfrac{6-4}{4-6}\right)^3}{2(2^2+0^3)}$

$$10^2-\frac{10+2\left(\frac{6-4}{4-6}\right)^3}{2(2^2+0^3)}=100-\frac{10+2\left(\frac{2}{-2}\right)^3}{2(4+0)}=100-\frac{10+2(-1)^3}{2(4)}=100-\frac{10-2}{8}=100-\frac{8}{8}=100-1=99$$

(t) $(4^3 - 1) \div \dfrac{5^2 + 4 \div 2}{8^2 - 9^2 + 8}$

$$(4^3 - 1) \div \frac{5^2 + 4 \div 2}{8^2 - 9^2 + 8} = (64 - 1) \div \frac{25 + 4 \div 2}{64 - 81 + 8} = 63 \div \frac{25 + 2}{-17 + 8} = 63 \div \frac{27}{-9}$$

$$= 63 \div (-3) = -21$$

See supplementary problem 1.19.

1.22 Evaluate $7 * 2 - 5 * 4$.

Algebraic Entry: 7 × 2 − 5 × 4 =

RPN: 7 ENTER 2 × 5 ENTER 4 × −

Last Entry: 5 × 4 = STO 7 × 2 − RCL =

Answer: −6

1.23 Evaluate $(3 + 2 * 4)(9 * 2 - 5)$.

Algebraic Entry: (3 + 2 × 4) × (9 × 2 − 5) =

RPN: 3 ENTER 2 ENTER 4 × + 9 ENTER 2 × 5 − ×

Last Entry: 2 × 4 + 3 = STO 9 × 2 − 5 × RCL =

Answer: 143

1.24 Evaluate $(2 * 3^4 + 5) * 7$.

Algebraic Entry: (2 × 3 y^x 4 + 5) × 7 =

RPN: 2 ENTER 3 ENTER 4 y^x × 5 + 7 ×

Last Entry: 3 y^x 4 × 2 + 5 × 7 =

Answer: 1169

1.25 Evaluate $\dfrac{-3}{7} + 15\left(\dfrac{7}{11}\right)$.

Algebraic Entry: 3 +/− ÷ 7 + 1 5 × 7 ÷ 1 1 =

RPN: 3 +/− ENTER 7 ÷ 1 5 ENTER 7 × 1 1 ÷ +

Last Entry: 3 +/− ÷ 7 = STO 1 5 × 7 ÷ 1 1 + RCL =

Answer: 9.116883117

1.26 Evaluate $3(5^4 - 4^5)$.

Algebraic Entry: 3 × (5 y^x 4 − 4 y^x 5) =

RPN: 3 ENTER 5 ENTER 4 y^x 4 ENTER 5 y^x − ×

Last Entry: 4 y^x 5 = STO 5 y^x 4 − RCL × 3 =

Answer: −1197

1.27 Evaluate $\dfrac{5 + 7^3}{2^5 - 5}$.

Algebraic Entry: (5 + 7 y^x 3) ÷ (2 y^x 5 − 5) =

PN: 5 ENTER 7 ENTER 3 y^x + 2 ENTER 5 y^x 5 − ÷

Last Entry: 2 y^x 5 − 5 = STO 7 y^x 3 + 5 ÷ RCL =

Answer: 12.88888889

1.28 Evaluate $(2-3*5)^3(7-2*8)^5$.

Algebraic Entry: [(] [2] [−] [3] [×] [5] [)] [y^x] [3] [×] [(] [7] [−] [2] [×] [8] [)] [y^x] [5] [=]

RPN: [2] [ENTER] [3] [ENTER] [5] [×] [−] [3] [y^x] [7] [ENTER] [2] [ENTER] [8] [×] [−] [5] [y^x] [×]

Last Entry: [3] [×] [5] [=] [STO] [2] [−] [RCL] [y^x] [3] [=] [*Write answer*] [2] [×] [8] [=] [STO] [7] [−] [RCL] [y^x] [5] [×] [*Re-enter first answer*] [=]

Answer: 129730653 or 1.29731×10^8 if 6 digits are displayed in Scientific Notation.

This last problem illustrates one of the greatest difficulties with the Last Entry calculator: the final calculation cannot always be conveniently arrived at without writing down an intermediate result and then reentering it later. The Last Entry calculator type is considered inferior to the other types. It is not generally referenced in calculator problems; either the Algebraic Entry or RPN calculator type is assumed.

See supplementary problem 1.20 for additional calculator practice.

1.29 Translate each of the following into mathematical expressions:

(*a*) **"Fifteen less than the sum of two numbers"**—use "x" for the first number and "y" for the second.

The translation

becomes "fifteen less than the sum of x and y" which in turn

becomes "fifteen less than $(x + y)$" and finally

"$(x + y) - 15$"

or "$x + y - 15$" note: parentheses not needed.

(*b*) **"The product of eight, and 12 more than a number"**—use "p" as the number.

The translation

becomes "the product of 8, and 12 more than p" and that

becomes "the product of 8 and $(p + 12)$" which gives

"$8(p + 12)$"

since the product requires 8 to be multiplied by the value of p increased by 12.

(*c*) **"The sum of 25 and the product of seven and a number"**—use "x" as the number.

The translation becomes (in ever better resolution)

"the sum of 25 and the product of 7 and x"

"the sum of 25 and $7x$"

"$25 + 7x$"

since the sum requires 25 and the product be added.

(*d*) **"Three times one number minus twice another number"**—use "x" for one number and "y" for the other.

Translating

"3 times x minus twice y"

"3 times x minus $2y$"

"$3x - 2y$."

(*e*) **"Five times a number is decreased by 7"**—use "s" for the number.

"five times s decreased by 7"

"$5s - 7$."

(*f*) **"The ratio of 4 to a quantity"**—use "q" for the quantity.

"the ratio of 4 to q"

"$4 \div q$" or "$\frac{4}{q}$."

(*g*) **"A number decreased by eleven, divided by twice the number"**—use "x" for the number.

"x decreased by 11, divided by twice x"

"(x decreased by 11) divided by ($2x$)"

"$(x-11) \div (2x)$" or "$\frac{x-11}{2x}$."

1.30 Translate each of the following into an equation or an inequality:

(*a*) **"A number is not the same as 20"**—use "n" for the number.

"n is not the same as 20"

"$n \neq 20$."

(*b*) **"Six more than one number is eight less than one-half another number"**—use "x" and "y" for the numbers.

"6 more than x is 8 less than $\frac{1}{2}y$"

"$x+6$ is $\frac{1}{2}y-8$"

"$x+6=\frac{1}{2}y-8$." or "$x+6=\frac{y}{2}-8$."

(*c*) **"The product of the sum of a number and three, and seven less than another number is more than ten"**—use "s" and "t" for the numbers.

"the product of the sum of s and 3, and 7 less than t is more than 10"

"the product of (the sum of s and 3) and (7 less than t) is greater than 10"

"(the sum of s and 3)(7 less than t) > 10"

"$(s+3)(t-7) > 10$."

(*d*) **"The difference of 50 and the product of three and a number is no more than 20"** —use "n" for the number.

"the difference of 50 and the product of 3 and n is no more than 20"

"the difference of 50 and (3 times n) is less than or equal to 20"

"$50 - 3n \leq 20$."

Number Problems

1.31 Translate each into a representative equation:

(*a*) **"Four more than twice a number is one less than three times the number."**

This is similar to previous problems. Let "x" represent the number. Read the statement carefully to convince yourself that only one number is indicated. Then the problem can be translated in the following stages:

"4 more than twice x = 1 less than 3 times x"

"4 more than $2x$ = 1 less than $3x$"

"$2x + 4 = 3x - 1$."

(*b*) **"The sum of two consecutive integers is 95."**

This problem requires a look at the meaning of consecutive integers. Examples of consecutive integers are 13 & 14, 79 & 80, and -134 & -133; the larger is simply one more than the smaller. If the first (or smaller) is identified as "x," then the next (or larger) would be identified as "$x + 1$." Using "x" as the first integer and "$x + 1$" as the next, the statement could be translated by steps as

$$\text{"(first)} + \text{(next)} = 95\text{"}$$
$$\text{"}(x) + (x + 1) = 95\text{"}$$

(The parentheses are included only to help show the respective integers involved).

Note: For consecutive even (e.g., 32 & 34) or consecutive odd (e.g., 119 & 121) integers, if the first number is designated as "x," the next would be designated as "$x + 2$." They differ by two whether they are consecutive even or consecutive odd. So, to obtain the next, simply add two.

(*c*) **"Seven times a number plus three times its reciprocal is 22."**

Since the reciprocal of any number is its multiplicative inverse, if "n" is designated as the number, then "$1/n$" would designate its reciprocal. Therefore, the statement could be translated in stages as follows:

$$\text{"7 times } n \text{ plus 3 times } (1/n) = 22\text{"}$$
$$\text{"}7n \text{ plus } 3(1/n) = 22\text{"}$$
$$\text{"}7n + 3(1/n) = 22.\text{"}$$

Coin Problems

1.32 Translate each of the following into an equation representing the coins indicated:

(*a*) **"A collection of nickels, dimes, and quarters is worth \$1.65. If there are 2 more dimes than nickels but only one more quarter than the number of nickels, how many of each coin are there?"**

Notice immediately the following:

"(value of all nickels) + (value of all dimes) + (value of all quarters) = \$1.65."

So, the value of nickels, dimes, and quarters must be symbolized. Also, all values must be in the same units so 165 cents would probably be easier to use than \$1.65 (no decimals). All values are then in cents, not dollars.

The question being asked in the problem normally identifies what the variable should represent. Here, the number of each coin denomination (value) is requested. A choice, however, is required for determining on *which* coin to focus first—to determine the number of that coin. Frequently in this type of problem, one coin is easier to "key on" than the others. In this specific problem it's the number of nickels. Letting "x" represent the number of nickels identifies the number of dimes and quarters rather easily from the second sentence. Since there are two more dimes than nickels, the number of dimes would be designated as "$x + 2$"; there is one more quarter than nickels so their quantity would be "$x + 1$." Be certain to note that each designation is for the number of coins; nothing involving a value occurs yet!

Another step in the process requires the values of the different denominations to be identified. Since the value of each coin is well known and the number of each coin is symbolized, they can be combined to give the total value of all coins of that type (denomination). The total value for all of one type of coin is the product of its quantity and its denominational value. The translation can be stepped through by

$$\begin{pmatrix}\text{Value of}\\ \text{all nickels}\end{pmatrix}+\begin{pmatrix}\text{Value of}\\ \text{all dimes}\end{pmatrix}+\begin{pmatrix}\text{Value of}\\ \text{all quarters}\end{pmatrix}=165$$

$$\begin{pmatrix}\text{Nickel}\\ \text{value}\end{pmatrix}\begin{pmatrix}\text{Number of}\\ \text{nickels}\end{pmatrix}+\begin{pmatrix}\text{Dime}\\ \text{value}\end{pmatrix}\begin{pmatrix}\text{Number of}\\ \text{dimes}\end{pmatrix}+\begin{pmatrix}\text{Quarter}\\ \text{value}\end{pmatrix}\begin{pmatrix}\text{Number of}\\ \text{quarters}\end{pmatrix}=165$$

"$(5)(x)+(10)(x+2)+(25)(x+1)=165$."

(*b*) **"An auditorium seating 1200 people sold out a performance and grossed $9000. If the general admission tickets were $6 and the reserved seating tickets were $10, how many general admission tickets were sold?"**

This problem doesn't have anything to do with coins so may seem misplaced. It does, however, involve two tickets with their own denominational value. That makes this a coin-type problem. Treating the tickets as coins puts it in perspective.

This problem is a bit trickier than the last in that there is seemingly no relationship between the number of tickets of the two types – the denominational quantities. The only relationship given is the fact that they combine for 1200 tickets; that is enough. Let "n" represent the number of general admission tickets sold. If you were to guess "n" to be 375, how many reserved seating tickets would there have to be? That number must "take up" the difference between 375 and 1200, i.e., it must be 1200 − 375 or 825. The 375 was just a guess to help think through the setup so simply replace it by "n" and you have the number of reserved seating tickets sold: "$1200 - n$." Using all of this and refining the translation might give the following sequence:

$$\begin{pmatrix}\text{Value of general}\\ \text{admission tickets}\end{pmatrix}+\begin{pmatrix}\text{Value of reserved}\\ \text{seating tickets}\end{pmatrix}=9000$$

$$\begin{pmatrix}\text{General admission}\\ \text{ticket value}\end{pmatrix}\begin{pmatrix}\text{Number of general}\\ \text{admission tickets}\end{pmatrix}+\begin{pmatrix}\text{Reserved seating}\\ \text{ticket value}\end{pmatrix}\begin{pmatrix}\text{Number of reserved}\\ \text{seating tickets}\end{pmatrix}=9000$$

"$(6)(n)+(10)(1200-n)=9000$."

Geometry Problems

1.33 Translate each of the following into an equation or inequality representing the figure:

(*a*) **"Find the length and width of a rectangle given that the perimeter is no more than 800 inches and the length is three times the width."**

Geometry problems require knowing some information about the geometric figure involved. Often that information is a formula included in the problem itself. When, as here, the author of the problem thinks the information to be reasonably standard knowledge at the level of the student, the formula is omitted. After working a few exercises, you will probably agree with that assessment for this figure.

This specific problem requires a formula for the relationship between the perimeter of a rectangle and its length and width. The perimeter of any figure is the distance around the figure, so the perimeter of a rectangle is

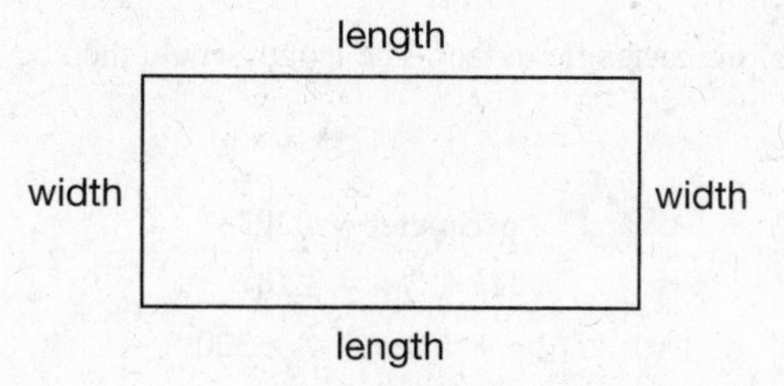

Figure 1.16

length + width + length + width or 2(length) + 2(width); so $P = 2l + 2w$. Now the problem can be translated into an algebraic form. Letting "w" represent the width (in inches for agreement of units with the perimeter), the length can then be represented by "$3w$" since it is three times the width. Using this information, the translation steps might proceed as

$$\text{"perimeter} \leq 800\text{"}$$

$$\text{"2(length) + 2(width)} \leq 800\text{"}$$

$$\text{"}2(3w) + 2(w) \leq 800.\text{"}$$

(*b*) **"How long is the hypotenuse in a right triangle with one leg 5 inches and the other 12 inches? (The Pythagorean Theorem states that the sum of the squares of the lengths of the two legs is equal to the square of the length of the hypotenuse.)"**

This problem has a formula embedded (in the stated Pythagorean Theorem) but requires understanding of the terminology. In this problem, a right triangle, its hypotenuse, and the legs are referenced. A right triangle is one in which the two shorter sides (called legs) intersect at a right angle or corner. The longer side (called the hypotenuse) is always opposite that right angle.

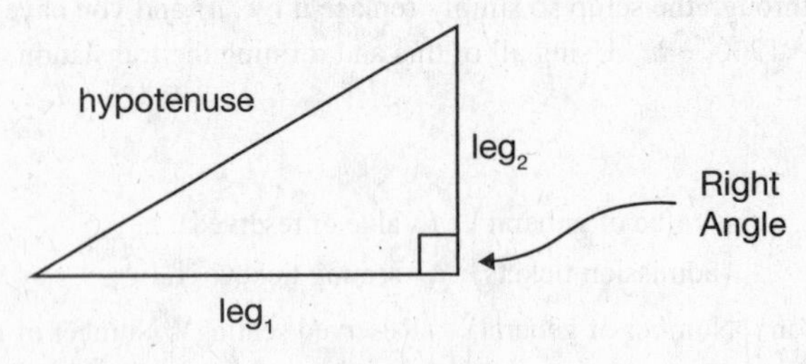

Figure 1.17

Since the problem asks for the length of the hypotenuse, it is logical to associate that with the variable; let's identify it as "x." Now the Pythagorean Theorem must be used. Since it states

$$\text{"}(leg_1)^2 + (leg_2)^2 = (hypotenuse)^2\text{"}$$

and one leg is 12, say leg_1, and the other, leg_2, is 5, then the theorem can be written for this problem now as

$$\text{"}(12)^2 + (5)^2 = (x)^2\text{"}$$

which is a valid algebraic translation of the original problem. Nothing more is required for the setup of this problem! Frequently when a formula is given in the problem, it is the basis for the setup.

(*c*) **"The length of a rectangle is 10 more than twice its width. If the perimeter is 320 feet, how wide is the rectangle?"**

Let "w" be the width of the rectangle in feet. The length would then be 10 more than twice w or $l = 2w + 10$. So,

$$\text{"perimeter} = 320\text{"}$$

$$\text{"}2l + 2w = 320\text{"}$$

$$\text{"}2(2w + 10) + 2w = 320.\text{"}$$

Refer to supplementary problems 1.21 through 1.49 for similar problems.

SUPPLEMENTARY PROBLEMS

1.1 Specify each set by listing its elements within braces:

(*a*) {integers between -2 and 5}
(*b*) {even rational numbers between -2 and 5}
(*c*) {first three natural numbers}
(*d*) {whole numbers less than five}

1.2 Let $S = \left\{-\frac{13}{4}, -\sqrt{4}, -1.787878\ldots, 2, 2.74, \sqrt{11}\right\}$. Specify each set :

(*a*) The members of S that represent integers.
(*b*) $\{x \mid x \in S \text{ and } x \in Q\}$
(*c*) $\{x \mid x \in W \text{ and } x \text{ is negative}\}$
(*d*) The set of irrational elements of S.

1.3 $A = \{1, 4, 7\}, B = \{1, 2, 5, 7\}$, and $C = \{-1, 2, 6\}$. Find the following:

(*a*) $B \cap C$
(*b*) $A \cup B$
(*c*) $A \cap C$
(*d*) $C \cap N$
(*e*) $B \cup W$
(*f*) $B \cap A$
(*g*) $J \cup C$

1.4 Name the property that is illustrated in each case.

(*a*) If $x + y = 8$ and $y = 2$, then $x + 2 = 8$.
(*b*) If $8 = x + y$, then $x + y = 8$.
(*c*) If $s + 3 = t$ and $t = 9$, then $s + 3 = 9$.
(*d*) If $bh = 20$ and $b = 4$, then $4h = 20$.
(*e*) $p = p$
(*f*) If $x = y$ and $y = t + 4$, then $x = t + 4$.

1.5 Express each relation using appropriate symbols.

(*a*) 4 is greater than -2
(*b*) x is nonpositive
(*c*) s is between 4 and 7
(*d*) -3.5 is less than -3.4
(*e*) 10 is greater than or equal to 10

1.6 Replace each question mark with the appropriate order symbol to form a true statement.

(*a*) $-3 ? -4$
(*b*) $-4 ? -3$
(*c*) $0 ? -1$
(*d*) $5 ? 5$
(*e*) $-99 ? -98$

1.7 Graph the indicated intervals on a number line.

(*a*) $[-2, 4)$
(*b*) $(-\infty, 1]$
(*c*) $[0, 4]$
(*d*) $(-3, \infty)$

1.8 Use set-builder notation to describe the intervals in problem 1.7.

1.9 Graph each of the indicated sets of real numbers on a separate number line.

(*a*) $\{x \mid x > -2\}$
(*b*) $\{t \mid -3 < t \leq 2\}$
(*c*) $\{y \mid y < 20\}$
(*d*) $\{s \mid -5 \leq s \leq -2\}$

1.10 State the name of the property illustrated in each of the following.

(*a*) $yx = xy$
(*b*) $2x + (-2x) = 0$
(*c*) $6x + 2x = (6 + 2)x$
(*d*) $4 + 0 = 4$
(*e*) $1t = t$
(*f*) $(1/2)2w = 1w$
(*g*) $4s + (2s + t) = (4s + 2s) + t$
(*h*) $1w = w$
(*i*) $(x - 3)5 = 5(x - 3)$
(*j*) $0 + (s - t) = (s - t) + 0$

1.11 Replace each question mark so that the given statement illustrates the stated property.

(a) $(3w + x) + 2y = ? + (x + 2y)$	Associative property of addition
(b) $4t? = 4st$	Commutative property of multiplication
(c) $0 + ? = 5$	Identity property of addition
(d) $1(s + 2t) = ?$	Identity property of multiplication
(e) $(1/4)4x = ?$	Multiplicative inverse property
(f) $3 + s + 5t = s + 5t + ?$	Commutative property of addition
(g) $0 = y + ?$	Additive inverse property
(h) $2s(t) = (?)t$	Associative property of multiplication
(i) $3x + 12y = ?(x + 4y)$	Distributive property

1.12 Find the following sums.

(a) $3 + 8$ (b) $7 + (-4)$ (c) $12 + (-5)$
(d) $-44 + (+55)$ (e) $-23 + (-15)$ (f) $0 + 22 + (-27)$

1.13 Find the following differences.

(a) $9 - 3$ (b) $12 - (-4)$ (c) $-14 - 8$
(d) $-36 - (-29)$ (e) $8 - 15$ (f) $-23 - (-33)$

1.14 Perform the indicated operations.

(a) $9 + 4$ (b) $9 - 4$ (c) $4 + (-9)$
(d) $-4 - 9$ (e) $12 + (-7) - 8$ (f) $-22 + 9$
(g) $9 - 15$ (h) $-11 + 8$ (i) $-8 - (-4) - 7$

1.15 Find the indicated products.

(a) $5(-7)$ (b) $12(3)$ (c) $0(5)$
(d) $(-8)(-6)$ (e) $(7)(-3)(0)$ (f) $4(-6)(-3)(-2)$

1.16 Find the indicated quotients.

(a) $\frac{-18}{3}$ (b) $24/(-4)$ (c) $15/0$ (d) $0/11$ (e) $\frac{-30}{-6}$

1.17 Perform the indicated operations.

(a) $3 + (-4)$ (b) $(-5)(9)$ (c) $48/12$ (d) $-9 - 22$
(e) $\frac{24}{0}$ (f) $14 - (-5)$ (g) $(-7)(-9)$ (h) $0(13)$
(i) $-2 + 44$ (j) $8 - 11$ (k) $56 \div 7$ (l) $0 - (-99)$

1.18 Write the following without exponents and evaluate.

(a) 5^2 (b) $(-2)^3$ (c) 0^6 (d) 1^4 (e) 3^4

1.19 Perform the indicated operations.

(a) $\frac{5(3-5)}{-2} - \frac{18}{-2}$ (b) $\frac{7+4(2)}{5} - 7$

(*c*) $6(-2) - 4(-7) + 5$

(*d*) $27 \div [3(9 - 3(4 - 2))]$

(*e*) $(3 + 2)^2$

(*f*) $(3 \cdot 2)^2$

(*g*) $(7 - 5)(1 + 13)$

(*h*) $2 \cdot 3 - 4(-4) + 8$

(*i*) $\dfrac{3(2) - 5(2)}{5(-3) + 2 \cdot 3^2 + 1}$

(*j*) $3^2 + 2^2$

(*k*) $\dfrac{1 + 36 \div (-4) - (-2)}{-2 - 1}$

(*l*) $-20 \div (-4)2 + 4 - 3(-2)$

(*m*) $\dfrac{8^2}{4} - (5 - 4^2)$

(*n*) $\dfrac{2 \cdot 6^2 \div 9 + 8}{4 - 8}$

(*o*) $\dfrac{3(2 - 4^2 \cdot 3 + 6)}{7^2 - (4 - 7)^2} + 7$

(*p*) $2^3 - 2[(-3)^2 - 2^3] + 3[3(-4) - 6 \div 2]$

1.20 Evaluate using a calculator:

(*a*) $8 * 7 - 6 * 9$

(*b*) $53 + 62 * 86$

(*c*) $(12 - 17 * 21)(13 * 19 - 12 * 20)$

(*d*) $(12 \div 3 * 5 - 4)^2$

(*e*) $(2 + 12)^2 - (2^2 + 12^2)$

(*f*) $\dfrac{3 * 4 + 6}{3 * 3}$

(*g*) $(-23 + 7 * 12)^2 (9 * 13)^3$

(*h*) $45 + 17\left(62 - \dfrac{512}{32}\right)(94 - 68)$

(*i*) $(537 * 453 - 398^2)^2$

1.21 Identify the mathematical symbol indicated by each of the following key words and phrases:

(*a*) quotient
(*b*) more than
(*c*) decreased by
(*d*) times
(*e*) increased by
(*f*) less
(*g*) subtracted from
(*h*) product
(*i*) is

Translate each of the following into a mathematical expression, equation, or inequality and identify its type; first define what the variable(s) represent(s) (if not given).
DO NOT solve the problem

1.22 15 more than a number — use "*n*" for the number.

1.23 Three times one number minus twice another number—use "*x*" for one number and "*y*" for the other number.

1.24 Four times the sum of six and a number—use "*t*" for the number.

1.25 A collection of dimes and quarters has a value of \$6.05. How many quarters are in the collection if there are 15 more dimes than quarters?

1.26 The sum of two consecutive even integers is 142.

1.27 Four times a number is 29 more than the number.

1.28 A collection of dimes and nickels is worth $2.85. If there are 40 coins, how many nickels are in the collection?

1.29 The area of a rectangle is 1462 square inches. The length is 43 inches. What is the width? [This requires the formula for the area of a rectangle: (*length*)(*width*) = *area*.]

1.30 The sum of three consecutive odd integers is 117.

1.31 First-class fare on an airline flight is $165 and coach fare is $135. There were 90 passengers on a flight for which receipts totaled $12,510. How many coach-fare passengers were on the flight?

1.32 The perimeter of a triangle is 71. If the longest side is 29 and another side is twice the length of the shortest side, how long is the shortest side?

1.33 Five more than twice a number is 33.

1.34 The symphony performance made $12,500 by selling 1125 tickets. Balcony tickets sold for $7 and main floor tickets sold for $12. How many balcony tickets were sold?

1.35 Six times one number minus the sum of four and another number.

1.36 The length of a rectangle is 10 more than twice its width. If the perimeter is 320 feet, how wide is the rectangle?

1.37 A collection of dimes and quarters is valued at $7.70. If the collection included 18 quarters, how many dimes were included?

1.38 Three times the sum of two and a number is three less than four times the number.

1.39 The perimeter of a rectangle is twelve times the width. The length is 10 inches. What is the width?

1.40 Fifteen subtracted from the sum of ten and a number.

1.41 Four times the sum of seven and a number is at least six times the number.

1.42 The perimeter of a triangle is 56 inches. The longest side is four times as long as the shortest side; the third side is twice as long as the shortest side. How long is the shortest side?

1.43 In a collection of pennies, nickels, and dimes, there are twice as many dimes as nickels and four more pennies than nickels. The collection is worth $2.90. How many of each coin were in the collection?

1.44 The length of a rectangle is 5 inches longer than the width. The area is less than 84 square inches. What is the width?

1.45 Take the sum of five and one number away from the product of three and another number.

1.46 The area of a triangle is one-half of the height times the base. If the height of a triangle is 5 and the area is 75, what is the length of the base?

1.47 John bought tablets and pencils for his algebra course. Each tablet cost 89 cents and each pencil cost 20 cents. Excluding taxes, the total cost was $8.34. How many tablets did John buy if he bought 15 pencils?

1.48 The shortest side of a triangle is 3 more than half as long as the medium side. The longest side is 7 more than triple the length of the medium side. What is the length of the medium side if the distance around the triangle is 190 feet?

1.49 A store sold 82 items for $1314. Item A sold for $18 each and item B sold for $15 each. Find the number of the type A items sold.

ANSWERS TO SUPPLEMENTARY PROBLEMS

1.1 (*a*) $\{-1, 0, 1, 2, 3, 4\}$ (*b*) $\{0, 2, 4\}$
(*c*) $\{1, 2, 3\}$ (*d*) $\{0, 1, 2, 3, 4\}$

1.2 (*a*) $\{-\sqrt{4}, 2\}$ (*b*) $\left\{-\frac{13}{4}, -\sqrt{4}, -1.787878\ldots, 2, 2.74\right\}$
(*c*) $\varnothing$ (*d*) $\{\sqrt{11}\}$

1.3 (*a*) $\{2\}$ (*b*) $\{1, 2, 4, 5, 7\}$ (*c*) $\varnothing$
(*d*) $\{2, 6\}$ (*e*) $\{0, 1, 2, \ldots\} = W$ (*f*) $\{1, 7\}$
(*g*) $\{\ldots, -2, -1, 0, 1, 2, \ldots\} = J$

1.4 (*a*) Substitution (*b*) Symmetric
(*c*) Transitive or substitution (*d*) Substitution
(*e*) Reflexive (*f*) Transitive or substitution

1.5 (*a*) $4 > -2$ (*b*) $x \leq 0$ (*c*) $4 < s < 7$
(*d*) $-3.5 < -3.4$ (*e*) $10 \geq 10$

1.6 (*a*) $>$ or $\geq$ (*b*) $<$ or $\leq$ (*c*) $>$ or $\geq$
(*d*) $\geq$ or $\leq$ or $=$ (*e*) $<$ or $\leq$

1.7 (*a*) See Figure 1.18. (*b*) See Figure 1.19. (*c*) See Figure 1.20. (*d*) See Figure 1.21.

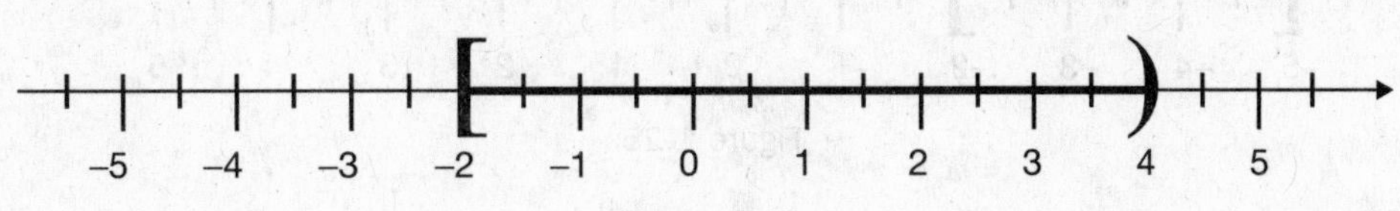

Figure 1.18

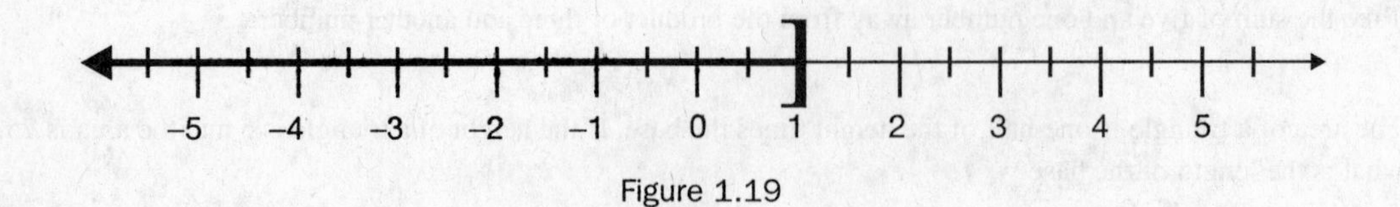

Figure 1.19

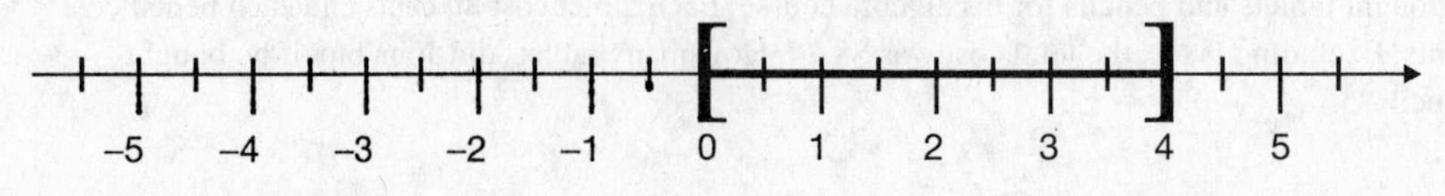

Figure 1.20

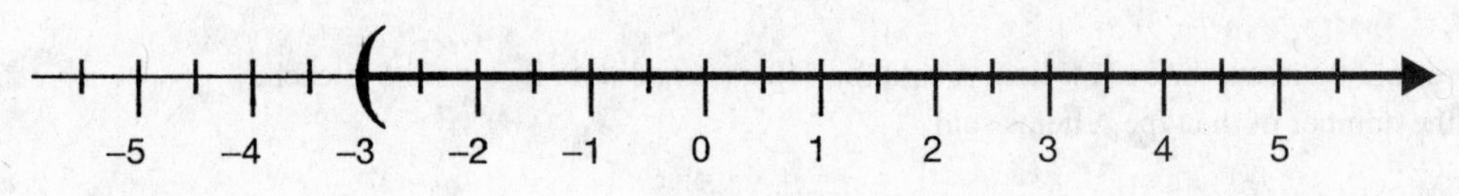

Figure 1.21

1.8 (*a*) $\{x \mid -2 \le x < 4\}$ (*b*) $\{x \mid x \le 1\}$
(*c*) $\{x \mid 0 \le x \le 4\}$ (*d*) $\{x \mid x > -3\}$

1.9 (*a*) See Figure 1.22. (*b*) See Figure 1.23. (*c*) See Figure 1.24. (*d*) See Figure 1.25.

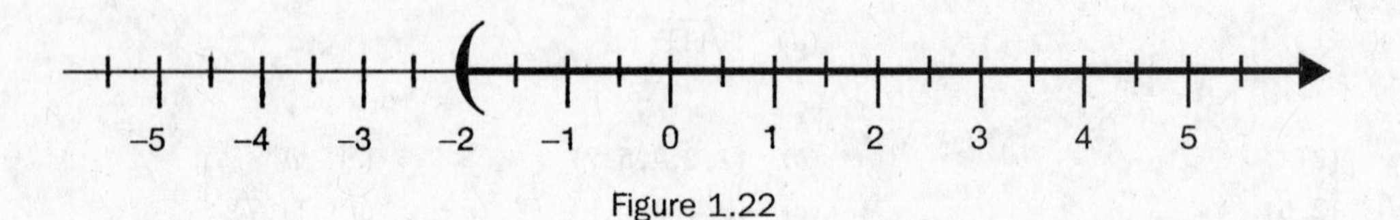

Figure 1.22

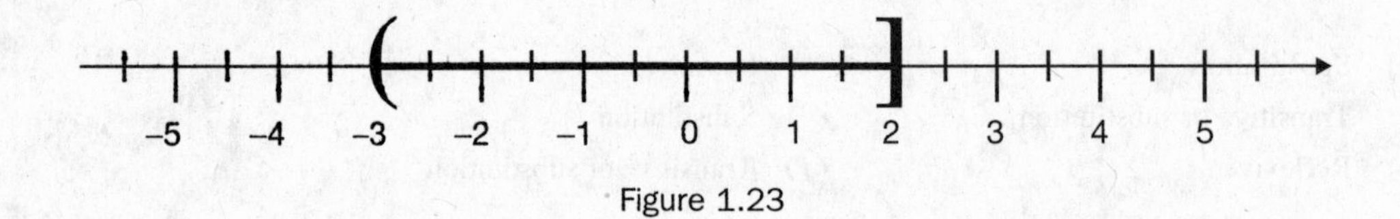

Figure 1.23

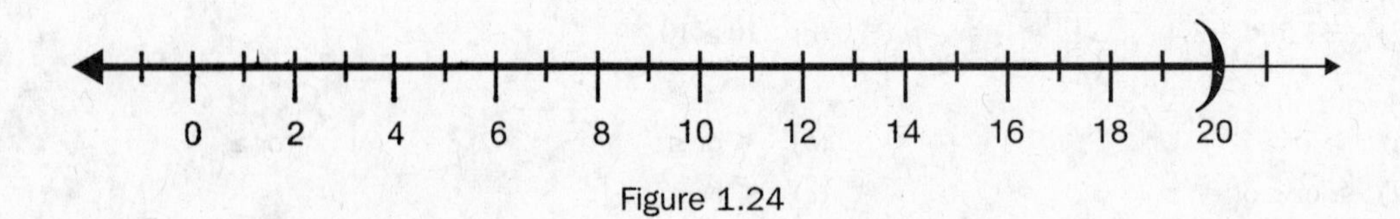

Figure 1.24

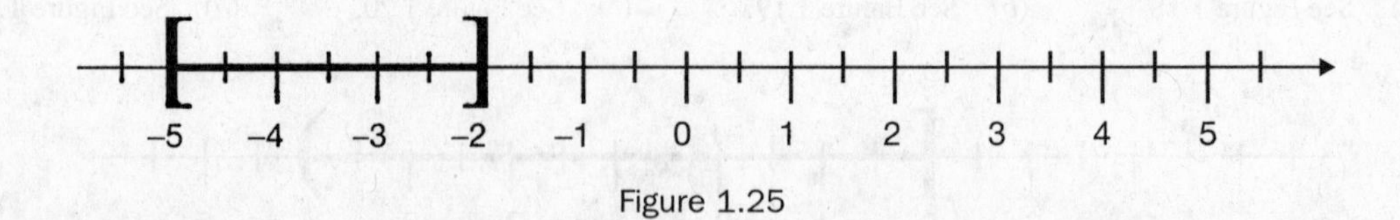

Figure 1.25

1.10 (*a*) Commutative property of multiplication (*b*) Additive inverse property
(*c*) Distributive property (*d*) Additive identity property
(*e*) Multiplicative identity property (*f*) Multiplicative inverse property
(*g*) Associative property of addition (*h*) Identity property of multiplication
(*i*) Commutative property of multiplication (*j*) Commutative property of addition

1.11 (*a*) $3w$ (*b*) s (*c*) 5
(*d*) $s + 2t$ (*e*) $1x$ (*f*) 3
(*g*) $-y$ (*h*) $2s$ (*i*) 3

1.12 (*a*) 11 (*b*) 3 (*c*) 7
(*d*) 11 (*e*) -38 (*f*) -5

1.13 (*a*) 6 (*b*) 16 (*c*) -22
(*d*) -7 (*e*) -7 (*f*) 10

1.14 (*a*) 13 (*b*) 5 (*c*) -5
(*d*) -13 (*e*) -3 (*f*) -13
(*g*) -6 (*h*) -3 (*i*) -11

1.15 (*a*) -35 (*b*) 36 (*c*) 0
(*d*) 48 (*e*) 0 (*f*) -144

1.16 (*a*) -6 (*b*) -6 (*c*) undefined (*d*) 0 (*e*) 5

1.17 (*a*) -1 (*b*) -45 (*c*) 4 (*d*) -31
(*e*) undefined (*f*) 19 (*g*) 63 (*h*) 0
(*i*) 42 (*j*) -3 (*k*) 8 (*l*) 99

1.18 (*a*) $5 \cdot 5 = 25$ (*b*) $(-2)(-2)(-2) = -8$ (*c*) $0 \cdot 0 \cdot 0 \cdot 0 \cdot 0 \cdot 0 = 0$
(*d*) $1 \cdot 1 \cdot 1 \cdot 1 = 1$ (*e*) $3 \cdot 3 \cdot 3 \cdot 3 = 81$

1.19 (*a*) 14 (*b*) -4 (*c*) 21 (*d*) 3 (*e*) 25 (*f*) 36
(*g*) 28 (*h*) 30 (*i*) -1 (*j*) 13 (*k*) 2 (*l*) 20
(*m*) 27 (*n*) -4 (*o*) 4 (*p*) -39

1.20 (*a*) 2 (*b*) 5385 (*c*) -2415
(*d*) 256 (*e*) 48 (*f*) 2
(*g*) 5959601973 (5.95960×10^9 if 6 digits are displayed and scientific notation is used). (*h*) 20377 (*i*) 7200710449 (7.20071×10^9 if 6 digits are displayed and scientific notation is is used).

1.21 (*a*) $\div$ (*b*) $+$ (*c*) $-$ (*d*) $\times$ or $*$ (*e*) $+$
(*f*) $-$ (*g*) $-$ (*h*) $\times$ or $*$ (*i*) $=$

Note: The choice of variable names (when not given) is arbitrary and any may replace those used in these answers. Other answers may be correct, dependent upon algebraic equivalence and representation for the variable. All parentheses shown in these answers are necessary (for the form given).

1.22 Expression; $n + 15$, or $15 + n$.

1.23 Expression; $3x - 2y$ (the order of operations coming from the words just as in an algebraic expression—multiplication takes precedence over subtraction).

1.24 Expression; $4(t + 6)$ note that 4 multiplies the sum (not just the 6).

1.25 Coin problem; let "q" represent the number of quarters in the collection: $25q + 10(q + 15) = 605$.

1.26 Number problem; let "x" be the smaller integer, so "$x + 2$" would be the larger: $x + x + 2 = 142$.

1.27 Expression; let "x" represent the number: $4x = x + 29$.

1.28 Coin problem; let "n" represent the number of nickels in the collection: $5n + 10(40 - n) = 285$.

1.29 Geometry problem; let the width be represented by w: $43w = 1462$.

1.30 Expression; let "n" represent the first integer: $n + n + 2 + n + 4 = 117$.

1.31 Coin problem; let "c" represent the number of coach-fare passengers on the flight: $135c + 165(90 - c) = 12510$.

1.32 Geometry problem; let the shortest side length be represented by s: $s + 2s + 29 = 71$.

1.33 Number problem; let "n" be the number: $2n + 5 = 33$.

1.34 Coin problem; let "x" represent the number of balcony tickets sold, so "$1125 - x$" represents the number of main floor tickets sold: $7x + 12(1125 - x) = 12{,}500$.

1.35 Expression; let "x" be the first number, "y" be the second number: $6x - (y + 4)$.

1.36 Geometry problem; let "w" be the width of the rectangle (in feet): $2w + 2(2w + 10) = 320$.

1.37 Coin problem; let "d" be the number of dimes in the collection: $10d + 25(18) = 770$.

1.38 Number problem; let "x" represent the number: $3(2 + x) = 4x - 3$.

1.39 Geometry problem; let "w" represent the width (in inches): $2(10) + 2w = 12w$.

1.40 Expression; let "n" be the number: $n + 10 - 15$.

1.41 Number problem; let "y" represent the number: $4(y + 7) \geq 6y$.

1.42 Geometry problem; let "s" represent the shortest side (in inches): $4s + 2s + s = 56$.

1.43 Coin problem; let "N" represent the number of nickels, so "$2N$" is the number of dimes and "$N + 4$" is the number of pennies: $1(N + 4) + 5N + 10(2N) = 290$.

1.44 Geometry problem; let "x" represent the width: $(x + 5)\,x < 84$.

1.45 Expression; let "p" be one number and "q" be the other number: $3q - (p + 5)$.

1.46 Geometry problem; let "b" represent the base: $\frac{1}{2}(5b) = 75$.

1.47 Coin problem; let "t" be the number of tablets John bought: $89t + 20(15) = 834$.

1.48 Geometry problem; let "m" be the length of the medium side (in feet):
$\frac{1}{2}m + 3 + m + 3m + 7 = 190$.

1.49 Coin problem; let "a" represent the number of type A items sold:
$18a + 15(82 - a) = 1314$.

Polynomials

2.1 Definitions

An *algebraic expression*, or simply an *expression*, is any meaningful collection of numerals, variables, and signs of operations. Several examples of expressions follow.

$$3s^2 - 4t; 5x^2 - 2x + 8; (a - b)^3; \text{and } \frac{2s^2 - 4t}{3s + t}.$$

In an expression of the form $P + Q + R + S + \ldots$, P, Q, R, and S are called the *terms* of the expression. Terms are separated by addition symbols, but may be separated by subtraction symbols also.

We now illustrate with several examples.

Expression	Number of terms
$4r + 5st$	2
$3x - 5y + 4w$	3
$5(4s - 6t)$	1

The last expression, $5(4s - 6t)$, contains two factors, but only one term. The second factor of the expression contains two terms, however. The first expression, $4r + 5st$, contains two terms. The first term of the expression has two factors, while the second term of the expression contains three factors.

A *coefficient* consists of any factor or group of factors of the product of the remaining factors in a term. In the expression $4r + 5st$ above, 4 is the coefficient of r in the first term while 5 is the coefficient of st in the second term. In addition, $5s$ is the coefficient of t, $5t$ is the coefficient of s, and st is the coefficient of 5 in the second term. Normally the word coefficient refers to the numerical coefficient in a term. In an expression such as $s - t$, the coefficient of s is 1 and the coefficient of t is -1, since $s - t = s + (-t) = 1s + (-1)t$.

A *monomial* is an algebraic expression of the form cx^n, or a product of such expressions, where c is a real number and n is a whole number. Some examples of monomials are

$$5x^2, 2st^3, w, \text{and } 9.$$

A *polynomial* is an algebraic expression whose terms are monomials. A *binomial* is a polynomial that contains two terms, and a *trinomial* is a polynomial that contains three terms. A polynomial contains a finite number of terms, and it may contain more than one variable. Some examples of binomials are

$$4x - 8; x^2 + y^2; 3r^2s - 4; \text{and } (x + y)^3 + 27.$$

Trinomials look like the following:

$$4x^2 - 7x + 9;\ 3r + 7s - 8t;\ 3x^2y + 5yz - 2z^2;\ \text{and } 4(a + b)^2 - 5(a + b) + 1.$$

The following expressions are not polynomials, since each expression contains a quantity with an exponent that is not a whole number.

$$4x^{-2} + 5x + 8;\ \frac{2}{x} + 3y^2 = 2x^{-1} + 3y^2;\ \text{and } 8s^{\frac{1}{2}} + 9s - 9s^2.$$

The *degree* of a monomial is the sum of the exponents of the variables it contains. The degree of a nonzero constant is zero, and the degree of zero is undefined.

The following monomials have the indicated degree: $4x^2$ has degree 2; $-4s^2t$ has degree 3; $7s^2y^3z^4$ has degree 9; and 7 has degree 0.

The degree of a polynomial is the highest degree of any of its terms.

A *polynomial* in one variable is an expression of the form

$$a_nx^n + a_{n-1}x^{n-1} + a_{n-2}x^{n-2} + \cdots + a_2x^2 + a_1x + a_0$$

where $a_n \neq 0$ and each a_i is a real number, n is a nonnegative integer (whole number), and x is a variable. The degree of the polynomial is n.

In the definition above, the expressions to the lower right of the a's are called *subscripts*. In a polynomial that contains many terms, subscripts are used to distinguish one coefficient from another. Note that the subscripts of the a's are the same as the exponents of x in the various terms.

When a polynomial is written with powers of the variable decreasing, it is said to be in *standard form*. The standard form of $8 - 5x^2 + 3x$ is $-5x^2 + 3x + 8$. In this illustration, $a_2 = -5$, $a_1 = 3$, $a_0 = 8$, and $n = 2$.

See solved problems 2.1–2.4.

2.2 Sums and Differences

In Section 1.3 the distributive property was stated as $a(b + c) = ab + ac$. The symmetric property allows us to rewrite the same property in the form $ab + ac = a(b + c)$. We can employ the distributive property in the latter form to simplify certain polynomials.

The following examples illustrate the idea.

EXAMPLE 1.

(*a*) $3x + 5x = (3 + 5)x = 8x$

(*b*) $4t^2 + 9t^2 = (4 + 9)t^2 = 13t^2$

(*c*) $2s + 4s + 8s = (2 + 4 + 8)s = 14s$

(*d*) $3x + x + 8y + 3y = (3 + 1)x + (8 + 3)y = 4x + 11y$

Terms which involve the same powers of the variables are called *like terms*. The process illustrated in the examples above is called *combining like terms*. In each case, the result represents the same real number as the original expression for all real-numbered replacements of the variable or variables. Unlike terms cannot be combined.

Expressions that represent the same real number for all replacements of the variable(s) are called *equivalent expressions*. Hence, equivalent expressions were obtained when we simplified the expressions in the above examples.

If two expressions represent different real numbers for some combination of replacements for the variable(s) involved, the expressions are not equivalent expressions. For example, $-(s + t)$ and $-s + t$ are not equivalent expressions. If we replace s by 4 and t by 3, we find that $-(s + t) = -(4 + 3) = -(7) = -7$, but $-s + t = -4 + 3 = -1$. We have employed what is known as a *counterexample* to show that the two expressions are not equivalent.

In Section 1.4 we defined the difference of a and b as $a - b = a + (-b)$. It is useful to think of the signs in an expression as being signs of the coefficients, and the operation as being addition. We shall illustrate with some problems.

See solved problems 2.5–2.7.

2.3 Products

In Section 1.4 we defined positive integer exponents. Recall that the exponent tells us how many times to use the base as a factor in an exponential expression. For instance, 4^3 means $4 \cdot 4 \cdot 4$. In general, $b^m b^n$ means m factors of b times n factors of b. There are then $m + n$ factors of b, which may be written as b^{m+n}. This result is stated formally as an important law of exponents.

LAW 1: For all positive integer exponents, $b^m b^n = b^{m+n}$.

The commutative and associative laws may be employed along with our first law of exponents to multiply monomials.

See solved problem 2.8.

Let us now consider other expressions that involve exponential factors. Specifically, let's discuss $(b^m)^n$ where m and n are natural numbers. The n means to use b^m as a factor n times and the m tells us to use b as a factor m times. Therefore, we have

$$\begin{aligned}(b^m)^n &= b^m \cdot b^m \cdot b^m \cdot \cdots \cdot b^m \ (n \text{ factors of } b^m)\\ &= b^{m+m+m+\cdots+m} \ (n \text{ terms of } m)\\ &= b^{mn}\end{aligned}$$

The result is merely mn factors of b. We restate this result formally as the second law of exponents.

LAW 2: For all positive integer exponents, $(b^m)^n = b^{mn}$.

Also observe that

$$\begin{aligned}(ab)^n &= (ab)(ab)(ab) \cdots (ab) \ (n \text{ factors of } ab)\\ &= (a \cdot a \cdot a \cdot \cdots \cdot a)(b \cdot b \cdot b \cdot \cdots \cdot b) \ (n \text{ factors of each})\\ &= a^n b^n\end{aligned}$$

We restate this result formally as the third law of exponents.

LAW 3: For all positive integer exponents, $(ab)^n = a^n b^n$.

The second law merely states that if some power of an expression is raised to a power, the result is that expression raised to the product of those powers. Similarly, the third law states that a power of a product of two expressions is the product of the same power of the expressions.

See solved problem 2.9.

We can generalize the distributive property using the associative property. The result is shown below.

$$\begin{aligned}a(b + c + d) &= a[(b + c) + d]\\ &= a(b + c) + ad\\ &= ab + ac + ad\end{aligned}$$

Observe that the factor a is multiplied by each term within parentheses. We can generalize this result when there is any finite number of terms within parentheses.

The same properties can be applied to find products of polynomials containing more than one term.

$$\begin{aligned}(a + b)(c + d) &= (a + b)c + (a + b)d \\ &= ac + bc + ad + bd\end{aligned}$$

The result simply says that every term in the first factor is multiplied by every term in the second factor. The same principle can be applied to products of all polynomials.

See solved problems 2.10–2.11.

SPECIAL PRODUCTS

1. $(a + b)^2 = (a + b)(a + b) = a^2 + ab + ab + b^2 = a^2 + 2ab + b^2$.
2. $(a - b)^2 = (a - b)(a - b) = a^2 - ab - ab + b^2 = a^2 - 2ab + b^2$.
3. $(a + b)(a - b) = a^2 - ab + ab - b^2 = a^2 - b^2$.
4. $(a + b)^3 = a^3 + 3a^2b + 3ab^2 + b^3$.
5. $(a - b)^3 = a^3 - 3a^2b + 3ab^2 - b^3$.

Take time to learn these special products immediately. Learning them now will save us considerable time in the future. We will employ them on a regular basis in subsequent sections when factoring. For your benefit in those sections, be sure to memorize from right to left as well.

It is helpful to state the above relationships in words in order to readily recognize and remember them. Form 1 tells us to square a binomial sum, square the first term, add two times the product of the terms, then add the square of the last term. Form 2, the square of a binomial difference, is stated similarly. State it to yourself now. Form 3 states that the product of the sum and difference of two quantities is the difference of the squares of those quantities. Forms 4 and 5 are the cubes of a binomial sum and difference, respectively. Observe the coefficients and exponents on a and b in successive terms on the right side of each relationship. The coefficients are the same magnitude in both expressions, although their signs differ. The exponents on a decrease by one in successive terms, while the exponents on b increase by one in successive terms. We now illustrate the use of the above forms.

See solved problem 2.12.

We also use the distributive property to find products of polynomials which contain more than two terms. We simply multiply each term of one factor by each of the terms in the other factor. The process is facilitated by employing a vertical form similar to the one used in arithmetic. This format allows us to align like terms vertically, thus facilitating the combining of those terms. We now illustrate the process.

See solved problem 2.13.

2.4 Factoring

In Section 1.3 we stated the distributive properties as

$$a(b + c) = ab + ac \text{ and } (b + c)a = ba + ca.$$

Essentially we are rewriting products as sums in both instances. We now reverse the process by rewriting sums as products of factors. This process is called *factoring*. We will begin with expressions which look like the right sides of the above equations and rewrite them in the form of the left sides.

The first type of factoring we shall consider is that of factoring expressions which contain common monomial factors in their terms. Recall that monomials consist of constants or a product of a constant and one or more variables raised to positive integer powers. In general, we shall consider only integer factors when factoring numbers for the time being.

The first step entails identifying factors which occur in every term of the expression. These common factors will be factored out of each term, thus obtaining a product of factors. We say that the expression is factored completely when there are no common factors remaining in the terms other than the number one. (One is a factor of every expression.) An expression which is factored completely is said to be *prime*.

See solved problem 2.14.

Factoring by Grouping

The concept of a common monomial factor can be extended to common binomial factors. Some polynomials with an even number of terms can frequently be factored by grouping terms with a common factor. For example, the polynomial $ax + bx + 3a + 3b$ or $(ax + bx) + (3a + 3b)$ may be written as $(a + b)x + (a + b)3$ using the distributive law twice; once for the first pair of factors and once for the last pair. Now observe that the two groups contain the common factor $(a + b)$. This common factor can be factored out to obtain $(a + b)(x + 3)$. Thus, the factored form of the original expression has been obtained.

Sometimes it may be helpful to substitute a single letter for a common binomial expression in order to facilitate the factoring process. In the above example, let p represent the polynomial $(a + b)$. It follows that

$$\begin{aligned} ax + bx + 3a + 3b &= (a + b)x + (a + b)3 & \\ &= px + p3 & \text{Substitute } p \text{ for } a + b \\ &= p(x + 3) & \text{Factor} \\ &= (a + b)(x + 3) & \text{Substitute } a + b \text{ for } p \end{aligned}$$

Feel free to employ a similar substitution in order to simplify the factoring of polynomials which may be factored by grouping or by associating certain terms. *Remember that the resulting groups we form must all contain the same common factor.*

See solved problem 2.15.

Factoring $x^2 + px + q$

Recall that $(x + a)(x + b) = x^2 + (a + b)x + ab$. Note that the constant term is the product of a and b, while the coefficient of x is the sum of a and b. Thus, to factor $x^2 + px + q$, we must find two factors of q whose sum is p. The process is relatively simple if q is not large, or if q has a small number of factors. Be aware that if q is positive, then a and b must have like signs. On the other hand, if q is negative, a and b must have unlike signs. Recall that multiplication is commutative; therefore, the order of the factors is immaterial. We now illustrate.

See solved problem 2.16.

Factoring $Ax^2 + Bx + C$ by Trial and Error

We now consider a more complicated form. Observe the product

$$(ax + b)(cx + d) = acx^2 + adx + bcx + bd = acx^2 + (ad + bc)x + bd.$$

The expression is complicated by the coefficients of the x terms in the binomials. Now think of $acx^2 + (ad + bc)x + bc$ as $Ax^2 + Bx + C$. If we wish to factor $Ax^2 + Bx + C$, we observe that the factors (if it is factorable) have the form $(ax + b)(cx + d)$. Furthermore $A = ac$, $B = ad + bc$, and $C = bd$. Hence, the process entails the determination of factors of A and C such that the sum of certain combinations of products of those factors is B. The process is easier than it may appear at this point, although we employ trial and error to obtain the result. Study the following problems carefully.

See solved problem 2.17.

Factoring $Ax^2 + Bx + C$ by Grouping

The trial-and-error method can be quite laborious; particularly when A and C have numerous factors. We now discuss an alternative method which employs factoring by grouping. We summarize the method now.

Factoring $Ax^2 + Bx + C$ by Grouping

1. Find the product AC.
2. Find two factors of AC which have a sum of B.
3. Rewrite the middle term, Bx, as a sum of terms whose coefficients are the factors found in step 2.
4. Factor by grouping.

Read the illustrated problems carefully.

See solved problem 2.18.

Factoring Using Special Products

We introduced several special products in Section 2.3. We shall now use them to factor expressions that fit those special product forms. They are the following:

1. $a^2 + 2ab + b^2 = (a + b)^2$
2. $a^2 - 2ab + b^2 = (a - b)^2$
3. $a^2 - b^2 = (a + b)(a - b)$

Observe that the first and last terms in each form are perfect square terms. When factoring trinomials, we should always ask ourselves if the first and last terms in our expression are perfect squares. If they are perfect square terms, we then look at the middle term. Ask if it is two times the product of the square roots of the first and last terms. If that is true, look at the signs of the coefficients to determine if they fit any of our special product forms. If so, we apply the appropriate pattern to write the factors. If the expression in question does not fit a particular pattern, we attempt to factor it using the methods we described previously in this section.

If the expression under consideration is a binomial, we ask ourselves if both terms are perfect squares. If both terms are perfect squares, are they subtracted? If they are, form 3 gives us the appropriate factors. If the terms are added, the expression is prime. That is, the sum of two perfect square terms does not factor.

It is now appropriate to illustrate the aforementioned concepts.

See solved problem 2.19.

There are two additional special product forms that we have introduced. These additional forms involve the third power, or cubes, of expressions. We should also be able to recognize and apply these relationships. They are the following:

4. $a^3 + b^3 = (a + b)(a^2 - ab + b^2)$
5. $a^3 - b^3 = (a - b)(a^2 + ab + b^2)$

The above forms are referred to as "the sum and difference of two cubes." In words, form 4 tells us that the first factor of the sum of two cubes is the sum of the cube roots of the terms. The terms in the second factor are related to the terms of the first factor (a and b) in the following manner. The first and last terms are the square of a and b, respectively. The middle term in the second factor is the opposite of the product of a and b. Form 5 can be described similarly.

See solved problem 2.20.

There are some general guidelines we should adhere to when factoring expressions in general. They are the following:

1. Identify and remove, that is, factor out, all common monomial factors first.
2. If the expression contains two terms, it may be the difference of two squares, or a sum or difference of cube terms. If so, apply the appropriate pattern. Recall that the sum of two squares is prime.
3. If the expression contains three terms, determine if two of those terms are perfect squares. If that is the case, the expression may be a perfect square binomial. Otherwise, it may be a general form.
4. If the expression contains four terms, determine if two of those terms are perfect cubes. If that is the case, the expression may be a perfect cube binomial. Otherwise, it may factor by grouping.

We wish to remind you that factoring is an extremely important skill that will be used in many situations. Learn the patterns we have introduced, and practice on supplementary problem 2.17. Try to determine the form each problem fits in order to determine its factors.

2.5 Division

Recall that b^n means that b is used as a factor n times when n is a positive integer. There are some additional laws of exponents which we have not yet introduced. Consider the expression b^m/b^n.

If the exponents m and n are positive integers, the numerator represents m factors of b, and the denominator represents n factors of b. Therefore, if $b \neq 0$ and $m > n$,

$$\begin{aligned}
\frac{b^m}{b^n} &= \frac{\overbrace{b \cdot b \cdot b \cdot \cdots \cdot b}^{m \text{ factors}}}{\underbrace{b \cdot b \cdot b \cdot \cdots \cdot b}_{n \text{ factors}}} \\
&= \frac{\left(\overbrace{b \cdot b \cdot b \cdot \cdots \cdot b}^{n \text{ factors}}\right)\left(\overbrace{b \cdot b \cdot b \cdot \cdots \cdot b}^{m-n \text{ factors}}\right)}{\underbrace{b \cdot b \cdot b \cdot \cdots \cdot b}_{n \text{ factors}}} \\
&= \left(\underbrace{1 \cdot 1 \cdot 1 \cdot \cdots \cdot 1}_{n \text{ factors}}\right)\left(\underbrace{b \cdot b \cdot b \cdot \cdots \cdot b}_{m-n \text{ factors}}\right) \\
&= b^{m-n}
\end{aligned}$$

The last lines follow since $b/b = 1$ and $1 \cdot b = b$

Similarly, if $n > m$,

$$\begin{aligned}
\frac{b^m}{b^n} = \frac{\overbrace{b \cdot b \cdot b \cdot \cdots \cdot b}^{m \text{ factors}}}{\underbrace{b \cdot b \cdot b \cdot \cdots \cdot b}_{n \text{ factors}}} &= \frac{\overbrace{b \cdot b \cdot b \cdot \cdots \cdot b}^{m \text{ factors}}}{\left(\underbrace{b \cdot b \cdot b \cdot \cdots \cdot b}_{m \text{ factors}}\right)\left(\underbrace{b \cdot b \cdot b \cdot \cdots \cdot b}_{n-m \text{ factors}}\right)} \\
&= \frac{1}{\left(\underbrace{b \cdot b \cdot b \cdot \cdots \cdot b}_{n-m \text{ factors}}\right)} = \frac{1}{b^{n-m}}
\end{aligned}$$

We are merely dividing out common factors that occur in both numerator and denominator in the above illustrations. The result in both instances is simply b raised to the positive difference of the powers m and n written in the numerator or denominator, whichever results in a positive exponent on b.

We'll now consider $(a/b)^n$ where $b \neq 0$ and n is a positive integer. We observe that

$$\left(\frac{a}{b}\right)^n = \overbrace{\left(\frac{a}{b}\right)\left(\frac{a}{b}\right)\left(\frac{a}{b}\right)\cdots\left(\frac{a}{b}\right)}^{n \text{ factors}} = \frac{\overbrace{a\cdot a\cdot a\cdot\cdots\cdot a}^{n \text{ factors}}}{\underbrace{b\cdot b\cdot b\cdot\cdots\cdot b}_{n \text{ factors}}} = \frac{a^n}{b^n}.$$

We have shown that if we are raising a quotient to some power n, the result is the numerator raised to that power divided by the denominator raised to the same power.

The results of all of the laws of exponents we have discussed are now restated for your reference. They are employed in expressions involving positive, integer exponents.

Laws of Exponents

1. $b^m \cdot b^n = b^{m+n}$
2. $(b^m)^n = b^{m \cdot n}$
3. $(ab)^n = a^n b^n$

4a. $\dfrac{b^m}{b^n} = b^{m-n}$ if $b \neq 0$ and $m > n$

4b. $\dfrac{b^m}{b^n} = \dfrac{1}{b^{n-m}}$ if $b \neq 0$ and $n > m$

5. $\left(\dfrac{a}{b}\right)^n = \dfrac{a^n}{b^n}$ if $b \neq 0$.

See solved problem 2.21.

The procedure for dividing a polynomial of more than one term by a monomial is a consequence of the distributive property. Recall that division by a denominator, d, is accomplished by multiplying by its multiplicative inverse (reciprocal) $1/d$. Hence, to divide $a + b$ by d, where $d \neq 0$, we write

$$\frac{a+b}{d} = (a+b)\cdot\frac{1}{d}$$

$$= a\cdot\frac{1}{d} + b\cdot\frac{1}{d}$$

$$= \frac{a}{d} + \frac{b}{d}$$

In words, the above sequence tells us that to divide a binomial sum by a monomial, we must divide each term in the binomial by the denominator and add the results. We can also view the procedure as the reverse of adding fractions that have the same denominator. An identical process may be employed regardless of the number of terms in the numerator.

See solved problem 2.22.

Recall that we check division by multiplying the quotient by the divisor and add the remainder. The result should equal the dividend. That is, (quotient)(divisor) + remainder = dividend. If we divide both sides of this equation by the divisor and apply the symmetric property of equality, we obtain the equivalent equation.

$$\frac{\text{dividend}}{\text{divisor}} = \text{quotient} + \frac{\text{remainder}}{\text{divisor}}$$

If we wish to divide a polynomial by a divisor that contains more than one term, we employ a process that is analogous to long division in arithmetic. We illustrate by considering the division of 158 by 12.

$$\begin{array}{rll} & 13 & \leftarrow \text{quotient} \\ \text{divisor} \rightarrow 12 & \overline{)158} & \leftarrow \text{dividend} \\ & \underline{12} & \leftarrow \text{subtract} \\ & 38 & \leftarrow \text{new dividend} \\ & \underline{36} & \leftarrow \text{subtract} \\ & 2 & \leftarrow \text{remainder (less than the divisor)} \end{array}$$

Dividing 158 by 12 = 10 + 2 is analogous to dividing $x^2 + 5x + 8$ by $x + 2$, where x represents 10. Compare the steps below with those above as we proceed through the problem. The polynomials must first be written in descending powers.

$$x+2\overline{)\,x^2+5x+8}$$

Divide the first term of the dividend by the first term of the divisor to obtain the first term of the quotient.

$$\begin{array}{rl} & x \\ x+2 & \overline{)x^2+5x+8} \end{array}$$

Now multiply the first term of the quotient by the entire divisor and subtract the product from the dividend. A good method to use when subtracting is to add the opposite; this technique is used in the far right display below.

$$\begin{array}{rl} & x \\ x+2 & \overline{)x^2+5x+8} \\ & \underline{x^2+2x} \\ & \quad 3x+8 \end{array} \quad \text{or} \quad \begin{array}{rl} & x \\ x+2 & \overline{)x^2+5x+8} \\ & \underline{-x^2-2x} \\ & \quad 3x+8 \end{array}$$

Next use the remainder as the new dividend and repeat the above procedure until the remainder is of <u>lower degree</u> than the divisor.

$$\begin{array}{rll} & x+3 & \leftarrow \text{quotient} \\ \text{divisor} \rightarrow x+2 & \overline{)x^2+5x+8} & \leftarrow \text{dividend} \\ & \underline{x^2+2x} & \leftarrow \text{subtract} \\ & 3x+8 & \leftarrow \text{new dividend} \\ & \underline{3x+6} & \leftarrow \text{subtract} \\ & 2 & \leftarrow \text{remainder} \end{array}$$

or, equivalently,

$$\begin{array}{rll} & x+3 & \leftarrow \text{quotient} \\ \text{divisor} \rightarrow x+2 & \overline{)x^2+5x+8} & \leftarrow \text{dividend} \\ & \underline{-x^2-2x} & \leftarrow \text{add} \\ & 3x+8 & \leftarrow \text{new dividend} \\ & \underline{-3x-6} & \leftarrow \text{add} \\ & 2 & \leftarrow \text{remainder} \end{array}$$

The above process may be used to divide any polynomial by another polynomial whose degree is less than or equal to that of the dividend. The procedure is summarized below.

1. Write the terms in the dividend and the divisor in descending powers. If a power of the dividend is absent, write zero as its coefficient.
2. Divide the first term of the dividend by the first term of the divisor to obtain the first term of the quotient.
3. Multiply the first term of the quotient by the entire divisor and subtract the product from (or add the opposite to) the dividend.
4. Use the difference as the new dividend and repeat the procedure until the difference is of lower degree than the divisor. The last difference obtained is called the remainder. Use only the most recent term of the quotient when multiplying by the divisor in each case.

See solved problem 2.23.

SYNTHETIC DIVISION

The procedure for dividing a polynomial by a binomial of the form $x - c$, where c is a constant, can be accomplished very efficiently by a method called *synthetic division*. To illustrate, return to the example discussed previously in which we divided $x^2 + 5x + 8$ by $x + 2$.

The result is shown below for reference.

$$
\begin{array}{rr@{\,}ll}
 & & x+3 & \leftarrow \text{quotient} \\
\text{divisor} \rightarrow & x+2 & \overline{)\,x^2+5x+8} & \leftarrow \text{dividend} \\
 & & \underline{x^2+2x} & \leftarrow \text{subtract} \\
 & & \quad 3x+8 & \leftarrow \text{new dividend} \\
 & & \quad \underline{3x+6} & \leftarrow \text{subtract} \\
 & & \qquad 2 & \leftarrow \text{remainder}
\end{array}
$$

All essential data are retained if we omit the variables, since the position of the term indicates the power of the term.

$$
\begin{array}{r|rrr}
 & 1 & 3 & \\ \hline
2) & 1 & 5 & 8 \\
 & 1 & 2 & \\ \hline
 & & 3 & 8 \\
 & & 3 & 6 \\ \hline
 & & & 2
\end{array}
$$

Now eliminate the coefficients which are duplicates of those directly above them. The array is condensed to

$$
\begin{array}{r|rrr}
 & 1 & 3 & \\ \hline
2) & 1 & 5 & 8 \\
 & & 2 & 6 \\ \hline
 & & 3 & 2
\end{array}
$$

Next, eliminate the three in the top row (quotient row), since it is a duplicate of the three in the bottom row. The leading coefficient in the top row could be written in the bottom row to provide additional vertical condensing. We now have

$$
\begin{array}{r|rrr}
2 & 1 & 5 & 8 \\
 & & 2 & 6 \\ \hline
 & 1 & 3 & \multicolumn{1}{|r}{2}
\end{array}
$$

Recall that subtraction is best accomplished by adding the opposite. Our subtraction step becomes an addition step if we change the sign of the divisor. The signs of the coefficients in the second row change also. The usual array then becomes

$$\begin{array}{r|rrr} -2 & 1 & 5 & 8 \\ & & -2 & -6 \\ \hline & 1 & 3 & 2 \end{array}$$

Note that the numbers in the middle row are the product of the divisor and the number in the bottom row of the previous column. The coefficients of the quotient are in the bottom row. The last number in the bottom row is the remainder. The usual form of the solution is written

$$\frac{x^2 + 5x + 8}{x + 2} = x + 3 + \frac{2}{x + 2}.$$

Another example is given following the summary of the procedure below.

To employ synthetic division to divide a polynomial by a divisor of the form $x - c$, where c is a constant:

1. Form the top row by writing the coefficients of the dividend. The dividend must be written in descending powers and 0 entered as the coefficient of all absent terms. Write the value of c to the left of these coefficients. (In the previous example, c was -2 since $x + 2$ is the same as $x - (-2)$.)
2. Bring down the first dividend entry as the first coefficient of the quotient in the bottom row.
3. Multiply this coefficient by c. Place the result in the second row beneath the next coefficient of the dividend. Now add and place the sum in the bottom row.
4. Repeat the procedure in step 3 until all entries in row one have been used.
5. The numbers in the bottom row are, from left to right, the coefficients of the quotient in descending powers. The quotient has degree one less than the degree of the dividend. The last number in the bottom row is the remainder.

See solved problem 2.24.

Synthetic division is extremely useful when factoring and finding roots of polynomials. These are topics that are addressed in detail in college algebra.

SOLVED PROBLEMS

2.1 State the degree of each of the following.

(*a*) $8a^2b$ (*b*) $x - 8$ (*c*) 10 (*d*) $5x^2 - 7x + 3$ (*e*) $5s^3t + 2s^2t^3 - 6$

(*a*) 3 (*b*) 1 (*c*) 0 (*d*) 2 (*e*) 5

2.2 Determine the number of terms in each of the following and state the degree of each polynomial.

(*a*) $7x^2 + 4x - 9$ (*b*) $3s^2 + 4t^2$ (*c*) $-7y^2z + 2yz^3$
(*d*) $(4s^3 - t)^0$ (*e*) 100 (*f*) $-3s + 4t + 2u - 7v + 11w$

(*a*) 3 terms; degree 2 (*b*) 2 terms; degree 2 (*c*) 2 terms; degree 4
(*d*) 1 term; degree 0 (*e*) 100 has 1 term; degree 0 (*f*) 5 terms; degree 1

See supplementary problem 2.1.

2.3 Identify the part which corresponds to the expressions in solved problem 2.2 that represent monomials, binomials, or trinomials.

Monomials: parts (*d*) and (*e*); binomials: parts (*b*) and (*c*); Trinomials: part (*a*). Note that part (*f*) represents none of the requested forms.

Refer to supplementary problem 2.2.

2.4 Write each of the following in standard form, and state the degree of each.

(*a*) $5s - 4 + 8s^2$

$8s^2 + 5s - 4$, degree 2;

(*b*) $x^4 + 7x^7 - x^2 + 2$

$7x^7 + x^4 - x^2 + 2$, degree 7;

(*c*) $2 + 8t^3 - 7t + 5t^2$

$8t^3 + 5t^2 - 7t + 2$, degree 3;

(*d*) $y - 7$

$y - 7$ has degree 1.

See supplementary problem 2.3 for similar exercises.

2.5 Perform the indicated operations.

(*a*) $4x - 2x + 7x$

$4x - 2x + 7x = 4x + (-2x) + 7x$	Definition of subtraction
$= (4 - 2 + 7)\,x$	Distributive property
$= 9x$	Number fact

(*b*) $5x - 8x - 3x^2 + 7x^2$

$5x - 8x - 3x^2 + 7x^2 = 5x + (-8x) + (-3x^2) + 7x^2$	Definition of subtraction
$= (5 - 8)x + (-3 + 7)x^2$	Distributive property
$= -3x + 4x^2$	Number fact
$= 4x^2 - 3x$	Commutative property

(*c*) $7 - 3t^2 + 4t - 4 + 5t^2$

$7 - 3t^2 + 4t - 4 + 5t^2 = 7 + (-3t^2) + 4t + (-4) + 5t^2$	Definition of subtraction
$= -3t^2 + 5t^2 + 4t + 7 + (-4)$	Commutative property
$= (-3 + 5)t^2 + 4t + 7 + (-4)$	Distributive property
$= 2t^2 + 4t + 3$	Number fact

Note that we wrote the results as polynomials in standard form. Ordinarily one should follow that practice unless the coefficient of the leading term is negative. In the latter case we often write the polynomial in the order of increasing powers of the variable. We find that negative signs are less likely to be overlooked when they are not associated with the leading coefficient.

An expression of the form $a + (b + c)$ may be written as $a + b + c$. However, we must be more careful when an expression of the form $a - (b + c)$ is encountered. Follow the steps below.

$a - (b + c) = a + [-(b + c)]$	Definition of subtraction
$= a + [-1(b + c)]$	$-q = -1q$
$= a + [-b - c]$	Distributive property
$= a - b - c$	Definition of subtraction

Hence, an expression in parentheses preceded by a negative sign is written equivalently without parentheses by replacing each term within parentheses by its additive inverse (negative or opposite). We illustrate in solved problem 2.6 below.

2.6 Perform the indicated operations.

(*a*) $10x - (4x + 8)$

$$\begin{aligned} 10x - (4x + 8) &= 10x - 4x - 8 && a - (b + c) = a - b - c \\ &= (10 - 4)x - 8 && \text{Distributive property} \\ &= 6x - 8 && \text{Number fact} \end{aligned}$$

(*b*) $(6y^2 + 2y - 7) - (4y - 10)$

$$\begin{aligned} (6y^2 + 2y - 7) - (4y - 10) &= 6y^2 + 2y - 7 - 4y + 10 && a - (b + c) = a - b - c \\ &= 6y^2 + 2y - 4y - 7 + 10 && \text{Commutative property} \\ &= 6y^2 + (2 - 4)y - 7 + 10 && \text{Distributive property} \\ &= 6y^2 - 2y + 3 && \text{Number fact} \end{aligned}$$

It is cumbersome and time consuming to include the details of each of the steps involved in the above process. As we acquire experience, we combine steps in order to conserve time and space. This combination of steps is often referred to as "*combining like terms*."

When an expression contains more than one set of grouping symbols, we begin by removing the innermost symbols and work outward. We illustrate once again.

2.7 Perform the indicated operations.

(*a*) $9x - [5 - (5x + 4)]$

$$\begin{aligned} 9x - [5 - (5x + 4)] &= 9x - [5 - 5x - 4] && a - (b + c) = a - b - c \\ &= 9x - [1 - 5x] && \text{Combine like terms} \\ &= 9x - 1 + 5x && a - (b + c) = a - b - c \\ &= 14x - 1 && \text{Combine like terms} \end{aligned}$$

(*b*) $3x^2 - [8x + (x^2 - 9)]$

$$\begin{aligned} 3x^2 - [8x + (x^2 - 9)] &= 3x^2 - [8x + x^2 - 9] && \text{Distributive property} \\ &= 3x^2 - 8x - x^2 + 9 && a - (b + c) = a - b - c \\ &= 2x^2 - 8x + 9 && \text{Combine like terms} \end{aligned}$$

(*c*) $(4a^2 - 3ab + 2b^2) + [(2a^2 - 4b^2) - ab]$

$$\begin{aligned} (4a^2 - 3ab + 2b^2) + [(2a^2 - 4b^2) - ab] &= 4a^2 - 3ab + 2b^2 + 2a^2 - 4b^2 - ab \\ & && \text{Distributive property} \\ &= 6a^2 - 4ab - 2b^2 && \text{Combine like terms} \end{aligned}$$

Note that like terms were combined within the bracket grouping prior to the removal of the brackets in problem 2.7(*a*). One should combine like terms whenever possible.

The process of rewriting polynomials by combining like terms is called *simplifying,* since the result is a polynomial which contains fewer terms than the original. It is simpler than the original. In general, to simplify means to perform the indicated operations when possible. The simple form of a polynomial is one which contains no like terms or grouping symbols. Thus, the results we obtained in solved problems 2.5 through 2.7 above are the simple form of the original polynomials.

See supplementary problems 2.4 through 2.6 for similar exercises.

2.8 Multiply.

(*a*) x^3x^2

$x^3x^2 = x^{3+2} = x^5$

(*b*) t^4t^2t

$t^4t^2t = t^{4+2+1} = t^7$

(c) $(2x^2)(4x)$

$(2x^2)\,(4x) = (2 \cdot 4)(x^2 \cdot x) = 8x^3$

(*d*) $(-3s^2t)(5st^5)$

$(-3s^2t)(5st^5) = (-3 \cdot 5)(s^2 \cdot s)(t \cdot t^5) = -15s^3t^6$

2.9 Multiply.

(*a*) $(b^3)^4$

$(b^3)^4 = b^{3 \cdot 4} = b^{12}$ Apply law 2

(*b*) $(x^2)^8$

$(x^2)^8 = x^{2 \cdot 8} = x^{16}$ Apply law 2

(c) $(st)^3$

$(st)^3 = s^3t^3$ Apply law 3

(*d*) $(3p)^2$

$(3p)^2 = 3^2p^2 = 9p^2$ Apply law 3

(*e*) $(4x^5y)^3$

$(4x^5y)^3 = 4^3(x^5)^3y^3$ Apply law 3

$= 64x^{15}y^3$ Apply law 2

See supplementary problem 2.7 for practice of the above concepts.

2.10 Multiply.

(*a*) $4(2x + y - 3z)$

$4(2x + y - 3z) = 4(2x) + 4y + 4(-3z) = 8x + 4y - 12z$

(*b*) $-2r^2s\,(3rs^2 - 5r^2s + r^3)$

$-2r^2s\,(3rs^2 - 5r^2s + r^3) = (-2r^2s)\,(3rs^2) + (-2r^2s)\,(-5r^2s) + (-2r^2s)r^3$

$= -6r^3s^3 + 10r^4s^2 - 2r^5s$

(c) $(2x + y)(x - y)$

$(2x + y)(x - y) = 2x(x - y) + y(x - y) = 2x \cdot x - 2x \cdot y + y \cdot x - y \cdot y = 2x^2 - xy - y^2$

(*d*) $(x + 3y)(2x + y)$

$(x + 3y)(2x + y) = x(2x + y) + 3y(2x + y) = x \cdot 2x + x \cdot y + 3y \cdot 2x + 3y \cdot y$

$= 2x^2 + xy + 6xy + 3y^2 = 2x^2 + 7xy + 3y^2$

Look at the expressions to the right of the second equal sign of the solution in problem 2.10, parts (*c*) and (*d*) above. In both cases the first terms represent the product of the **F**irst terms of the factors; the second terms are the product of the **O**utside terms; the third terms are the product of the **I**nside terms; and the last terms are the product of the **L**ast terms. We refer to the above process as the *FOIL* method of finding the product. The letters in the word *FOIL* remind us of which terms we must multiply when finding the product of two binomials. This technique is applicable only when the factors are binomials. We must remember, however, that we are simply applying a generalized form of the distributive property. See additional problems below.

2.11 Multiply.

(*a*) $(x - 5)(x - 2)$

$(x - 5)(x - 2) = x^2 - 2x - 5x + 10 = x^2 - 7x + 10$

(*b*) $(w + 8)(w - 3)$

$(w + 8)(w - 3) = w^2 - 3w + 8w - 24 = w^2 + 5w - 24$

(*c*) $(4s + 2t)(5s + t)$

$(4s + 2t)(5s + t) = 20s^2 + 4st + 10st + 2t^2 = 20s^2 + 14st + 2t^2$

Observe that the FOIL method normally positions like terms in juxtaposition in the product. This arrangement facilitates combining like terms. There are several special products which we encounter frequently. We should learn them and be able to recognize them when they occur. Think of them as patterns for which we should watch.

2.12 Employ the special product forms to find the following products.

(*a*) $(x + 4)^2$

It is form 1. Therefore, $(x + 4)^2 = x^2 + 2 \cdot 4x + 4^2 = x^2 + 8x + 16$.

(*b*) $(2x + y)^2$

It is form 1. Therefore, $(2x + y)^2 = (2x)^2 + 2(2x)y + y^2 = 4x^2 + 4xy + y^2$.

(*c*) $(x - 5)^2$

It is form 2. Therefore, $(x - 5)^2 = x^2 - 2 \cdot 5x + (-5)^2 = x^2 - 10x + 25$.

(*d*) $(s - 3t)^2$

It is form 2. Therefore, $(s - 3t)^2 = s^2 - 2s(3t) + (-3t)^2 = s^2 - 6st + 9t^2$.

(*e*) $(x + 4)(x - 4)$

It is form 3. Therefore, $(x + 4)(x - 4) = x^2 - 4^2 = x^2 - 16$.

(*f*) $(4x + y)(4x - y)$

It is form 3. Therefore, $(4x + y)(4x - y) = (4x)^2 - y^2 = 16x^2 - y^2$.

(*g*) $(a + 4)^3$

It is form 4. Therefore, $(a + 4)^3 = a^3 + 3a^2(4) + 3a(4^2) + 4^3 = a^3 + 12a^2 + 48a + 64$.

See supplementary problem 2.8 for similar problems.

2.13 Use the vertical format to find the following products. The result appears in the bottom line.

(*a*) $(x^2 - 3x + 1)(2x - 5)$

$$
\begin{array}{r}
x^2 - 3x + 1 \\
2x - 5 \\
\hline
2x^3 - 6x^2 + 2x \phantom{{}-5} \\
- 5x^2 + 15x - 5 \\
\hline
2x^3 - 11x^2 + 17x - 5
\end{array}
$$

(*b*) $(t^3 + 4t - 3)(t^2 - 7)$

$$\begin{array}{r} t^3 + 4t - 3 \\ t^2 - 7 \\ \hline t^5 + 4t^3 - 3t^2 \qquad\qquad \\ -7t^3 \qquad - 28t + 21 \\ \hline t^5 - 3t^3 - 3t^2 - 28t + 21 \end{array}$$

Do supplementary problem 2.9 for practice.

2.14 Factor completely.

(*a*) $4s + 4t$

$4s + 4t = 4(s + t)$

(*b*) $5x^2y - 50z$

$5x^2y - 50z = 5(x^2y - 10z)$

(*c*) $6x^2 + 8x$

$6x^2 + 8x = 2x(3x + 4)$

(*d*) $-27s^3 - 6s$

$-27s^3 - 6s = -3s(9s^2 + 2)$

(*e*) $8x^3y^2 + 2x^2y^2 - 6x^2y$

$8x^3y^2 + 2x^2y^2 - 6x^2y = 2x^2y(4xy + y - 3)$

(*f*) $-r^2s^2t - r^2st^2 - rs^2t^2$

$-r^2s^2t - r^2st^2 - rs^2t^2 = -rst(rs + rt + st)$

Refer to supplementary problem 2.10 for similar problems.

2.15 Factor completely.

(*a*) $xs + xt + 3s + 3t$

$xs + xt + 3s + 3t = (xs + xt) + (3s + 3t)$

$= x(s + t) + 3(s + t)$

$= xp + 3p$ — Replace $s + t$ by p

$= (x + 3)p$ — Factor

$= (x + 3)(s + t)$ — Replace p by $s + t$

(*b*) $2x^2 + 3x + 4x + 6$

$2x^2 + 3x + 4x + 6 = (2x^2 + 3x) + (4x + 6)$

$= x(2x + 3) + 2(2x + 3)$ — Factor

$= xp + 2p = (x + 2)p$ — Replace $2x + 3$ by p and factor

$= (x + 2)(2x + 3)$ — Replace p by $2x + 3$

(*c*) $10x^2 + 2xy - 25x - 5y$

$$10x^2 + 2xy - 25x - 5y = (10x^2 + 2xy) - (25x + 5y)$$
$$= 2x(5x + y) - 5(5x + y) = 2xp - 5p$$

Factor and replace $5x + y$ by p

$$= (2x - 5)p = (2x - 5)(5x + y)$$

Factor and replace p by $5x + y$

(*d*) $(2x + 1)(y - 1) + 4(y - 1)$

$$(2x + 1)(y - 1) + 4(y - 1) = (y - 1)[(2x + 1) + 4]$$
$$= (y - 1)(2x + 5)$$

(*e*) $3a(a + 1)^2 + (a + 1)^2$

$$3a(a + 1)^2 + (a + 1)^2 = (a + 1)^2(3a + 1)$$

Do supplementary problem 2.11 for additional practice of the above concepts.

2.16 Factor.

(*a*) $x^2 + 7x + 10$.

Since $+10$ is positive, its factors must have like signs. We choose positive factors of 10 since their sum must equal 7. The factors of 10 whose sum is 7 are 2 and 5. Hence, the required factors are $(x + 2)$ and $(x + 5)$. We write

$$x^2 + 7x + 10 = (x + 2)(x + 5).$$

(*b*) $x^2 - x - 6$.

Since -6 is negative, its factors must have unlike signs. The possibilities are -1 and $+6$; $+1$ and -6; $+2$ and -3; and -2 and $+3$. We must choose the pair whose sum is -1, since the coefficient of x is -1. The correct pair is $+2$ and -3. Hence, the required factors are $(x + 2)$ and $(x - 3)$. We write

$$x^2 - x - 6 = (x + 2)(x - 3).$$

(*c*) $t^2 - 6t + 5$

Determine factors of $+5$ whose sum is -6.

$$t^2 - 6t + 5 = (t - 1)(t - 5)$$

(*d*) $s^2 + 13s + 42$

Choose factors of $+42$ whose sum is $+13$.

$$s^2 + 13s + 42 = (s + 6)(s + 7)$$

The student should now attempt supplementary problem 2.12.

2.17 Factor completely.

(*a*) $2x^2 + 11x + 15$

We first consider the various possibilities. $A = ac = 1 \cdot 2$ or $2 \cdot 1$ and $C = bd = 1 \cdot 15, 15 \cdot 1, 3 \cdot 5$ or $5 \cdot 3$. We need not consider negative factors since all coefficients are positive. Now $B = ad + bc = 11$. By trial and error we observe that $1 \cdot 5 + 3 \cdot 2 = 5 + 6 = 11$. Thus, $ad = 1 \cdot 5$ and $bc = 3 \cdot 2$. We now write the factored form of the polynomial. That is,

$$2x^2 + 11x + 15 = (x + 3)(2x + 5).$$

We should always multiply the factors we obtain in order to check the result.

(*b*) $6x^2 + 5x - 4$

The possible factors of 6 are $1 \cdot 6$ and $2 \cdot 3$. The factors of -4 have unlike signs. The possibilities are $-1 \cdot 4$, $-4 \cdot 1$, and $-2 \cdot 2$. We must find the correct combination by trial and error. Checking various combinations leads to the conclusion that $(2x - 1)$ and $(3x + 4)$ are the correct factors. Hence,

$$6x^2 + 5x - 4 = (2x - 1)(3x + 4).$$

(*c*) $3x^2 - 16x + 16$

In this case A = 3; therefore, the possible factors are 1 and 3. Also $B = -16$ and $C = 16$. We must use factors of C which have like signs, since it is positive. Observe that $B = -16$; therefore, we must employ negative factors of 16. The alternatives are $(-1)(-16)$, $(-2)(-8)$, and $(-4)(-4)$. We determine by trial error that $(x - 4)$ and $(3x - 4)$ are the correct factors. Hence,

$$3x^2 - 16x + 16 = (x - 4)(3x - 4).$$

(*d*) $4x^2 - 8x - 5$

Observe that $A = 4$, $B = -8$, and $C = -5$. The possible factors of $A = 4$ are $1 \cdot 4$ and $2 \cdot 2$. Choose factors of $C = -5$ with unlike signs. They are $-1 \cdot 5$ and $-5 \cdot 1$. We know that the factors of our trinomial are binomial in form and that the sum of the products of the outer and inner terms in those factors must equal $-8x$. That tells us that the largest magnitude product should be negative. The number of trials in our process can be reduced if we keep that fact in mind. Ultimately we find that $(2x + 1)$ and $(2x - 5)$ are the correct factors. Hence, our expression becomes

$$4x^2 - 8x - 5 = (2x + 1)(2x - 5).$$

The trial-and-error method of factoring must be employed many times in order to become proficient using it. We can develop a good intuitive sense about finding appropriate factors if we use it often. Be sure to check the potential factors in a systematic manner in order to avoid overlooking a combination which may be correct.

The reader should now try supplementary problem 2.13.

2.18 Factor using the grouping method.

(*a*) $8x^2 + 6x - 27$

We first find the product $AC = 8(-27) = -216$. Next, find two factors of -216 whose sum is $B = 6$. One factor must be positive, the other must be negative, and they are approximately the same magnitude, since their sum is 6. The positive factor has the larger magnitude also, since their sum is positive. We eventually conclude that the factors are -12 and 18. Now rewrite $6x$ as $-12x + 18x$. That is, $8x^2 + 6x - 27 = 8x^2 - 12x + 18x - 27$. Finally factor the last expression by grouping as shown below.

$$\begin{aligned} 8x^2 - 12x + 18x - 27 &= (8x^2 - 12x) + (18x - 27) \\ &= 4x(2x - 3) + 9(2x - 3) \\ &= (4x + 9)(2x - 3) \end{aligned}$$

(*b*) $15x^2 - 37x + 20$

The product $AC = 15(20) = 300$. We now find factors of 300 whose sum is $B = -37$. In this case, we will need two negative factors since $AC = 300$ is positive and $B = -37$ is negative. After several attempts, we find that -12 and -25 work. Therefore,

$$\begin{aligned} 15x^2 - 37x + 20 &= 15x^2 - 12x - 25x + 20 \\ &= (15x^2 - 12x) - (25x - 20) \\ &= 3x(5x - 4) - 5(5x - 4) \\ &= (3x - 5)(5x - 4) \end{aligned}$$

(*c*) $5s^2 - 2s + 10$

$AC = 50$. Negative factors are once again required. The negative factors of 50 are $(-1)(-50)$, $(-2)(-25)$, and $(-5)(-10)$. None of these pairs has a sum of -2. This fact tells us that the expression is not factorable; it is prime.

Although we employ a process of elimination to find the correct combination of factors of AC, factoring by grouping is usually more efficient than factoring by trial and error. Either method may be used, although factoring by grouping is best in most cases. We must practice both procedures many times in order to become proficient at factoring. Factoring is a very important skill in algebra and in subsequent courses in which you may enroll. Our advice is this: practice, practice, practice! You should now practice by working supplementary problem 2.14.

2.19 Employ the special products to factor the following.

(*a*) $x^2 + 10x + 25$

It is a trinomial whose first and last terms are perfect squares and the coefficients are all positive. We next ask if the middle term is twice the product of the square roots of the first and last terms. The square root of x squared is x and the square root of 25 is 5. (Technically, the square root of x squared is $+x$ or $-x$ and 25 has $+5$ and -5 as square roots. We use the roots with positive signs in form 1 stated above.) The middle term is $10x$, which is indeed twice the product of the square roots of the first and last terms. The trinomial is a form 1 expression; hence, the factors are both $x + 5$. We write

$$x^2 + 10x + 25 = (x + 5)^2.$$

(*b*) $x^2 - 10x + 25$

This expression is identical to the expression in part (*a*) above except for the sign of the coefficient of the middle term. (We use the negative square root of 25 this time). In this instance, the trinomial is a form 2 expression. Therefore, the factors are both $x - 5$. We write

$$x^2 - 10x + 25 = (x - 5)^2.$$

(*c*) $t^2 - 4$

We note that both terms are perfect squares and that they are subtracted. Hence, the expression is the difference of two squares and therefore fits form 3. We write

$$t^2 - 4 = (t + 2)(t - 2).$$

The reader should think through the rationale in the remaining problems.

(*d*) $4s^2 + 28s + 49$

$$\begin{aligned} 4s^2 + 28s + 49 &= (2s)^2 + 2(7)(2s) + 7^2 \\ &= (2s + 7)^2 \end{aligned}$$

(*e*) $9y^2 - 42y + 49$

$$\begin{aligned} 9y^2 - 42y + 49 &= (3y)^2 - 2(7)(3y) + 7^2 \\ &= (3y - 7)^2 \end{aligned}$$

(*f*) $25x^2 - 64y^2$

$$\begin{aligned} 25x^2 - 64y^2 &= (5x)^2 - (8y)^2 \\ &= (5x + 8y)(5x - 8y) \end{aligned}$$

See supplementary problem 2.15 for practice problems.

2.20 Factor.

(*a*) $x^3 + 27$

$$\begin{aligned} x^3 + 27 &= x^3 + 3^3 \\ &= (x + 3)(x^2 - 3x + 9) \end{aligned}$$

(*b*) $8t^3 - 1$

$$\begin{aligned} 8t^3 - 1 &= (2t)^3 - 1^3 \\ &= (2t - 1)(4t^2 + 2t + 1) \end{aligned}$$

(*c*) $27s^3 - 1000t^3$

$$\begin{aligned} 27s^3 - 1000t^3 &= (3s)^3 - (10t)^3 \\ &= (3s - 10t)(9s^2 + 30st + 100t^2) \end{aligned}$$

Refer to supplementary problem 2.16 for additional practice.

2.21 Write each expression as a product or quotient, whichever is applicable, in which each variable occurs only once and all exponents are positive. That is, simplify.

(*a*) $\dfrac{x^6}{x^4}$

$$\frac{x^6}{x^4} = x^{6-4} = x^2$$

(*b*) $\dfrac{3x^3}{12x^5}$

$$\frac{3x^3}{12x^5} = \frac{1}{4x^{5-3}} = \frac{1}{4x^2}$$

(*c*) $\dfrac{(y^2)^3}{(3y)^2}$

$$\frac{(y^2)^3}{(3y)^2} = \frac{y^6}{9y^2} = \frac{y^{6-2}}{9} = \frac{y^4}{9}$$

(*d*) $\dfrac{(st^2)^3}{(s^2t)^2}$

$$\frac{(st^2)^3}{(s^2t)^2} = \frac{s^3(t^2)^3}{(s^2)^2t^2} = \frac{s^3t^6}{s^4t^2} = \frac{t^{6-2}}{s^{4-3}} = \frac{t^4}{s}$$

(*e*) $\left(\frac{a^2c}{3}\right)^2\left(\frac{-3}{abc}\right)^3$

$$\left(\frac{a^2c}{3}\right)^2\left(\frac{-3}{abc}\right)^3=\left(\frac{(a^2)^2c^2}{3^2}\right)\cdot\frac{(-3)^3}{a^3b^3c^3}=\frac{a^4c^2}{9}\cdot\frac{-27}{a^3b^3c^3}=\frac{-27a^4c^2}{9a^3b^3c^3}=\frac{-3a}{b^3c}$$

(*f*) $\frac{c^{2n}}{c^n}$

$$\frac{c^{2n}}{c^n}=c^{2n-n}=c^n$$

Refer to supplementary problem 2.18 for similar problems.

2.22 Divide.

(*a*) $\frac{10x^4+20x^3}{5x^2}$

$$\frac{10x^4+20x^3}{5x^2}=\frac{10x^4}{5x^2}+\frac{20x^3}{5x^2}=2x^2+4x$$

(*b*) $\frac{2t^3+10t^4}{2t^3}$

$$\frac{2t^3+10t^4}{2t^3}=\frac{2t^3}{2t^3}+\frac{10t^4}{2t^3}=1+5t$$

(*c*) $\frac{5x^3y-7x^2y^2+12xy^3}{xy}$

$$\frac{5x^3y-7x^2y^2+12xy^3}{xy}=\frac{5x^3y}{xy}-\frac{7x^2y^2}{xy}+\frac{12xy^3}{xy}=5x^2-7xy+12y^2$$

(*d*) $\frac{-24s^5+36s^3-12s^2}{-4s^2}$

$$\frac{-24s^5+36s^3-12s^2}{-4s^2}=\frac{-24s^5}{-4s^2}+\frac{36s^3}{-4s^2}-\frac{12s^2}{-4s^2}=6s^3-9s+3$$

See supplementary problem 2.19 for similar problems.

2.23 Divide.

(*a*) $x+3\overline{)2x^2+x-5}$

Divide $2x^2$ by x to obtain $2x$. Subtract the product of $2x$ and $x+3$ from (or add the opposite to) the dividend.

$$\begin{array}{r} 2x \\ x+3\overline{)2x^2+x-5} \\ 2x^2+6x \\ \hline -5x-5 \end{array} \quad \text{or} \quad \begin{array}{r} 2x \\ x+3\overline{)2x^2+x-5} \\ -2x^2-6x \\ \hline -5x-5 \end{array}$$

Now divide $-5x$ by x to obtain -5. Write -5 in the quotient line and multiply it by the divisor. Subtract the product from (or add the opposite to) the new dividend acquired in the first portion of the procedure.

$$\begin{array}{r} 2x-5 \\ x+3\overline{)2x^2+x-5} \\ \underline{2x^2+6x} \\ -5x-5 \\ \underline{-5x-15} \\ 10 \end{array} \quad \text{or} \quad \begin{array}{r} 2x-5 \\ x+3\overline{)2x^2+x-5} \\ \underline{-2x^2-6x} \\ -5x-5 \\ \underline{5x+15} \\ 10 \end{array}$$

The quotient is $2x - 5$ and the remainder is 10. We often write

$$\frac{2x^2+x-5}{x+3} = 2x-5+\frac{10}{x+3}$$

(*b*) $\dfrac{4a^4-2a^3+8a-5}{2a^2+1}$

Note that the coefficient of a^2 in the dividend is 0. Follow the four-step procedure stated previously to obtain

$$\begin{array}{r} 2a^2-a-1 \\ 2a^2+1\overline{)4a^4-2a^3+0a^2+8a-5} \\ \underline{4a^4+2a^2} \\ -2a^3-2a^2+8a-5 \\ \underline{-2a^3-a} \\ -2a^2+9a-5 \\ \underline{-2a^2-1} \\ 9a-4 \end{array}$$

or, equivalently,

$$\begin{array}{r} 2a^2-a-1 \\ 2a^2+1\overline{)4a^4-2a^3+0a^2+8a-5} \\ \underline{-4a^4-2a^2} \\ -2a^3-2a^2+8a-5 \\ \underline{2a^3+a} \\ -2a^2+9a-5 \\ \underline{2a^2+1} \\ 9a-4 \end{array}$$

Therefore, $\dfrac{4a^4-2a^3+8a-5}{2a^2+1} = 2a^2-a-1+\dfrac{9a-4}{2a^2+1}$

Work supplementary problem 2.20 to practice the above procedure.

2.24 Use synthetic division to divide the following.

(*a*) $x^4 - 3x^3 + 5x - 8$ by $x - 2$.

The associated array follows. Refer to the five-step procedure given above as you analyze the process. Observe that $c = 2$.

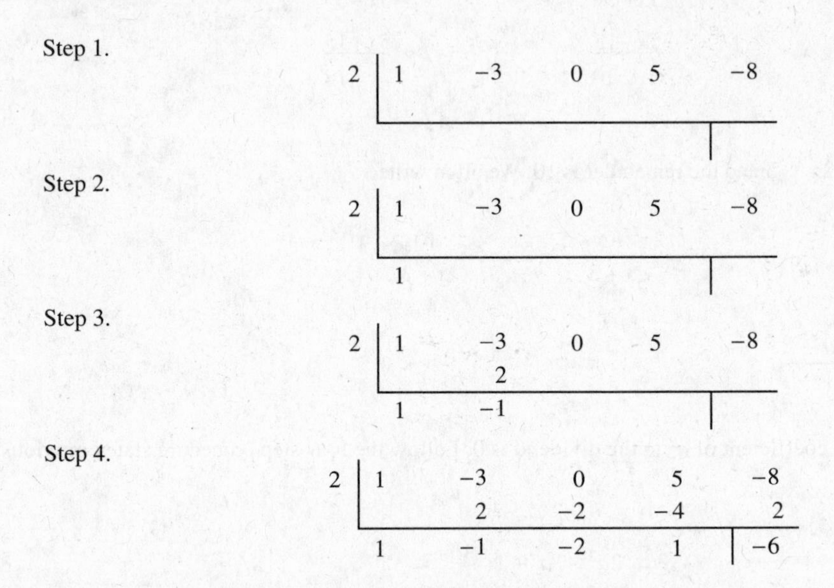

Step 1.

$$\begin{array}{r|rrrr|r} 2 & 1 & -3 & 0 & 5 & -8 \\ \hline & & & & & \end{array}$$

Step 2.

$$\begin{array}{r|rrrr|r} 2 & 1 & -3 & 0 & 5 & -8 \\ \hline & 1 & & & & \end{array}$$

Step 3.

$$\begin{array}{r|rrrr|r} 2 & 1 & -3 & 0 & 5 & -8 \\ & & 2 & & & \\ \hline & 1 & -1 & & & \end{array}$$

Step 4.

$$\begin{array}{r|rrrr|r} 2 & 1 & -3 & 0 & 5 & -8 \\ & & 2 & -2 & -4 & 2 \\ \hline & 1 & -1 & -2 & 1 & -6 \end{array}$$

Step 5. The coefficients of the terms in the quotient are $1, -1, -2$, and 1. The remainder is -6. The quotient is $x^3 - x^2 - 2x + 1$ and the remainder is -6. We write

$$\frac{x^4 - 3x^3 + 5x - 8}{x - 2} = x^3 - x^2 - 2x + 1 - \frac{6}{x - 2}.$$

(*b*) $3t^4 - 7t^3 - t^2 + 5$ by $t + 4$

The associated array follows. Refer to the five-step procedure given above as you analyze the process. Observe that $c = -4$.

Step 1.

$$\begin{array}{r|rrrr|r} -4 & 3 & -7 & -1 & 0 & 5 \\ \hline & & & & & \end{array}$$

Step 2.

$$\begin{array}{r|rrrr|r} -4 & 3 & -7 & -1 & 0 & 5 \\ \hline & 3 & & & & \end{array}$$

Step 3.

$$\begin{array}{r|rrrr|r} -4 & 3 & -7 & -1 & 0 & 5 \\ & & -12 & & & \\ \hline & 3 & -19 & & & \end{array}$$

Step 4.

$$\begin{array}{r|rrrr|r} -4 & 3 & -7 & -1 & 0 & 5 \\ & & -12 & 76 & -300 & 1200 \\ \hline & 3 & -19 & 75 & -300 & 1205 \end{array}$$

Step 5. The coefficients of the terms in the quotient are $3, -19, 75$, and -300. The remainder is 1205. The quotient is $3t^3 - 19t^2 + 75t - 300$ and the remainder is 1205. We write

$$\frac{3t^4 - 7t^3 - t^2 + 5}{t+4} = 3t^3 - 19t^2 + 75t - 300 + \frac{1205}{t+4}.$$

Remember that synthetic division is applicable only when the divisor has the form $x - c$. The degree of the quotient is always one less than the degree of the dividend, since the divisor is a first-degree polynomial.

See supplementary problem 2.21.

SUPPLEMENTARY PROBLEMS

2.1 Determine the number of terms and the degree of each of the following.

(*a*) $6x^3y$ (*b*) $1 - 3t^2 + 8t^3$ (*c*) 7

(*d*) $s + 10$ (*e*) $x^2 - 4^3$ (*f*) $5ab^3 - 6ab^2 + 2a^3b^2$

2.2 Identify the part which corresponds to the polynomials in problem 2.1 which represent monomials, binomials, or trinomials.

2.3 Write each of the following polynomials in standard form and state the degree of each.

(*a*) $4x^2 + 3 - 2x$ (*b*) $8 + t^3$

(*c*) $5s^2 - 3s + 9s^3 - 2$ (*d*) $2y - 3y^2 + 10 + y^5$

(*e*) $x^4 - 3x^2 + 2x^6 - 4x$

2.4 Simplify each expression.

(*a*) $(2x + 6) + (4 - 5x)$ (*b*) $(8x + 5) - (2x - 7)$

(*c*) $4 - [5 - (3p + 4) + 7]$ (*d*) $(5t^2 - 3t + 4) + (2t^2 - 6t - 8)$

(*e*) $[(5y^2 - 11y + 5 - (4y + 8)] + (8 - 3y)$ (*f*) $[(6a - 2) + (a + 4)] - (3a + 7)$

(*g*) $(3s^2 + 2s - 5) - (6s^2 - 8s - 1)$ (*h*) $7s - 4 - (s^2 - 4s - 12)$

(*i*) $3s - (4s + 2s^2) - [s^2 - (4s + 5) + 8]$

2.5 Perform the indicated operations.

(*a*) Add $2x^2 + 3x - 5$ and $4x^2 + 4x - 3$

(*b*) Add $3t + 1$ to $4t^2 - 4t + 5$

(*c*) Subtract $3t + 1$ from $4t^2 - 4t + 5$

(*d*) Find the difference of $5y - 2$ and $2y - 5$

(*e*) Subtract $2s^2 + s - 4$ from the sum of $3s^2 + s - 5$ and $3s - 11$.

2.6 Show by counterexample that the following are not equivalent.

(*a*) $-(x + 2)$ and $-x + 2$ (*b*) $-(x - y)$ and $-x - y$.

Hint: Choose a particular value for each of the variables and evaluate each expression. Are the results the same?

2.7 Perform the indicated operations, that is, simplify.

(*a*) $(5r)^2$ (*b*) $(-3s^2)(2s^3)$ (*c*) $(-4x^2y)(-5xy^2z)$

(*d*) $(2t^3)^3$ (*e*) $(-3s^2t^3)^2$ (*f*) $(-4b^6)(-5b^{10})(3b^7)$

(*g*) $(x^2)^3(x^3)^2$ (*h*) $(-x^3)^2$ (*i*) $(3x^2y)^2(-2xy^3)^3$

(*j*) $-(2bc)^2(3b^2c)^2$ (*k*) $ab^2(-a^2bc)^2$ (*l*) $(6b^n)^2$

(*m*) $(-4b^{2n})^3(b^4)$

2.8 Employ the special product forms to find the following products.

(*a*) $(x-2)^2$ (*b*) $(2x+5)^2$ (*c*) $(2x-y)^2$

(*d*) $(x+2)(x-2)$ (*e*) $(s+2t)^2$ (*f*) $(a-3)(a+3)$

(*g*) $(3r-7s)^2$ (*h*) $(3p+2t)^2$ (*i*) $(3x-2y)(3x+2y)$

(*j*) $(x^n-1)^2$ (*k*) $(2a-3b)^3$ (*l*) $(4x+y)^3$

(*m*) $(5-2t)^3$

2.9 Use the vertical to find the following products.

(*a*) $(x+4)(3x^2+5x-8)$ (*b*) $[(x-3)(2x+7)](3x-1)$

(*c*) $(4t^2-5t+1)(2t^2+3t-6)$ (*d*) $(x^3+5x+2)(x^2-3)$

2.10 Factor completely.

(*a*) $9x+3y$ (*b*) $21s+14t$

(*c*) $15a^2-18ab$ (*d*) $-2x^2y-7xy^2$

(*e*) $10x^3y^2-20x^2y^2+35x^2y$ (*f*) $-20s^3t^4-16s^2t^3-12st^2$

2.11 Factor completely.

(*a*) $ax+bx+7a+7b$ (*b*) $5y^2+10y+6y+12$

(*c*) $12n^2-10n+18n-15$ (*d*) $12s^2-9st-16s+12t$

(*e*) $(3x-2)(t+1)+6(t+1)$ (*f*) $5a(b-2)^2-(b-2)^2$

2.12 Factor.

(*a*) x^2+4x+3 (*b*) x^2-2x-3 (*c*) $t^2+5t-14$

(*d*) $s^2+10s+24$ (*e*) $s^2-9s+20$ (*f*) $y^2-5y-24$

2.13 Factor the following using the trial-and-error method.

(*a*) $3x^2+11x+10$ (*b*) $2t^2-7t-4$ (*c*) $4y^2+8y-21$

(*d*) $6x^2+29x+20$ (*e*) $9s^2-15s+4$ (*f*) $5p^2+8p-4$

(*g*) $8x^2-2x-3$ (*h*) $12y^2+13y-35$

2.14 Factor the following using the grouping method.

(*a*) $6x^2+17x+5$ (*b*) $9t^2+9t-10$ (*c*) $2s^2-9s-56$

(*d*) $12s^2-32s+21$ (*e*) $2y^2+y+5$ (*f*) $2x^2+5x-42$

(*g*) $25t^2+35t+12$ (*h*) $90x^2-67x+12$

2.15 Use special product forms to factor the following.

(a) $16x^2 - 24x + 9$ (b) $9t^2 + 48t + 64$ (c) $25s^2 + 70st + 49t^2$

(d) $121x^2 - 64y^2$ (e) $121x^2 + 64y^2$ (f) $25a^2 - 60ab + 36b^2$

(g) $p^2 + 8pq + 16q^2$ (h) $x^{2n} - y^{4n}$

2.16 Factor.

(a) $64y^3 + 27$ (b) $1000 - x^3$ (c) $64s^3 + 27t^3$ (d) $125p^3 - 729$

2.17 Factor completely. These problems involve the various forms that have been introduced.

(a) $s^2 - 6s + 9$ (b) $s^2 - 4st + 4t^2$ (c) $81x^2y^2 - 25$

(d) $4x^2 + 4xy^2 + y^4$ (e) $5a^3b - 45ab^3$ (f) $27 - 8t^3$

(g) $(x + y)^2 - 9$ (h) $16x^3 + 54$ (i) $a^4 + 28a^2b + 27b^2$

(j) $35s^2 + 10s - 50$ (k) $8x^3 - 16x^2 - x + 2$

2.18 Write each expression as a product or quotient, whichever is applicable, in which each variable occurs only once and all exponents are positive. That is, simplify.

(a) $\dfrac{4y^8}{y^3}$ (b) $\dfrac{(3s)^2(6t^3)^2}{(18st)(t^6)}$ (c) $\left(\dfrac{5x^2y}{10z}\right)^2(2x^2yz^2)^2$

(d) $\dfrac{(-x^2y)^2}{(x^3y)^3}$ (e) $\dfrac{(a^3b)^3}{ab^2}$ (f) $\dfrac{(st^2)^2(s^2t)^3}{(-s^3t)^3}$

(g) $\left[\dfrac{3xy^2}{(6x^2y^2)^2}\right]^2$ (h) $\left(\dfrac{s^2t}{r}\right)^3\left(\dfrac{r^2}{s^4t^2}\right)^2$ (i) $\dfrac{(a^{n+1}b^{n-1})^2}{a^nb^n}$

2.19 Divide.

(a) $\dfrac{x^3 + x^2}{x}$ (b) $\dfrac{2a^4 + 6a^2 + 8a}{2a}$

(c) $\dfrac{6t - 3t^2 - 12t^3}{3t}$ (d) $\dfrac{-4s^3t + 16s^2t^2 - 20st^3}{-2st}$

2.20 Divide.

(a) $\dfrac{t^2 + 2t + 3}{t + 3}$ (b) $\dfrac{-3x^2 + 13x + 10}{x - 3}$

(c) $\dfrac{4a^2 + 9}{2a + 3}$ (d) $\dfrac{n^3 - 1}{n - 1}$

(e) $\dfrac{6x^4 - 3x^3 - 11x^2 + 2x + 4}{3x^2 - 1}$

2.21 Use synthetic division to divide.

(a) $\dfrac{x^2 - 3x + 4}{x + 2}$ (b) $\dfrac{x^2 - 7x + 5}{x + 3}$

(c) $\dfrac{x^3 - 5x^2 + 12x - 27}{x - 3}$ (d) $\dfrac{2x^4 - 13x^3 + 17x^2 + 18x - 24}{x - 4}$

(e) $\dfrac{x^5 - 25x^3 - 3x^2 + 20x - 15}{x - 5}$ (f) $\dfrac{x^3 + 27}{x + 3}$

ANSWERS TO SUPPLEMENTARY PROBLEMS

2.1 (*a*) 1 term, degree 4 (*b*) 3 terms, degree 3 (*c*) 1 term, degree 0
(*d*) 2 terms, degree 1 (*e*) 2 terms, degree 2 (*f*) 3 terms, degree 5

2.2 Monomials: parts (*a*) and (*c*).
Binomials: parts (*d*) and (*e*).
Trinomials: parts (*b*) and (*f*).

2.3 (*a*) $4x^2 - 2x + 3$; degree 2. (*b*) $t^3 + 8$; degree 3.
(*c*) $9s^3 + 5s^2 - 3s - 2$; degree 3. (*d*) $y^5 - 3y^2 + 2y + 10$; degree 5.
(*e*) $2x^6 + x^4 - 3x^2 - 4x$; degree 6.

2.4 (*a*) $10 - 3x$ (*b*) $6x + 12$ (*c*) $3p - 4$
(*d*) $7t^2 - 9t - 4$ (*e*) $5y^2 - 18y + 5$ (*f*) $4a - 5$
(*g*) $-3s^2 + 10s - 4$ (*h*) $8 + 11s - s^2$ (*i*) $-3s^2 + 3s - 3$

2.5 (*a*) $6x^2 + 7x - 8$ (*b*) $4t^2 - t + 6$ (*c*) $4t^2 - 7t + 4$
(*d*) $3y + 3$ (*e*) $s^2 + 3s - 12$.

2.6 (*a*) For $x = -6$: $-(x + 2) = -(-6 + 2) = 4$, but $-x + 2 = -(-6) + 2 = 8$
(*b*) For $x = 8$ and $y = 3$: $-x - y = (8-3) = -5$, but $-x - y = -(8) - (3) = -11$

2.7 (*a*) $25r^2$ (*b*) $-6s^5$ (*c*) $20x^3y^3z$ (*d*) $8t^9$
(*e*) $9s^4t^6$ (*f*) $60b^{23}$ (*g*) x^{12} (*h*) x^6
(*i*) $-72x^7y^{11}$ (*j*) $-36b^6c^4$ (*k*) $a^5b^4c^2$ (*l*) $36b^{2n}$
(*m*) $-64b^{6n+4}$

2.8 (*a*) $x^2 - 4x + 4$ (*b*) $4x^2 + 20x + 25$
(*c*) $4x^2 - 4xy + y^2$ (*d*) $x^2 - 4$
(*e*) $s^2 + 4st + 4t^2$ (*f*) $a^2 - 9$
(*g*) $9r^2 - 42rs + 49s^2$ (*h*) $9p^2 + 12pt + 4t^2$
(*i*) $9x^2 - 4y^2$ (*j*) $x^{2n} - 2x^n + 1$
(*k*) $8a^3 - 36a^2b + 54ab^2 - 27b^3$ (*l*) $64x^3 + 48x^2y + 12xy^2 + y^3$
(*m*) $125 - 150t + 60t^2 - 8t^3$

2.9 (*a*) $3x^3 + 17x^2 + 12x - 32$ (*b*) $6x^3 + x^2 - 64x + 21$
(*c*) $8t^4 + 2t^3 - 37t^2 + 33t - 6$ (*d*) $x^5 + 2x^3 + 2x^2 - 15x - 6$

2.10 (*a*) $3(3x + y)$ (*b*) $7(3s + 2t)$ (*c*) $3a(5a - 6b)$
(*d*) $-xy(2x + 7y)$ (*e*) $5x^2y(2xy - 4y + 7)$ (*f*) $-4st^2(5s^2t^2 + 4st + 3)$

2.11 (*a*) $(a + b)(x + 7)$ (*b*) $(y + 2)(5y + 6)$ (*c*) $(6n - 5)(2n + 3)$
(*d*) $(4s - 3t)(3s - 4)$ (*e*) $(t + 1)(3x + 4)$ (*f*) $(b - 2)^2(5a - 1)$

2.12 (*a*) $(x + 1)(x + 3)$ (*b*) $(x + 1)(x - 3)$ (*c*) $(t + 7)(t - 2)$
(*d*) $(s + 4)(s + 6)$ (*e*) $(s - 5)(s - 4)$ (*f*) $(y - 8)(y + 3)$

2.13 (*a*) $(x + 2)(3x + 5)$ (*b*) $(t - 4)(2t + 1)$ (*c*) $(2y - 3)(2y + 7)$
(*d*) $(6x + 5)(x + 4)$ (*e*) $(3s - 4)(3s - 1)$ (*f*) $(5p - 2)(p + 2)$
(*g*) $(4x - 3)(2x + 1)$ (*h*) $(3y + 7)(4y - 5)$

2.14 (*a*) $(2x + 5)(3x + 1)$ (*b*) $(3t - 2)(3t + 5)$ (*c*) $(2s + 7)(s - 8)$
(*d*) $(6s - 7)(2s - 3)$ (*e*) prime (*f*) $(2x - 7)(x + 6)$
(*g*) $(5t + 4)(5t + 3)$ (*h*) $(10x - 3)(9x - 4)$

2.15 (*a*) $(4x - 3)^2$ (*b*) $(3t + 8)^2$ (*c*) $(5s + 7t)^2$
(*d*) $(11x + 8y)(11x - 8y)$ (*e*) prime (*f*) $(5a - 6b)^2$
(*g*) $(p + 4q)^2$ (*h*) $(x^n + y^{2n})(x^n - y^{2n})$

2.16 (*a*) $(4y + 3)(16y^2 - 12y + 9)$ (*b*) $(10 - x)(100 + 10x + x^2)$
(*c*) $(4s + 3t)(16s^2 - 12st + 9t^2)$ (*d*) $(5p - 9)(25p^2 + 45p + 81)$

2.17 (*a*) $(s - 3)^2$ (*b*) $(s - 2t)^2$
(*c*) $(9xy + 5)(9xy - 5)$ (*d*) $(2x + y^2)^2$
(*e*) $5ab(a + 3b)(a - 3b)$ (*f*) $(3 - 2t)(9 + 6t + 4t^2)$
(*g*) $(x + y + 3)(x + y - 3)$ (*h*) $2(2x + 3)(4x^2 - 6x + 9)$
(*i*) $(a^2 + 27b)(a^2 + b)$ (*j*) $5(7s^2 + 2s - 10)$
(*k*) $(8x^2 - 1)(x - 2)$

2.18 (*a*) $4y^5$ (*b*) $\dfrac{18s}{t}$ (*c*) $x^8y^4z^2$
(*d*) $\dfrac{1}{x^5y}$ (*e*) a^8b (*f*) $\dfrac{-t^4}{s}$
(*g*) $\dfrac{1}{144x^6y^4}$ (*h*) $\dfrac{r}{s^2t}$ (*i*) $a^{n+2}b^{n-2}$

2.19 (*a*) $x^2 + x$ (*b*) $a^3 + 3a + 4$ (*c*) $2 - t - 4t^2$ (*d*) $2s^2 - 8st + 10t^2$

2.20 (*a*) $t-1+\dfrac{6}{t+3}$ (*b*) $-3x+4+\dfrac{22}{x-3}$

(*c*) $2a-3+\dfrac{18}{2a+3}$ (*d*) n^2+n+1

(*e*) $2x^2-x-3+\dfrac{x+1}{3x^2-1}$

2.21 (*a*) $x-5+\dfrac{14}{x+2}$ (*b*) $x-10+\dfrac{35}{x+3}$

(*c*) $x^2-2x+6-\dfrac{9}{x-3}$ (*d*) $2x^3-5x^2-3x+6$

(*e*) $x^4+5x^3-3x+5+\dfrac{10}{x-5}$ (*f*) x^2-3x+9

CHAPTER 3

Rational Expressions

3.1 Basic Properties

Recall that a rational number has the form a/b, where a and b are integers and $b \neq 0$.

Similarly, a rational expression is the quotient (ratio) of two polynomials. It is an algebraic fraction defined for all real values of the variable(s) in the numerator and denominator provided the denominator is not equal to zero. Be reminded that polynomials are expressions whose variables have nonnegative integer exponents. In other words, rational expressions look like the following.

$$\frac{x^2}{x-1}, \frac{3x-2}{x^2+4x-7}, \frac{4}{t+5}, \frac{5s^3+3s^2+s-6}{t^4-3} \text{ and } 5a^2-3a+1.$$

The last expression is rational since it may be written as a quotient in which the denominator is one.

Any particular quotient (fraction) can be written in infinitely many forms. For example,

$$\frac{1}{2}=\frac{2}{4}=\frac{3}{6}=\cdots \text{ and } \frac{2}{7}=\frac{6}{21}=\frac{18}{63}\cdots.$$

In each case we obtain an equivalent fraction by multiplying the numerator and denominator by the same nonzero constant, say k. Equivalent expressions result since we are actually multiplying the fraction by one, $(k/k = 1)$, and one is the multiplicative identity element.

The principle we have been illustrating is called the *fundamental principle of fractions*. Symbolically we write the following.

Fundamental Principle of Fractions

$$\frac{a}{b}=\frac{a}{b}\cdot 1=\frac{a}{b}\cdot\frac{k}{k}=\frac{ak}{bk} \text{ where } b, k \neq 0.$$

In the above expression, a, b, and k each represent a polynomial. Some or all may also be constants, since constants are zero-degree polynomials.

We will employ the fundamental principle of fractions often in a later section in which we add and subtract fractions.

We shall now use the same principle to reduce rational expressions. Think of the equations above in reverse order. We then may write

$$\frac{ak}{bk} = \frac{a}{b} \text{ for } b, k \neq 0. \text{ Also, } \frac{ka}{kb} = \frac{a}{b} \text{ for } b, k \neq 0.$$

The preceding statements tell us that we may divide out the same factor(s) which occur in the numerator and denominator. Recall that factors are quantities which are multiplied, while terms are quantities which are added or subtracted. Equivalent expressions are not obtained when terms are divided out. DO NOT DIVIDE OUT (CANCEL) TERMS!

It would be advisable to review the special products and factoring before continuing this section.

See solved problem 3.1.

Sometimes the numerator and/or the denominator of a fraction are negative. We know that $(-1)(-1) = 1$, and $-1/-1 = 1$ and 1 is the identity element for multiplication. Hence, there are several equivalent forms of fractions obtained by multiplying the fraction times 1 written either as $(-1)(-1)$ or $-1/-1$. They are

$$\frac{a}{b} = \frac{-a}{-b} = -\frac{-a}{b} = -\frac{a}{-b}, b \neq 0 \text{ and } \frac{-a}{b} = \frac{a}{-b} = -\frac{-a}{-b} = -\frac{a}{b}, b \neq 0.$$

There are three signs associated with a fraction—the sign of the numerator, the sign of the denominator, and the sign in front of the fraction. Essentially the above equations tell us that we can change any combination of two of the three signs associated with a fraction without affecting the results. We call a/b and $-a/b$ the *standard forms* of the various quotients. If any two of the three signs of a fraction are negative, the standard form is a/b. If either one or three of the signs are negative, the standard form of the fraction is $-a/b$. The standard forms are the most appropriate forms to use. Always write your results in standard form when fractions are encountered.

See solved problem 3.2.

If the numerator or denominator of a fraction contain two terms, the best form or standard form is less apparent. Observe that

$$-(b - a) = -b + a = a - b.$$

We are actually distributing a factor of negative one and applying the commutative property of addition. If the expression is read from right to left, we can first apply the commutative property of addition and then factor out a negative one to obtain $-(b - a)$. Therefore,

$$\frac{-c}{a-b} = \frac{-c}{-(b-a)} = \frac{c}{b-a}, \text{ where } a \neq b. \text{ (The denominator cannot be zero.)}$$

Similarly,

$$-\frac{b-a}{c} = \frac{-(b-a)}{c} = \frac{a-b}{c}, \text{ where } c \neq 0.$$

In these instances, $c/(b - a)$ and $(a - b) / c$ are generally considered to be the best forms since those forms contain the fewest negative signs.

See solved problem 3.3.

3.2 Products and Quotients

We now possess the essential tools for performing operations on rational expressions. We shall begin with the operation of multiplication since it is the simplest. The definition of multiplying rational expressions is identical for arithmetic and algebraic quantities.

Definition of Multiplication $\frac{a}{b}\cdot\frac{c}{d}=\frac{ac}{bd}$, where $b,d\neq 0$.

In words, the product of two fractions is the product of the numerators divided by the product of the denominators, provided neither denominator is zero.

If there are common factors in any of the numerators and denominators, they should be divided out prior to multiplying the fractions.

See solved problem 3.4.

Our next objective is to learn how to divide rational expressions. We know that

$$18\div 6=\frac{18}{6}=3 \text{ since } 3\cdot 6=18$$

That is, quotient times divisor is equal to the dividend. Similarly,

$$\frac{a}{b}\div\frac{c}{d}=\frac{ad}{bc} \text{ since } \frac{ad}{bc}\cdot\frac{c}{d}=\frac{a}{b}\left(\frac{d}{c}\cdot\frac{c}{d}\right)=\frac{a}{b}\cdot 1=\frac{a}{b}$$

That is, quotient times divisor is equal to the dividend.

In Section 1.3 we learned that the reciprocal of a quantity is one divided by the quantity. Therefore,

$$\frac{1}{\frac{c}{d}}=1\div\frac{c}{d}=1\cdot\frac{d}{c}=\frac{d}{c}$$

In addition, the definition of multiplication tells us that

$$\frac{ad}{bc}=\frac{a}{b}\cdot\frac{d}{c}$$

We employ this equivalence to formally state the definition of division of rational expressions.

Definition of Division $\frac{a}{b}\div\frac{c}{d}=\frac{a}{b}\cdot\frac{d}{c}$, where $b,c,d\neq 0$.

In words, the quotient of two fractions is obtained by multiplying the dividend by the multiplicative inverse (reciprocal) of the divisor. Hence, a division problem is converted to a multiplication problem. We discussed multiplication in the first part of this section.

The operation of division is the inverse of multiplication, just as subtraction is the inverse of addition. Inverse operations undo each other.

Analyze the following problems and relate the steps to the previous statements.

See solved problem 3.5.

Now consider some exercises that entail both operations of multiplication and division. Watch for opportunities to apply the fundamental principle of fractions to divide out common factors that may occur in numerators and denominators of fractions which are multiplied.

See solved problem 3.6.

3.3 Sums and Differences

Same Denominator

We employ the definition of division from Section 3.2 and the distributive property to obtain expressions that demonstrate how to add and subtract rational expressions. Hence, if $c \neq 0$,

$$\frac{a}{c}+\frac{b}{c}=a\left(\frac{1}{c}\right)+b\left(\frac{1}{c}\right) \qquad \text{Definition of division}$$

$$=(a+b)\left(\frac{1}{c}\right) \qquad \text{Distributive property}$$

$$=\frac{a+b}{c} \qquad \text{Definition of division}$$

The above sequence of steps defines how fractions are added. We now employ that result and the definition of subtraction of real numbers (i.e., $a - b = a + (-b)$) to define subtraction of fractions.

$$\frac{a}{c}-\frac{b}{c}=\frac{a}{c}+\left(\frac{-b}{c}\right) \qquad \text{Definition of subtraction}$$

$$=\frac{a+(-b)}{c} \qquad \text{Definition of addition of fractions}$$

$$=\frac{a-b}{c} \qquad \text{Definition of subtraction}$$

We now state the definitions of addition and subtraction of rational expressions formally.

Definition of Addition $\frac{a}{c}+\frac{b}{c}=\frac{a+b}{c}, c \neq 0.$

Definition of Subtraction $\frac{a}{c}-\frac{b}{c}=\frac{a-b}{c}, c \neq 0.$

The above definitions may be stated verbally in the following manner. To add or subtract fractions which contain the same denominator, we merely add or subtract, respectively, the numerators and divide the result by the denominator. Note: Don't double, square, or alter the original denominator in any other way. Simply divide the new numerator by the original denominator.

See solved problem 3.7.

Unlike Denominators

If the fractions involved possess different denominators, we must rewrite the fractions as equivalent fractions that have the same denominator. We employ the fundamental principle of fractions to accomplish the task.

$$\frac{a}{b} = \frac{ak}{bk}, b, k \neq 0,$$

The primary task when finding equivalent fractions is that of determining the appropriate k. The denominator bk must be an expression that is exactly divisible by each of the original denominators. The expression desired for bk is the "smallest" quantity which each of the original denominators divides exactly. It is called the *least common denominator* or *LCD* of the fractions. The expression that k represents is found by first factoring all of the denominators into prime factors. Next, divide the factors of each individual denominator into the LCD. The quotient is the appropriate expression to use for k in each case. The k expression is the product of all of the factors of the denominator bk = LCD which are not common to the LCD. (We illustrate below.)

Be reminded that the denominator bk must be an expression that is exactly divisible by each of the original denominators. If the fractions $f_1 = 7/132$ and $f_2 = 5/63$ are the fractions which we require to have the same denominator, we first find the prime factors of the denominators. The results are below:

$132 = 2^2 \cdot 3 \cdot 11$ and $63 = 3^2 \cdot 7$. The appropriate bk = LCD is $2^2 \cdot 3^2 \cdot 7 \cdot 11 = 2772$.

We now return to the task of determining the appropriate k for each denominator. Since the LCD in our illustration is $2^2 \cdot 3^2 \cdot 7 \cdot 11$ the necessary k for f_1 is $\dfrac{2^2 \cdot 3^2 \cdot 7 \cdot 11}{2^2 \cdot 3 \cdot 11} = 3 \cdot 7$.

Similarly, the appropriate k for f_2 is $\dfrac{2^2 \cdot 3^2 \cdot 7 \cdot 11}{3^2 \cdot 7} = 2^2 \cdot 11$. In each case, the resulting denominator is $bk = \text{LCD} = 2^2 \cdot 3^2 \cdot 7 \cdot 11 = 2772$.

The process described above may be stated more concisely as follows. To find the LCD of two or more fractions:

1. Find the prime factors of each denominator.
2. The LCD is the product of the different factors which occur in the various denominators. Repeated factors are used the largest number of times they occur in any particular denominator.

See solved problem 3.8.

Now that we can rewrite fractions with different denominators as equivalent fractions having the same denominator, we are in a position to find sums and differences of all fractions. The process is summarized below.

1. Find the LCD.
2. Rewrite each fraction as an equivalent fraction with the LCD as its denominator.
3. Apply the definition of addition or subtraction of fractions with the same denominator.

See solved problem 3.9.

3.4 Mixed Operations and Complex Fractions

We sometimes are required to perform some combination of the four fundamental operations on rational expressions or fractions. We now possess the tools needed for the task. You should review the order of operations in Chapter 1 before you continue.

See solved problem 3.10.

A fraction in which the numerator and/or denominator contain fractions is called a *complex fraction*. We apply the definition of division of fractions in the usual way in order to divide one fraction by another. Symbolically, we write

$$\frac{\frac{a}{b}}{\frac{c}{d}} = \frac{a}{b} \div \frac{c}{d} = \frac{a}{b}\,\frac{d}{c} \text{ where } b, c, d \neq 0.$$

See solved problem 3.11.

If there is a sum or difference in the numerator and/or the denominator, there are two methods that may be employed successfully. Method one employs the fundamental principle of fractions to obtain a simple fraction. Method two involves simplifying the numerator and/or the denominator prior to performing the division. We illustrate both methods.

See solved problem 3.12

SOLVED PROBLEMS

3.1 Reduce to lowest terms. Assume no denominator is zero.

(a) $\dfrac{20}{42}$

$$\frac{20}{42} = \frac{2 \cdot 10}{1 \cdot 21} = \frac{10}{21}$$

(b) $\dfrac{4x+6}{14}$

$$\frac{4x+6}{14} = \frac{2(2x+3)}{2 \cdot 7} = \frac{2x+3}{7}$$

(c) $\dfrac{4s^2t^3}{40st^5}$

$$\frac{4s^2t^3}{40st^5} = \frac{4 \cdot s \cdot s \cdot t^3}{4 \cdot 10 \cdot s \cdot t^3 \cdot t^2} = \frac{s}{10t^2}$$

(d) $\dfrac{(p+t)^2}{2p+2t}$

$$\frac{(p+t)^2}{2p+2t} = \frac{(p+t)(p+t)}{2(p+t)} = \frac{p+t}{2}$$

(e) $\dfrac{(a+b)^2}{(a+b)^3}$

$$\frac{(a+b)^2}{(a+b)^3} = \frac{(a+b)^2}{(a+b)^2\,(a+b)} = \frac{1}{a+b}$$

(*f*) $\dfrac{x^2 - y^2}{3x^2 + 5xy + 2y^2}$

$$\frac{x^2 - y^2}{3x^2 + 5xy + 2y^2} = \frac{(x + y)(x - y)}{(x + y)(3x + 2y)} = \frac{x - y}{3x + 2y}$$

(*g*) $\dfrac{12x^2 + 60xy + 75y^2}{24x + 60y}$

$$\frac{12x^2 + 60xy + 75y^2}{24x + 60y} = \frac{3(4x^2 + 20xt + 25y^2)}{12(2x + 5y)} = \frac{3(2x + 5y)^2}{12(2x + 5y)} = \frac{2x + 5y}{4}$$

Remember! It is advisable to check your answers. See the following paragraph.

The results obtained when reducing fractions may be verified by substituting numerical values such as 2, 3, 4, etc. for the various variables involved. Substitute the values into the original expression and simplify. Next, substitute the corresponding values used in the original expression into the simplified version and compare the results. You should refrain from using 1 and 0 when checking, since they sometimes result in special outcomes. This technique is not fool-proof, although it usually allows you to determine if errors were made.

We illustrate the above procedure by checking solved problem 3.1(*g*). We wish to determine if

$\dfrac{12x^2 + 60xy + 75y^2}{24x + 60y} = \dfrac{2x + 5y}{4}$. Let $x = 3, y = 2$ and substitute into the left side to obtain

$\dfrac{12x^2 + 60xy + 75y^2}{24x + 60y} = \dfrac{12(3)^2 + 60(3)2 + 75(2)^2}{24(3) + 60(2)} = 4$. Now substitute the values for x and y into the

right side to obtain $\dfrac{2x + 5y}{4} = \dfrac{2(3) + 5(2)}{4} = 4$. The results are identical; hence, the simplified form of

the original expression is most likely correct.

Do supplementary problem 3.1 to improve your skills.

3.2 Write the following in standard form. Assume no denominator is zero.

(*a*) $-\dfrac{2}{-x}$ (*b*) $\dfrac{-3}{-s}$ (*c*) $-\dfrac{-6}{x^2}$ (*d*) $-\dfrac{-x}{-(x + 3)}$ (*e*) $-\dfrac{t}{-(5 - t)}$

(*a*) $\dfrac{2}{x}$ (*b*) $\dfrac{3}{s}$ (*c*) $\dfrac{6}{x^2}$ (*d*) $\dfrac{-x}{x + 3}$ (*e*) $\dfrac{t}{5 - t}$ or $\dfrac{-t}{t - 5}$

See supplementary problem 3.2.

3.3 Reduce to lowest terms and write in standard form. Assume no denominator is zero.

(*a*) $\dfrac{s - t}{t^2 - s^2}$

$$\frac{s - t}{t^2 - s^2} = \frac{s - t}{(t - s)(t + s)} = \frac{-(t - s)}{(t - s)(t + s)} = \frac{-1}{t + s} \text{ or } \frac{-1}{s + t}$$

(b) $\frac{2-x}{4-x^2}$

$$\frac{2-x}{4-x^2}=\frac{2-x}{(2-x)(2+x)}=\frac{1}{2+x} \text{ or } \frac{1}{x+2}$$

(c) $\frac{3-9a}{-3}$

$$\frac{3-9a}{-3}=\frac{3(1-3a)}{-3}=-1(1-3a)=3a-1$$

(d) $\frac{y-x}{x-y}$

$$\frac{y-x}{x-y}=\frac{-(x-y)}{x-y}=-1$$

(e) $\frac{-5(3-y)}{y^2+2y-15}$

$$\frac{-5(3-y)}{y^2+2y-15}=\frac{5(y-3)}{(y-3)(y+5)}=\frac{5}{y+5}$$

See supplementary problems 3.3 and 3.4 for additional practice.

3.4 Write each product as a single fraction in lowest terms.

(a) $\frac{3}{16}\cdot\frac{4}{15}$

$$\frac{3}{16}\cdot\frac{4}{15}=\frac{1}{4}\cdot\frac{1}{5}=\frac{1}{20}$$

(b) $\frac{5a^2}{b}\cdot\frac{b^3}{10a}$

$$\frac{5a^2}{b}\cdot\frac{b^3}{10a}=\frac{ab^2}{2}$$

(c) $\frac{35a^2b^4}{9ab^2}\cdot\frac{3a^3bc}{5b^2c^2}$

$$\frac{35a^2b^4}{9ab^2}\cdot\frac{3a^3bc}{5b^2c^2}=\frac{7ab^2}{3}\cdot\frac{a^3}{bc}=\frac{7a^4b}{3c}$$

(d) $\frac{s^2-t^2}{2s+4t}\cdot\frac{s+2t}{5s-5t}$

$$\frac{s^2-t^2}{2s+4t}\cdot\frac{s+2t}{5s-5t}=\frac{(s+t)(s-t)}{2(s+2t)}\cdot\frac{s+2t}{5(s-t)}=\frac{s+t}{10}$$

It is possible to perform some of the operations in different combinations in the above problems without affecting the result. As long as you employ the fundamental principle of fractions in an appropriate manner, the results will be correct.

See supplementary problem 3.5 for similar problems.

3.5 Divide and simplify. (Simplify means to reduce to lowest terms here.)

(a) $\frac{3}{4}\div\frac{25}{16}$

$$\frac{3}{4}\div\frac{25}{16}=\frac{3}{4}\cdot\frac{16}{25}=\frac{12}{25}$$

(b) $\frac{5}{8}\div 3$

$$\frac{5}{8}\div 3=\frac{5}{8}\cdot\frac{1}{3}=\frac{5}{24}$$

(*c*) $24xy^2 \div \dfrac{4y}{5}$

$$24xy^2 \div \frac{4y}{5} = \frac{24xy^2}{1} \cdot \frac{5}{4y} = 6xy \cdot 5 = 30xy$$

(*d*) $\dfrac{s^2t}{5r^3} \div \dfrac{st}{20r^3}$

$$\frac{s^2t}{5r^3} \div \frac{st}{20r^3} = \frac{s^2t}{5r^3} \cdot \frac{20r^3}{st} = 4s$$

(*e*) $\dfrac{6a-27}{5b} \div \dfrac{8a-36}{7b}$

$$\frac{6a-27}{5b} \div \frac{8a-36}{7b} = \frac{3(2a-9)}{5b} \cdot \frac{7b}{4(2a-9)} = \frac{21}{20}$$

(*f*) $\dfrac{10l+15w}{8} \div \dfrac{30l+45w}{4}$

$$\frac{10l+15w}{8} \div \frac{30l+45w}{4} = \frac{5(2l+3w)}{8} \cdot \frac{4}{15(2l+3w)} = \frac{1}{2 \cdot 3} = \frac{1}{6}$$

(*g*) $\dfrac{x^2+6x+5}{x^2-x-6} \div \dfrac{x^2+2x-15}{x^2-4}$

$$\frac{x^2+6x+5}{x^2-x-6} \div \frac{x^2+2x-15}{x^2-4} = \frac{(x+5)(x+1)}{(x-3)(x+2)} \cdot \frac{(x+2)(x-2)}{(x+5)(x-3)} = \frac{(x+1)(x-2)}{(x-3)^2}$$

(*h*) $\dfrac{3w^2+6w-24}{5w^2+30w+45} \div \dfrac{w^2-4w+4}{10w+30}$

$$\frac{3w^2+6w-24}{5w^2+30w+45} \div \frac{w^2-4w+4}{10w+30} = \frac{3(w^2+2w-8)}{5(w^2+6w+9)} \div \frac{w^2-4w+4}{10(w+3)}$$

$$= \frac{3(w+4)(w-2)}{5(w+3)^2} \cdot \frac{10(w+3)}{(w-2)^2} = \frac{3(w+4) \cdot 2}{(w+3)(w-2)} = \frac{6(w+4)}{(w+3)(w-2)}$$

Note that answers appear in factored form. If we were to distribute, it would be less apparent that the results are in lowest terms.

See supplementary problem 3.6 to practice the above concepts.

3.6 Perform the indicated operations and simplify.

(*a*) $\dfrac{15x^2y}{25a^2b^3} \cdot \left(\dfrac{2x}{5y} \div \dfrac{8bx^3}{25ay^3} \right)$

$$\frac{15x^2y}{25a^2b^3} \cdot \left(\frac{2x}{5y} \div \frac{8bx^3}{25ay^3} \right) = \frac{3x^2y}{5a^2b^3} \cdot \left(\frac{2x}{5y} \cdot \frac{25ay^3}{8bx^3} \right) = \frac{3x^2y}{5a^2b^3} \cdot \left(\frac{5ay^2}{4bx^2} \right) = \frac{3y^3}{4ab^4}$$

(*b*) $\dfrac{18s^2+18st+4t^2}{3s^2-2st-t^2} \div \left(\dfrac{s-t}{3s^2+4st+t^2} \cdot \dfrac{3s+2t}{s+t} \right)$

$$\frac{18s^2+18st+4t^2}{3s^2-2st-t^2} \div \left(\frac{s-t}{3s^2+4st+t^2} \cdot \frac{3s+2t}{s+t} \right)$$

$$= \frac{2(9s^2+9s+2t^2)}{(3s+t)(s-t)} \div \left[\frac{s-t}{(3s+t)(s+t)} \cdot \frac{3s+2t}{s+t} \right] = \frac{2(3s+2t)(3s+t)}{(3s+t)(s-t)} \div \left[\frac{(s-t)(3s+2t)}{(3s+t)(s+t)^2} \right]$$

$$= \frac{2(3s+2t)}{(s-t)} \cdot \frac{(3s+t)(s+t)^2}{(s-t)(3s+2t)} = \frac{2(3s+t)(s+t)^2}{(s-t)^2}$$

See supplementary problem 3.7 for additional practice.

3.7 Perform the indicated operation.

(*a*) $\frac{2}{7}+\frac{4}{7}$

$\frac{2}{7}+\frac{4}{7}=\frac{2+4}{7}=\frac{6}{7}$

(*b*) $\frac{x}{5}+\frac{y}{5}$

$\frac{x}{5}+\frac{y}{5}=\frac{x+y}{5}$

(*c*) $\frac{s}{3x}-\frac{2t}{3x}$

$\frac{s}{3x}-\frac{2t}{3x}=\frac{s-2t}{3x}$

(*d*) $\frac{3x-1}{x+2}-\frac{x-4}{x+2}$

$\frac{3x-1}{x+2}-\frac{x-4}{x+2}=\frac{3x-1-(x-4)}{x+2}=\frac{3x-1-x+4}{x+2}=\frac{2x+3}{x+2}$

(*e*) $\frac{x^2+15x-4}{x^2-x-2}+\frac{x^2-x-32}{x^2-x-2}$

$\frac{x^2+15x-4}{x^2-x-2}+\frac{x^2-x-32}{x^2-x-2}=\frac{x^2+15x-4+x^2-x-32}{x^2-x-2}=\frac{2x^2+14x-36}{x^2-x-2}$

$=\frac{2(x^2+7x-18)}{(x+1)(x-2)}=\frac{2(x-2)(x+9)}{(x+1)(x-2)}=\frac{2(x+9)}{x+1}$

Remember to reduce fractions when the numerator and denominator contain common factors. Also remember that no variable may have a value which would result in a zero denominator.

Refer to supplementary problem 3.8 for similar exercises.

3.8 Find the LCD for the following fractions.

(*a*) $\frac{6}{175}, \frac{5}{36}$, and $\frac{8}{135}$

$175 = 5^2 \cdot 7$; $36 = 2^2 \cdot 3^2$; and $135 = 5 \cdot 3^3$. The LCD $= 5^2 \cdot 7 \cdot 2^2 \cdot 3^3 = 2^2 \cdot 3^3 \cdot 5^2 \cdot 7 = 18{,}900$

(*b*) $\frac{2}{15x^2}$ and $\frac{5}{18xy}$

$15x^2 = 3 \cdot 5x^2$ and $18xy = 2 \cdot 3^2 \cdot xy$ so the LCD $= 2 \cdot 3^2 \cdot 5x^2y = 90x^2y$

(*c*) $\frac{3s+4}{s^2-4}$ and $\frac{2s-5}{s^2+4s+4}$

$s^2-4=(s+2)(s-2)$ and $s^2+4s+4=(s+2)^2$ so the LCD $=(s+2)^2(s-2)$

Note: In most circumstances, the factored form of the LCD is the most useful.

Supplementary problem 3.9 involves similar exercises.

3.9 Perform the indicated operation(s).

(*a*) $\frac{1}{x}+\frac{1}{2x}$

The LCD is $2x$. Therefore, $\frac{1}{x}+\frac{1}{2x}=\frac{1}{x}\cdot\frac{2}{2}+\frac{1}{2x}=\frac{2+1}{2x}=\frac{3}{2x}$

(*b*) $\frac{1}{t}+\frac{t-2}{t^2}-\frac{1}{2t}$

The LCD is $2t^2$. Therefore,

$$\frac{1}{t}+\frac{t-2}{t^2}-\frac{1}{2t}=\frac{1}{t}\cdot\frac{2t}{2t}+\frac{2}{2}\cdot\frac{t-2}{t^2}-\frac{1}{2t}\cdot\frac{t}{t}$$

$$=\frac{2t+2(t-2)-t}{2t^2}=\frac{2t+2t-4-t}{2t^2}=\frac{3t-4}{2t^2}$$

(*c*) $\frac{5}{18w}+\frac{3w}{4w-20}$

$18w=2\cdot 3^2 w$ and $4w-20=4(w-5)=2^2(w-5)$ so the LCD is $2^2\cdot 3^2 w(w-5)$ $=36w(w-5)$. Therefore,

$$\frac{5}{18w}+\frac{3w}{4w-20}=\frac{5}{2\cdot 3^2 w}\cdot\frac{2(w-5)}{2(w-5)}+\frac{3w}{2^2(w-5)}\cdot\frac{3^2 w}{3^2 w}$$

$$=\frac{10(w-5)}{2^2\cdot 3^2 w(w-5)}+\frac{3^2 w^2}{2^2\cdot 3^2 w(w-5)}=\frac{10w-50+27w^2}{2^2\cdot 3^2 w(w-5)}$$

$$=\frac{27w^2+10w-50}{36w(w-5)}$$

(*d*) $\frac{p}{p+2}-\frac{p-1}{p^2+7p+10}$

$p^2+7p+10=(p+2)(p+5)$ so the LCD is $(p+2)(p+5)$.

$$\frac{p}{p+2}-\frac{p-1}{p^2+7p+10}=\frac{p}{p+2}-\frac{p-1}{(p+2)(p+5)}=\frac{p}{p+2}\cdot\frac{p+5}{p+5}-\frac{p-1}{(p+2)(p+5)}$$

$$=\frac{p(p+5)-(p-1)}{(p+2)(p+5)}=\frac{p^2+5p-p+1}{(p+2)(p+5)}=\frac{p^2+4p+1}{(p+2)(p+5)}$$

(*e*) $\frac{1}{s}-\frac{2s}{s^2-4}+\frac{1}{s+2}$

$s^2-4=(s+2)(s-2)$ so the LCD is $s(s+2)(s-2)$.

$$\frac{1}{s}-\frac{2s}{s^2-4}+\frac{1}{s+2}=\frac{1}{s}\cdot\frac{(s+2)(s-2)}{(s+2)(s-2)}-\frac{2s}{(s+2)(s-2)}\cdot\frac{s}{s}+\frac{1}{s+2}\cdot\frac{s(s-2)}{s(s-2)}$$

$$=\frac{s^2-4-2s^2+s^2-2s}{s(s+2)(s-2)}=\frac{-2s-4}{s(s+2)(s-2)}$$

$$=\frac{-2(s+2)}{s(s+2)(s-2)}=\frac{-2}{s(s-2)}$$

(*f*) $\frac{2}{x^2+x-6}-\frac{5}{x^2+2x-3}$

$x^2+x-6=(x+3)(x-2)$ and $x^2+2x-3=(x+3)(x-1)$.

The LCD is $(x+3)(x-2)(x-1)$. Therefore,

$$\frac{2}{x^2+x-6}-\frac{5}{x^2+2x-3}=\frac{2}{(x+3)(x-2)}\cdot\frac{(x-1)}{(x-1)}-\frac{5}{(x+3)(x-1)}\cdot\frac{(x-2)}{(x-2)}$$

$$=\frac{2x-2}{(x+3)(x-2)(x-1)}-\frac{5x-10}{(x+3)(x-2)(x-1)}$$

$$=\frac{(2x-2)-(5x-10)}{(x+3)(x-2)(x-1)}=\frac{2x-2-5x+10}{(x+3)(x-2)(x-1)}$$

$$=\frac{-3x+8}{(x+3)(x-2)(x-1)}=\frac{8-3x}{(x+3)(x-2)(x-1)}$$

Recall that changing any combination of two of the three signs of a fraction results in an equivalent fraction. We employ the technique in the following problem.

(*g*) $2 + \frac{1}{x-y} - \frac{4}{y-x}$

$y - x$ may be written as $-(x - y)$. Rewrite the last fraction as an equivalent fraction with denominator $x - y$. The LCD is then $x - y$.

$$2 + \frac{1}{x-y} - \frac{4}{y-x} = 2 + \frac{1}{x-y} + \frac{4}{x-y} = 2 \cdot \frac{x-y}{x-y} + \frac{1}{x-y} + \frac{4}{x-y}$$

$$= \frac{2(x-y)+1+4}{x-y} = \frac{2x-2y+5}{x-y}$$

Practice by doing supplementary problem 3.10.

3.10 Perform the indicated operations and simplify.

(*a*) $\left(\frac{2}{x} + \frac{1}{y}\right) \div \frac{1}{x^2y}$

$$\left(\frac{2}{x} + \frac{1}{y}\right) \div \frac{1}{x^2y} = \left(\frac{2}{x} \cdot \frac{y}{y} + \frac{1}{y} \cdot \frac{x}{x}\right) \div \frac{1}{x^2y} = \frac{2y+x}{xy} \div \frac{1}{x^2y}$$

$$= \frac{2y+x}{xy} \cdot \frac{x^2y}{1} = (2y+x)\,x = x^2 + 2xy$$

(*b*) $\left(\frac{t+9}{t-3} - \frac{2}{t}\right) \div \frac{t^2-36}{t^2-9}$

$$\left(\frac{t+9}{t-3} - \frac{2}{t}\right) \div \frac{t^2-36}{t^2-9} = \left(\frac{t+9}{t-3} \cdot \frac{t}{t} - \frac{2}{t} \cdot \frac{t-3}{t-3}\right) \div \frac{t^2-36}{t^2-9}$$

$$= \left(\frac{t^2+9t-2t+6}{t(t-3)}\right) \div \frac{(t+6)(t-6)}{(t+3)(t-3)}$$

$$= \frac{t^2+7t+6}{t(t-3)} \cdot \frac{(t+3)(t-3)}{(t+6)(t-6)} = \frac{(t+6)(t+1)}{t} \cdot \frac{(t+3)}{(t+6)(t-6)}$$

$$= \frac{(t+1)(t+3)}{t(t-6)}$$

3.11 Express $\dfrac{\dfrac{2s^2t}{3x^2y}}{\dfrac{8st^3}{15x^3y}}$ as a simple fraction in lowest terms.

$$\frac{\dfrac{2s^2t}{3x^2y}}{\dfrac{8st^3}{15x^3y}} = \frac{2s^2t}{3x^2y} \cdot \frac{15x^3y}{8st^3} = \frac{5sx}{4t^2}$$

3.12 Simplify.

(*a*) $\dfrac{1 + \dfrac{2}{x}}{\dfrac{3}{y} + 1}$

Method 1: The fundamental principle of fractions says that we obtain an equivalent fraction if we multiply both numerator and denominator by the same quantity k. The appropriate k for our purpose is merely the LCD of the fractions which appear in the numerator and denominator. In our example, the LCD $= xy$. The entire sequence of essential steps is shown below.

$$\frac{1+\frac{2}{x}}{\frac{3}{y}+1}=\frac{\left(1+\frac{2}{x}\right)xy}{\left(\frac{3}{y}+1\right)xy}=\frac{1\cdot xy+\frac{2}{x}\cdot xy}{\frac{3}{y}\cdot xy+1\cdot xy}=\frac{xy+2y}{3x+xy}=\frac{y(x+2)}{x(y+3)}$$

Method 2: We begin by finding the sums in the numerator and denominator of the complex fraction. We next perform the division. The necessary steps are shown below.

$$\frac{1+\frac{2}{x}}{\frac{3}{y}+1}=\frac{1\cdot\frac{x}{x}+\frac{2}{x}}{\frac{3}{y}+1\cdot\frac{y}{y}}=\frac{\frac{x+2}{x}}{\frac{3+y}{y}}=\frac{x+2}{x}\cdot\frac{y}{3+y}=\frac{y(x+2)}{x(y+3)}$$

Observe that identical results were obtained by both methods. Method 1 is usually the more efficient method to employ since the distributive step can normally be done mentally.

We demonstrate another example below.

(*b*) $$\frac{\frac{s}{s-1}+\frac{2}{s+1}}{4+\frac{1}{s^2-1}}$$

Method 1: Recall that $s^2 - 1 = (s + 1)(s - 1)$. The LCD of $\frac{s}{s-1}, \frac{2}{s+1}$, and $\frac{1}{s^2-1}$ is $(s + 1)(s - 1)$.

We now multiply both numerator and denominator by the LCD and simplify.

$$\frac{\left(\frac{s}{s-1}+\frac{2}{s+1}\right)(s+1)(s-1)}{\left(4+\frac{1}{s^2-1}\right)(s+1)(s-1)}=\frac{\frac{s}{s-1}\cdot(s+1)(s-1)+\frac{2}{s+1}\cdot(s+1)(s-1)}{4\cdot(s+1)(s-1)+\frac{1}{s^2-1}\cdot(s+1)(s-1)}$$

$$=\frac{s(s+1)+2(s-1)}{4(s+1)(s-1)+1}=\frac{s^2+s+2s-2}{4(s^2-1)+1}=\frac{s^2+3s-2}{4s^2-3}$$

Method 2: The LCD of the fractions in the numerator as well as the denominator is $(s + 1)(s - 1)$. We now perform the indicated operations in the numerator and the denominator followed by the division of the results.

$$\frac{\frac{s}{s-1}+\frac{2}{s+1}}{4+\frac{1}{s^2-1}}=\frac{\frac{s}{s-1}\cdot\frac{s+1}{s+1}+\frac{2}{s+1}\cdot\frac{s-1}{s-1}}{4\cdot\frac{(s+1)(s-1)}{(s+1)(s-1)}+\frac{1}{(s+1)(s-1)}}$$

$$=\frac{\frac{s(s+1)+2(s-1)}{(s+1)(s-1)}}{\frac{4(s+1)(s-1)+1}{(s+1)(s-1)}}=\frac{s(s+1)+2(s-1)}{(s+1)(s-1)}\cdot\frac{(s+1)(s-1)}{4(s+1)(s-1)+1}$$

$$=\frac{s^2+s+2s-2}{4(s^2-1)+1}=\frac{s^2+3s-2}{4s^2-3}$$

Notice that both methods produce the same result once again. Practice the above concepts by doing supplementary problem 3.11.

SUPPLEMENTARY PROBLEMS

Assume no denominator is zero in each of the following problems.

3.1 Reduce to lowest terms.

(a) $\frac{18}{60}$ (b) $\frac{7x-21}{28}$ (c) $\frac{30a^4b^2}{40a^2b^5}$

(d) $\frac{3s-12t}{(s-4t)^2}$ (e) $\frac{(x-y)^2}{2x^2-5xy+3y^2}$ (f) $\frac{5x^{2n}y^k}{35x^ny^2}$

3.2 Write the following in standard form.

(a) $-\frac{-a}{8}$ (b) $\frac{-10}{-s}$ (c) $-\frac{x}{4-y}$

(d) $-\frac{7}{-b}$ (e) $-\frac{-s}{-t^3}$ (f) $-\frac{3}{-(x-3)}$

3.3 Reduce to lowest terms and write in standard form.

(a) $\frac{4-t}{t^2-16}$ (b) $\frac{49-x^2}{-2x-14}$ (c) $\frac{a^2-a-6}{18-3a-a^2}$

(d) $\frac{4}{-4s^2-36}$ (e) $\frac{40+12x-4x^2}{6y(x^2-3x-10)}$

3.4 Employ the fundamental principle of fractions to reduce to lowest terms. Write your answers in standard form.

(a) $-\frac{28a^2b}{35ab^3}$ (b) $\frac{30}{6x-18y}$ (c) $\frac{-5s^2t^4u^3}{-20st^4u^7}$

(d) $\frac{4k-16}{4-k}$ (e) $\frac{7a+7b}{a^2+b^2}$ (f) $\frac{2x^2y-4xy^2+14x^2y^2}{-2x}$

(g) $\frac{9x^4y^3z+3xy^2z}{12x^2y^2z-24x^3y^2z^2}$ (h) $\frac{(a+b)^3}{a^2-b^2}$ (i) $\frac{3s-3t}{t-s}$

(j) $\frac{x^2+2xy+y^2}{4x^2-xy-5y^2}$ (k) $\frac{s^3-t^3}{8s^2+8st+8t^2}$ (l) $\frac{a^{2n}+a^nb^n}{3a^{3n}+3a^{2n}b^n}$

3.5 Write each product as a single fraction in lowest terms.

(a) $\frac{12}{35}\cdot\frac{10}{27}$ (b) $\frac{14x}{15}\cdot\frac{5y^2}{7x^3}$

(c) $\frac{a^3}{x^2y}\cdot\frac{3x^3y}{ab^2}$ (d) $\frac{2y}{4xy-6y^2}\cdot\frac{2x-3y}{14x}$

(e) $\frac{4a^2-1}{b^2-16}\cdot\frac{b^2-4b}{2a+1}$ (f) $\frac{t^2-t-2}{t^2-3t-10}\cdot\frac{t^2-4t-5}{t^2+4t+3}$

(g) $\frac{(s-5)^2}{(s+3)^2}\cdot\frac{s^2+7s+12}{s^2-s-20}$ (h) $\frac{3x^2-7x-6}{3x^2-13x-10}\cdot\frac{4x^2-18x-10}{2x^2-x-1}$

3.6 Divide and simplify.

(*a*) $14 \div \dfrac{2}{3}$

(*b*) $\dfrac{4s^3}{7t^2} \div \dfrac{8s}{21t}$

(*c*) $30a^3b^3 \div \dfrac{15ab^2}{9}$

(*d*) $\dfrac{4x}{2x+y} \div \dfrac{8x^2-8xy}{2x+y}$

(*e*) $\dfrac{6x-15y}{x^2-2xy-3y^2} \div \dfrac{12x-30y}{x+y}$

(*f*) $\dfrac{p^2+p-12}{p^2+p-2} \div \dfrac{p^2-4p+3}{p^2+5p+6}$

(*g*) $\dfrac{s^3-27}{4s^2-4s} \div \dfrac{s-3}{4s}$

(*h*) $(a+2)^3 \div \dfrac{a^3+8}{a^2+4}$

3.7 Perform the indicated operations and simplify.

(*a*) $\dfrac{32ab^2}{25xy^3} \cdot \dfrac{15x^2y}{4a^3b} \cdot \dfrac{a^2b^2}{2xy}$

(*b*) $\dfrac{15s^2t^3}{16wx^2} \div \left(\dfrac{3st}{12w^3} \cdot \dfrac{10x}{21w}\right)$

(*c*) $\dfrac{x-y}{2x^2+3xy+y^2} \cdot \left(\dfrac{x^2+xy}{x^2-2xy+y^2} \div \dfrac{x^2+xy}{2x^2-xy-y^2}\right)$

(*d*) $\left(\dfrac{3s-2t}{s+t} \div \dfrac{3s^2-2st}{3st+2t^2}\right) \div \dfrac{5s+3t}{s^2-t^2}$

3.8 Perform the indicated operation.

(*a*) $\dfrac{p}{4} - \dfrac{3}{4}$

(*b*) $\dfrac{5}{t} + \dfrac{w}{t}$

(*c*) $\dfrac{3x}{x^2y} - \dfrac{2x-5}{x^2y}$

(*d*) $\dfrac{8s+5}{s-3} + \dfrac{4-5s}{s-3}$

(*e*) $\dfrac{2t^2-8t-11}{t^2+3t-4} + \dfrac{t^2+14t-13}{t^2+3t-4}$

3.9 Find the LCD for the following fractions.

(*a*) $\dfrac{4}{45}, \dfrac{7}{600}$, and $\dfrac{5}{42}$

(*b*) $\dfrac{x}{28s^2t}$ and $\dfrac{y}{30st^3}$

(*c*) $\dfrac{5a+2}{a^2-3a-10}$ and $\dfrac{3a}{a^2+a-30}$

3.10 Perform the indicated operation(s).

(*a*) $\dfrac{2}{3x} + \dfrac{5}{2x}$

(*b*) $\dfrac{x+4}{3} - \dfrac{x+1}{4}$

(*c*) $\dfrac{5}{a-b} + \dfrac{2}{b-a}$

(*d*) $5 + \dfrac{7}{t-2}$

(*e*) $\dfrac{3}{5} + \dfrac{1}{t} - \dfrac{t}{t+5}$

(*f*) $\dfrac{2s-1}{s+6} - \dfrac{s+3}{s-2}$

(*g*) $\dfrac{1}{x} + \dfrac{3}{x+3} - \dfrac{3}{x^2+5x+6}$

(*h*) $\dfrac{4c}{c^2-3c-4} - \dfrac{2c}{c^2-6c+8}$

3.11 Perform the indicated operations and simplify.

(a) $\frac{3}{x}-\frac{x}{4}\times\frac{5}{x^2}$ (b) $\frac{4}{x}+\frac{x}{3}\div\frac{x^2}{2}$

(c) $\frac{4}{x}\div\frac{x}{3}\times\frac{x^2}{2}$ (d) $\left(\frac{3}{x}-\frac{x}{4}\right)\frac{5}{x^2}$

(e) $\left(\frac{3}{s}-1\right)\left(\frac{3}{s}+1\right)$ (f) $a+b-\frac{1}{a-b}$

(g) $\dfrac{1-\frac{1}{c}}{1+\frac{1}{c}}$ (h) $\frac{t+2}{t^2+5t+6}+\frac{t-1}{t^2+7t+12}-\frac{2}{t+4}$

(i) $\dfrac{\frac{2}{x+y}}{\frac{8}{x-y}}$ (j) $\dfrac{\frac{8a^3b^2}{3cd^2}}{\frac{4ab^2}{15c^2d}}$

(k) $\dfrac{\frac{s-t}{s}-\frac{s+t}{t}}{\frac{s-t}{t}+\frac{s+t}{s}}$ (l) $\dfrac{\frac{3p}{7}+\frac{q}{14}}{\frac{3p}{28}-\frac{q}{2}}$

(m) $\dfrac{x-\frac{4}{x+6}}{\frac{1}{x+6}+x}$ (n) $\dfrac{a+b}{\frac{1}{a}+\frac{1}{b}}$

(o) $\dfrac{\frac{2x^2-6x}{x+2}-2x}{\frac{x}{x-3}-\frac{4x}{x^2-x-6}}$

ANSWERS TO SUPPLEMENTARY PROBLEMS

3.1 (a) $\frac{3}{10}$ (b) $\frac{x-3}{4}$ (c) $\frac{3a^2}{4b^3}$

(d) $\frac{3}{s-4t}$ (e) $\frac{x-y}{2x-3y}$ (f) $\frac{x^ny^{k-2}}{7}$

3.2 (a) $\frac{a}{8}$ (b) $\frac{10}{s}$ (c) $\frac{x}{y-4}$ or $\frac{-x}{4-y}$

(d) $\frac{7}{b}$ (e) $\frac{-s}{t^3}$ (f) $\frac{3}{x-3}$ or $\frac{-3}{3-x}$

3.3 (a) $\frac{-1}{t+4}$ (b) $\frac{x-7}{2}$ (c) $\frac{-(a-2)}{a+6}$

(d) $\frac{-1}{s^2+9}$ (e) $\frac{-2}{3y}$

3.4 (a) $\frac{-4a}{5b^2}$ (b) $\frac{5}{x-3y}$ (c) $\frac{s}{4u^4}$

(d) -4 (e) It is in lowest terms. (f) $-y(x-2y+7xy)$

(g) $\frac{3x^3y+1}{4x(1-2xz)}$ (h) $\frac{(a+b)^2}{a-b}$ (i) -3

(j) $\frac{x+y}{4x-5y}$ (k) $\frac{s-t}{8}$ (l) $\frac{1}{3a^n}$

3.5 (a) $\frac{8}{63}$ (b) $\frac{2y^2}{3x^2}$ (c) $\frac{3a^2x}{b^2}$

(d) $\frac{1}{14x}$ (e) $\frac{b(2a-1)}{b+4}$ (f) $\frac{(t-2)(t+1)}{(t+2)(t+3)}$

(g) $\frac{s-5}{s+3}$ (h) $\frac{2(x-3)}{x-1}$

3.6 (a) 21 (b) $\frac{3s^2}{2t}$ (c) $18a^2b$

(d) $\frac{1}{2(x-y)}$ (e) $\frac{1}{2(x-3y)}$ (f) $\frac{(p+4)(p+3)}{(p-1)^2}$

(g) $\frac{s^2+3s+9}{s-1}$ (h) $\frac{(a+2)^2(a^2+4)}{a^2-2a+4}$

3.7 (a) $\frac{12b^3}{5y^3}$ (b) $\frac{63st^2w^3}{8x^3}$ (c) $\frac{1}{x+y}$ (d) $\frac{t(3s+2t)(s-t)}{s(5s+3t)}$

3.8 (a) $\frac{p-3}{4}$ (b) $\frac{5+w}{t}$ or $\frac{w+5}{t}$ (c) $\frac{x+5}{x^2y}$

(d) $\frac{3s+9}{s-3}$ or $\frac{3(s+3)}{s-3}$ (e) $\frac{3(t-2)}{t-1}$

3.9 (a) $2^3 \cdot 3^2 \cdot 5^2 \cdot 7 = 12,600$ (b) $420s^2t^3$ (c) $(a+2)(a-5)(a+6)$

3.10 (a) $\frac{19}{6x}$ (b) $\frac{x+13}{12}$ (c) $\frac{3}{a-b}$

(d) $\frac{5t-3}{t-2}$ (e) $\frac{25+20t-2t^2}{5t(t+5)}$ (f) $\frac{s^2-14s-16}{(s+6)(s-2)}$

(g) $\frac{2(2x^2+4x+3)}{x(x+3)(x+2)}$ (h) $\frac{2c(c-5)}{(c+1)(c-4)(c-2)}$

3.11 (a) $\frac{7}{4x}$ (b) $\frac{14}{3x}$ (c) 6

(d) $\frac{5(12-x^2)}{4x^3}$ (e) $\frac{9-s^2}{s^2}$ (f) $\frac{a^2-b^2-1}{a-b}$

(g) $\frac{c-1}{c+1}$ (h) $\frac{-3}{(t+3)(t+4)}$ (i) $\frac{x-y}{4(x+y)}$

(j) $\frac{10a^2c}{d}$ (k) -1 (l) $\frac{2(6p+q)}{3p-14q}$

(m) $\frac{x^2+6x-4}{x^2+6x+1}$ (n) ab (o) $\frac{-10(x-3)}{x-2}$

First-Degree Equations and Inequalities

Solving various kinds of equations and inequalities is one of the most essential and useful skills in mathematics. In this chapter, we will develop techniques for solving first-degree equations and inequalities. We shall assume that variables may be replaced by any real number.

4.1 Solving First-Degree Equations

Definition 1. An *equation* is a mathematical statement that two expressions are equal.

Equations may be true always, sometimes, or never.

Definition 2. An equation that is always true is called an *identity*.

Some examples of identities are $x + 2 = 2 + x$ and $x^2 = (-x)^2$. Each is true for every real value of x.

Definition 3. An equation that is sometimes true is called a *conditional* equation.

Some examples of conditional equations are $x + 2 = 8$ and $x^2 - x - 6 = 0$. Each is true for specific values of x, 6 for the first equation and both 3 and -2 for the second, but false for all other values. The equations are true for certain "conditions" only.

Definition 4. An equation that is never true is called a *contradiction*.

Some examples of contradictions are $x + 2 = x$ and $3 = 5$. Neither is true for any value of x.

See solved problems 4.1–4.2.

Definition 5. A value of the variable that results in a true equation is called a *solution or root* of the equation. The value is said to "satisfy" the equation.

Definition 6. The set of all of the solutions of an equation is called the *solution set* of the equation.

Verify that $x = 8$ is a solution of each of the following equations. Replace x by 8 and observe that a true statement results in each case.

(*i*) $x = 8$
(*ii*) $3x = 24$
(*iii*) $3x - 3 = 21$
(*iv*) $3(x - 1) + 5 = 26$

Definition 7. Equations which have exactly the same solution set are called *equivalent equations*.

Hence, the equations which appear immediately prior to the preceding definition are all equivalent equations.

The solution to the first equation in the sequence above is obvious. The solution to each of the successive equations in the sequence is increasingly more difficult to verify.

In general, the process of solving an equation involves transforming the equation into a sequence of equivalent equations. The sequence should result in "simpler" equations as we proceed. Ultimately we hope to obtain an equation whose solution is obvious. We employ the axioms and properties of equality in an appropriate order to accomplish the task.

Our efforts in this section will primarily be devoted to finding solutions of conditional equations. Additionally, the equations will be restricted to first-degree equations.

Definition 8. A *first-degree equation* is an equation that can be written in the form $ax + b = 0$, where a and b are constants and $a \neq 0$.

A first-degree equation is an equation in which the variable appears to the first power only and is not part of a denominator. Some examples of first-degree equations are $4x + 9 = 0$; $5x - 9 = 2x$; and $2(x + 3) - 4(2 - x) = x$.

Review the axioms of equality which were presented in Chapter 1 before continuing.

The substitution and reflexive properties of equality were stated in Chapter 1. The substitution property allows us to substitute equal quantities for one another. The reflexive property states that a quantity equals itself.

Suppose $a = b$ and $a + 4 = a + 4$ are given. Since $a = b$, we may substitute b for a on the right side of $a + 4 = a + 4$ to obtain $a + 4 = b + 4$. Compare $a = b$ with $a + 4 = b + 4$. If we add 4 to both sides of $a = b$, we obtain $a + 4 = b + 4$. This sequence of statements illustrates that if two expressions are equal, the same quantity may be added to both sides to obtain equal expressions. This property is known as the *addition property of equality*.

Since subtraction is defined in terms of addition, the addition property of equality also allows us to subtract the same quantity from both sides of an equation to obtain equal expressions.

Analogous arguments may be made for multiplication and division by nonzero quantities also.

We now state the above properties formally. Assume the variables a, b, and q represent real number values or quantities.

Addition property of equality	If $a = b$, then $a + q = b + q$.
Multiplication property of equality	If $a = b$, then $aq = bq$ provided that $q \neq 0$.

Equivalent equations are obtained when the properties of equality are applied to an equation. Those properties allow us to obtain equivalent equations when we add or subtract the same quantity to or from both sides, or when we multiply or divide both sides by the same nonzero quantity.

See solved problem 4.3.

The process used to solve first-degree equations not involving fractions is summarized below.

1. Use the distributive property and simplify each side of the equation if possible.
2. Use the addition property of equality to obtain an equation that contains the variable term on one side and the constant term on the other side.
3. Use the multiplication property of equality to make the coefficient of the variable one. In other words, isolate the variable.
4. Check the result in the original equation.

See solved problem 4.4.

4.2 Graphs of First-Degree Equations

We discussed solving first-degree equations in one variable in Section 4.1. We now consider first-degree equations in two variables. A solution to an equation in two variables consists of a pair of numbers. A value for each variable is needed in order for the equation to be identified as true or false.

Consider the equation $y = 3x - 4$. If $x = 1$, then $y = 3(1) - 4 = -1$. So, if x is replaced by 1 and y is replaced by -1, the equation is satisfied.

The reader can verify that each of the pairs of numbers given below also satisfy the equation.

If $x = 0, y = -4$.
If $x = 3, y = 5$.
If $x = -2, y = -10$.

It is customary to express the pairs of numbers as an *ordered pair*. We usually write the pair in alphabetical order and use parentheses to express the fact that a specific order is intended. Hence, (x, y) is the notation used to express ordered pairs for the equation $y = 3x - 4$. The specific ordered pairs from the values above which satisfy the equation are $(1, -1)$, $(0, -4)$, $(3, 5)$, and $(-2, -10)$. In fact, there are infinitely many ordered pairs which satisfy the given equation as well as all other first-degree equations in two variables. Each ordered pair found above is a solution of the equation $y = 3x - 4$.

In the ordered pair (x, y), x is the *first component* or *abscissa* and y is the *second component* or *ordinate*. The solution set consists of all the ordered pairs which satisfy the equation.

Since there are infinitely many elements in the solution set, we must employ a way to display it other than listing the members of the set. Back in the 17th century, the French mathematician/philosopher René Descartes developed a method to display infinite sets of this type. He created what is now referred to as the *rectangular*, or *Cartesian, coordinate system*. It associates each ordered pair of real numbers with a unique point in the plane. He placed a horizontal number line and a vertical number line perpendicular to one another at a point called the *origin*. The origin is located at the point which corresponds to zero on each number line. The horizontal line is commonly called the x-axis, and the vertical line is called the y-axis. Together they are referred to as the coordinate axes. The coordinate axes separate the plane into four regions called *quadrants*. We use Roman numerals to designate the quadrants.

The first component or *x-coordinate* of the ordered pair is the directed distance of the point from the vertical axis. The second component or *y-coordinate* of the ordered pair is the directed distance of the point from the horizontal axis. The components of the ordered pair that corresponds to a particular point are called the *coordinates* of the point, and the point is called the *graph* of the ordered pair. If a point lies to the right of the vertical axis, its x-coordinate is positive. On the other hand, the x-coordinate is negative if the point lies to the left of the vertical axis. Similarly the y-coordinate is positive if the point lies above the horizontal axis and negative if the point lies below the horizontal axis. If a point lies on one of the axes, its directed distance from that axis is zero and the corresponding coordinate is zero. The coordinates of the origin are $(0, 0)$.

A rectangular coordinate system is shown in Figure 4.1.

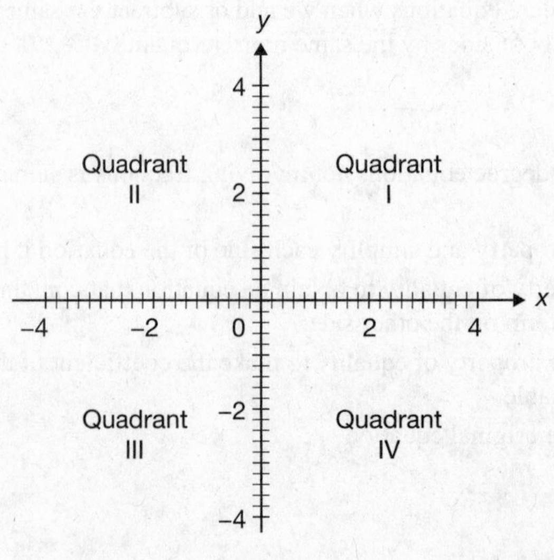

Figure 4.1

See solved problem 4.5.

We now return to the idea of displaying a solution set which contains infinitely many elements.

Consider the solution set of the equation $y = x - 2$. Since x and y represent any real numbers, we can choose arbitrary values for x, then evaluate the right side of the equation to find the corresponding y value. We find several ordered pairs in the solution set and observe that a pattern emerges when we graph the points on a rectangular coordinate system. The results are shown below.

If $x = -4$, $y = -6$. The point with coordinates $(-4, -6)$ is on the graph.

If $x = -2$, $y = -4$. The point with coordinates $(-2, -4)$ is on the graph.

If $x = 0$, $y = -2$. The point with coordinates $(0, -2)$ is on the graph.

If $x = 2$, $y = 0$. The point with coordinates $(2, 0)$ is on the graph.

If $x = 4$, $y = 2$. The point with coordinates $(4, 2)$ is on the graph.

Now graph the points which correspond to the ordered pairs we found. See Figure 4.2 (*a*).

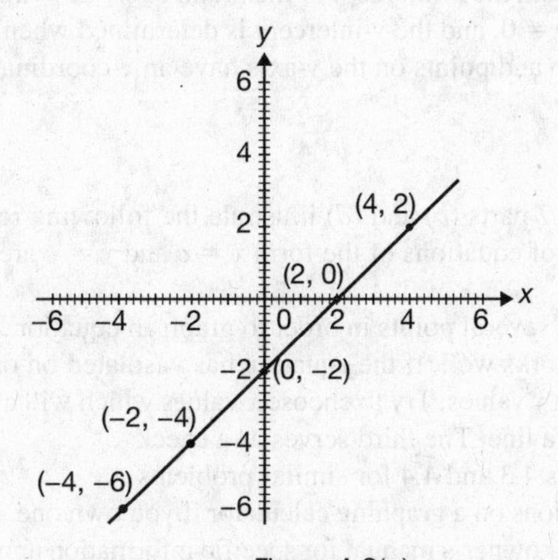

Figure 4.2(*a*)

The plotted points appear to be on a line. The line which passes through the points graphed in Figure 4.2 (*a*) is shown in Figure 4.2 (*b*). The line actually displays the solution set of the equation $y = x - 2$. In other words, every ordered pair which satisfies the equation $y = x - 2$ is associated with a point on the line and every point on the line is associated with an ordered pair which satisfies the equation. We have displayed infinitely many ordered pairs. The line is referred to as the graph of the solution set of the equation or, more simply, as the graph of the equation.

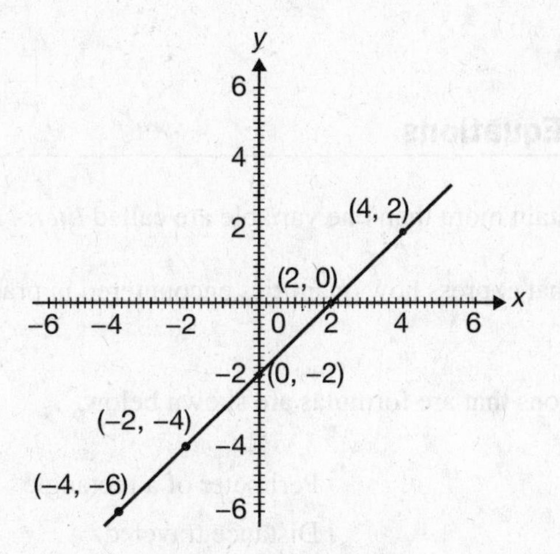

Figure 4.2(*b*)

Even though it is not shown on the graph, the line will continue indefinitely in both directions. Sometimes an arrow is shown at each end to indicate this; we will omit the arrows.

The following theorem can be proven. We omit the proof.

Theorem 1. The graph of an equation of the form $ax + by = c$, where a and b are not both zero, is a straight line. Conversely, every straight line is the graph of an equation of the form $ax + by = c$.

First-degree equations in the form $ax + by = c$ are called *linear equations* for this reason. The form $ax + by = c$ is commonly called the *standard form* of the equation. We shall introduce other forms in a subsequent section.

The above theorem is very helpful, since we only need to find two points on the line in order to graph it. It is advisable to find a third point as a check. The points where the graph crosses the axes are often significant and easy to find.

Definition 9. The *x-intercepts* of a graph are the x values of the points where the graph intersects the x-axis. The *y-intercepts* of a graph are the y values of the points where the graph intersects the y-axis.

Refer to Figure 4.2 (*b*) and note that the x-intercept of the graph of $y = x - 2$ is 2 and the y-intercept is -2. The x-intercept is determined when $y = 0$, and the y-intercept is determined when $x = 0$. In all cases, points on the x-axis have a y-coordinate of zero and points on the y-axis have an x-coordinate of zero.

See solved problems 4.6–4.7.

The results of solved problem 4.7 parts (*c*) and (*d*) illustrate the following relationships. In a two-dimensional coordinate system, the graphs of equations of the form $x = c$ and $y = c$ are vertical lines and horizontal lines, respectively.

Remember, we must always plot several points in order to graph an equation. If the equation is in standard form, the intercept method usually works well. If the equation has y isolated on one side, pick arbitrary values for x and determine the corresponding y values. Try to choose x values which will avoid fractional y-coordinates. It is advisable to plot three points on a line. The third serves as a check.

Refer to supplementary problems 4.3 and 4.4 for similar problems.

You can graph first-degree equations on a graphing calculator if you own one. We shall describe the process in a general manner here. Consult your owner's manual for specific information about the calculator you possess.

In general, one must set the appropriate "window" or domain and range for the equation to be graphed. The domain consists of the x values to be utilized, and the range consists of the y values to be used. The default settings on the calculator may be satisfactory, but normally you must determine the appropriate window for the equation at hand. This can be done if you determine the intercepts of the graph.

The next task is that of entering the equation to be graphed. Your calculator most likely accepts only equations with y isolated on one side of the equation and all remaining terms on the other. Rewrite the equation if necessary. Be careful to manipulate the terms correctly. Clear graphs done previously, enter the expression, and graph it.

See solved problem 4.8.

4.3 Formulas and Literal Equations

Definition 10. Equations that contain more than one variable are called *literal equations*.

Definition 11. Literal equations that express how quantities encountered in practical applications are related are called *formulas*.

Some examples of literal equations that are formulas are shown below.

$P = 2l + 2w$	Perimeter of a rectangle
$d = rt$	Distance traveled

$A = \frac{1}{2}bh$	Area of a triangle
$I = Prt$	Simple interest

We shall apply the techniques discussed in Section 4.1 to solve various literal equations and formulas.

Definition 12. A literal equation is said to be *solved explicitly* for a variable if that variable is isolated on one side of the equation (and does not occur on the other side).

See solved problem 4.9.

4.4 Applications

Solving real-world problems very often entails translating verbal statements into appropriate mathematical statements. Unfortunately, there is no simple method available to accomplish this. The reader should review Section 1.7 for assistance with the task. We offer the following suggestions as an orderly approach to apply.

1. Read the problem carefully! You may need to read it several times in order to understand what is being said and what you are asked to find.
2. Draw diagrams or figures whenever possible. This will help you analyze the problem. Your figure should be drawn and labeled as accurately as possible in order to avoid wrong conclusions.
3. Identify the unknown quantity (or quantities) and use a variable to label it (them). Your variable(s) will frequently represent the value(s) requested in the problem statement. The first letter of a key word for your unknown quantity may be a good letter to use. Write a complete sentence that states explicitly what your variable represents. Don't be ambiguous or vague.
4. Determine how the known and unknown quantities are related. The words of the problem may tell you. If not, there may be a particular formula which is relevant.
5. Write an equation that relates the known and unknown quantities. Be careful when doing this! Ask yourself if your equation translates the words of the problem accurately. Your equation must seem reasonable and make sense.
6. Solve the equation.
7. Answer the question that was asked. The value of the variable you used may not be the answer to the question. It depends upon how you defined the variable you are using.
8. Check the answer in the statement of the problem.

Proportions can quite often be employed to answer a variety of questions.

Definition 13. The quotient of two quantities, a/b, is called a *ratio*.

Definition 14. A statement that two ratios are equal, $a/b = c/d$, is called a *proportion*.

Write ratios of related quantities when you construct proportions. There is often more than one way to do this. We will illustrate.

See solved problems 4.10–4.20.

4.5 Solving First-Degree Inequalities

Definition 15. Expressions that utilize the relations $<, \leq, >$, or $\geq$ are called *inequalities*.

Some examples of first-degree inequalities are $3t + 5 > 10$ and $\frac{x-5}{3} \leq 0$.

Definition 16. Any element of the replacement set (domain) of the variable for which the inequality is true is called a *solution*.

Definition 17. The set that contains all of the solutions of an inequality is called the *solution set* of the inequality.

Definition 18. *Equivalent inequalities* are inequalities that have the same solution set.

To solve inequalities, we shall employ a technique that parallels our approach to solving equations. We shall obtain a sequence of equivalent inequalities until we arrive at one whose solution set is obvious.

The solution set usually includes one or more intervals when graphed on a number line. Graphing the solution set helps us visualize it. Review Section 1.2 for graphs of various intervals on the number line.

Properties of Inequalities

If $a < b$, then for real numbers a, b, and q:

(1) $a + q < b + q$ and $a - q < b - q$. (Adding or subtracting the same quantity on each side of an inequality preserves the order (sense) of the inequality.)

(2) If $q > 0$, $aq < bq$ and $a/q < b/q$. (Multiplying or dividing both sides of an inequality by a positive quantity preserves the order (sense) of the inequality.)

(3) If $q < 0$, $aq > bq$ and $a/q > b/q$. (Multiplying or dividing both sides of an inequality by a negative quantity reverses the order (sense) of the inequality.)

The properties above were stated for the "$<$" relation. The above properties are also valid for the $\leq$, $>$, and $\geq$ relations.

EXAMPLE 1. Given that $2 < 5$, then

$$2 + 3 < 5 + 3 \quad \text{or } 5 < 8 \qquad 2 - 3 < 5 - 3 \quad \text{or } -1 < 2 \qquad 2(3) < 5(3) \quad \text{or } 6 < 15$$

$$\frac{2}{3} < \frac{5}{3} \qquad 2(-3) > 5(-3) \quad \text{or } -6 > -15 \qquad \frac{2}{-3} > \frac{5}{-3} \quad \text{or } \frac{-2}{3} > \frac{-5}{-3}$$

See solved problems 4.21.

4.6 Graphs of First-Degree Inequalities

In Section 4.2, we graphed equations of the form $ax + by = c$. The graphs were determined to be straight lines. We now direct our attention to graphs of first-degree inequalities. The process is merely an extension of the techniques employed in Section 4.2. The graph is a visual display of the solution set of the inequality being discussed. The graph is a region of the coordinate plane called a half-plane. A *half-plane* is the region on either side of a line in a plane.

See solved problem 4.22.

The process illustrated in the above problems is summarized below. To graph linear inequalities:

1. Graph the equation which represents the boundary of the solution set. Draw a solid line if the relation involves $\leq$ or $\geq$. Draw a dashed line if the relation involves $<$ or $>$.
2. Choose a convenient test point not on the line, and substitute its coordinates into the inequality. Choose a point which has at least one zero coordinate for convenience.

3. If the coordinates of the test point satisfy the inequality, shade the region which contains the point. If the coordinates of the test point do not satisfy the inequality, shade the region on the other side of the line.

4.7 Applications Involving Inequalities

Statements that describe inequalities are set up in a manner analogous to those that describe equations. The appropriate relation symbol will be one of the inequality symbols rather than the equals symbol.

See solved problems 4.23–4.29.

4.8 Absolute-Value Equations and Inequalities

In Chapter 1 we defined the absolute value of a real number as the distance between that number and zero on the number line. For example, $|-2| = 2$, since the distance between the points associated with -2 and 0 on the number line are 2 units apart.

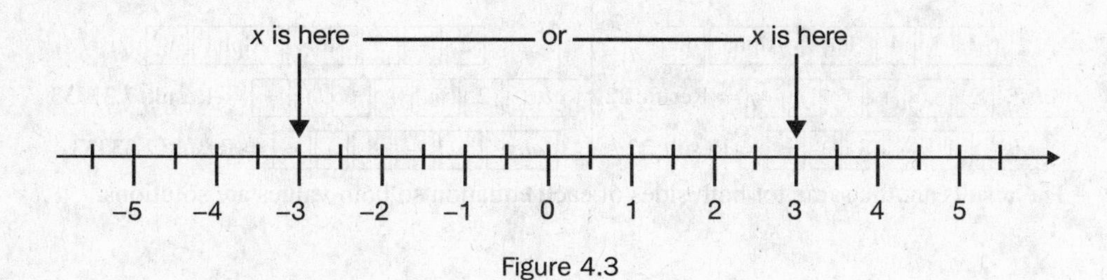

Figure 4.3

The equation $|x| = 3$ means that the distance between the points associated with x and 0 on the number line is 3. Hence, x is either -3 or 3 as illustrated in Figure 4.3.

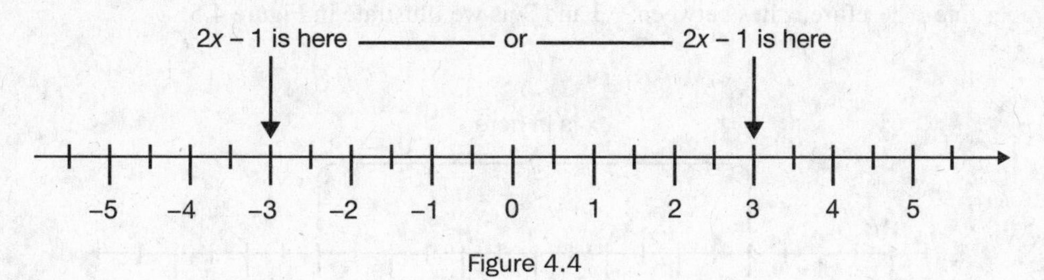

Figure 4.4

Therefore, $x = -3$ and $x = 3$ are solutions to $|x| = 3$. The solution set is $\{-3, 3\}$.

Similarly, $|2x - 1| = 3$ means that the distance between the points associated with $2x - 1$ and 0 is 3. Refer to Figure 4.4.

We conclude that the solutions to $2x - 1 = -3$ and $2x - 1 = 3$ both satisfy $|2x - 1| = 3$.

$$2x - 1 = -3 \quad \text{or} \quad 2x - 1 = 3$$
$$2x = -2 \qquad\qquad 2x = 4$$
$$x = -1 \qquad\qquad x = 2$$

Therefore, $x = -1$ and $x = 2$ are solutions to $|2x - 1| = 3$. The solution set is $\{-1, 2\}$.

Both examples above illustrate that there are two solutions to equations which involve the absolute value of linear expressions that are equal to a positive number. Both solutions are obtained by solving appropriate related equations. We generalize our observations below.

Property 1: If $c \geq 0$, $|q| = c$ is equivalent to $q = -c$ or $q = c$.

Verbally we say that if c is a nonnegative real number, the solutions to the absolute value of a linear quantity q that is equal to c are the solutions to the equations $q = -c$ and $q = c$. The application of property 1 permits us to write equations equivalent to the original equation, free of absolute values, which we can solve.

See solved problem 4.30.

To check an absolute value equation using the calculator method we employed for solved problem 4.4(*e*), you will need to find the absolute value function for your calculator. Most graphing calculators use a built-in function, *abs*(). (You will most likely have to find it as a function on your calculator, since typing "abs" probably will not be understood by the calculator.) Now we can check whether the left and right sides of the equation give the same result for each of the values we obtained as solutions. The check then for solved problem 4.30(*h*) follows.

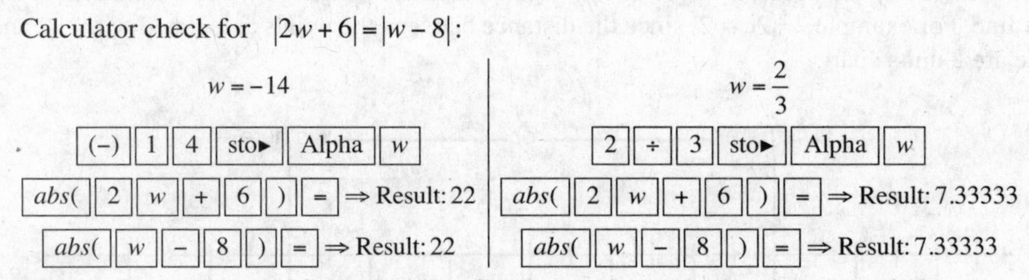

The results are the same for both sides of each equation so both values are solutions.

We employ a similar technique to solve absolute value inequalities. The distance interpretation of absolute value allows us to replace the original absolute value inequality with equivalent inequalities which are free of absolute values.

Consider the statement $|x| < 3$. The statement means that the distance between x and 0 is less than three units on the number line. Therefore, x lies between -3 and 3 as we illustrate in Figure 4.5.

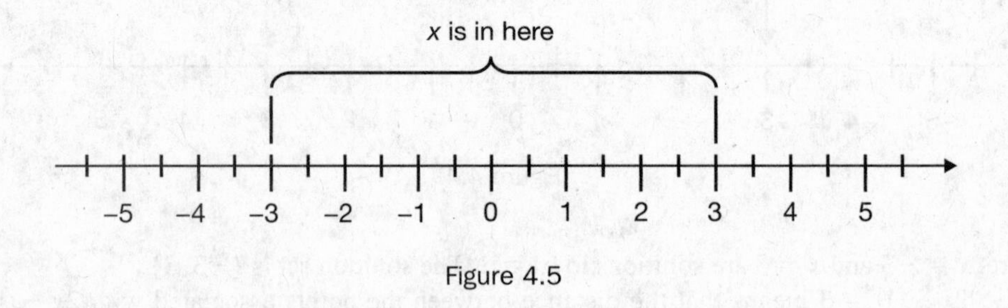

Figure 4.5

The set-builder notation of the solution set is $\{x|-3 < x < 3, x \in R\}$. The solution set consists of all real numbers in the interval $(-3, 3)$. $|x| < 3$ is equivalent to the double inequality $-3 < x < 3$. We shall primarily employ the interval notation for solution sets of inequalities for simplicity. Refer to Section 1.2 to review interval notation.

Similarly, $|x + 2| < 3$ means that $x + 2$ is less than three units from 0 on the number line. Refer to Figure 4.6.

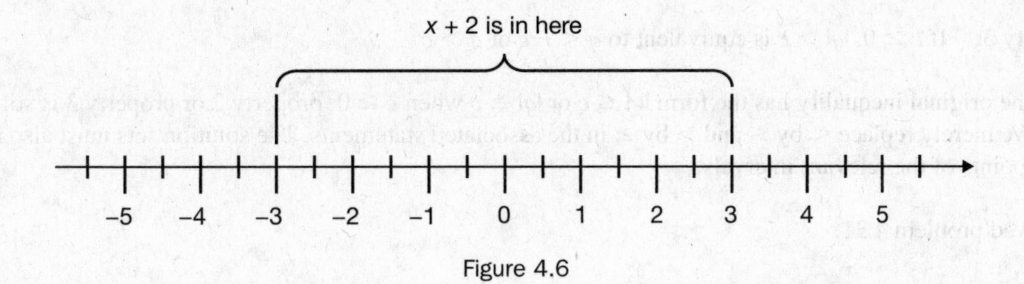

Figure 4.6

In other words, the double inequality $-3 < x + 2 < 3$ must be true. Subtract 2 from each of the three members of the inequality to obtain $-5 < x < 1$. The solution set is $\{x | -5 < x < 1, x \in R\}$. In interval notation, the solution set is $(-5, 1)$.

The above examples illustrate the general relation that we state as property 2.

Property 2: If $c \geq 0$, $|q| < c$ is equivalent to $-c < q < c$.

Now consider the statement $|x| > 3$. The statement means that the distance from x to 0 is more than three units. Therefore, x lies to the left of -3 or to the right of 3 on the number line. Refer to Figure 4.7.

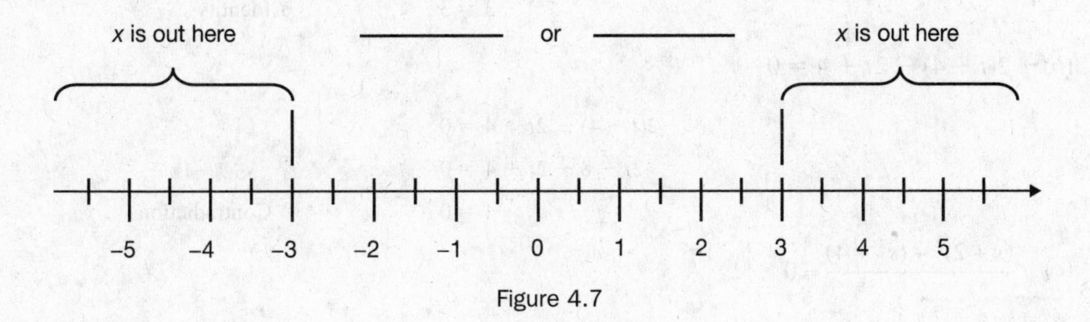

Figure 4.7

In other words, $x < -3$ or $x > 3$. The solution set is $\{x|x < -3, x \in R\} \cup \{x|x > 3, x \in R\}$ in set-builder notation. We state the solution as $(-\infty, -3) \cup (3, \infty)$ in interval notation.

Now we shall solve $|x - 1| > 3$. The distance from $x - 1$ to 0 is more than three units. Hence, $x - 1$ lies to the left of -3 or to the right of 3 on the number line. Refer to Figure 4.8.

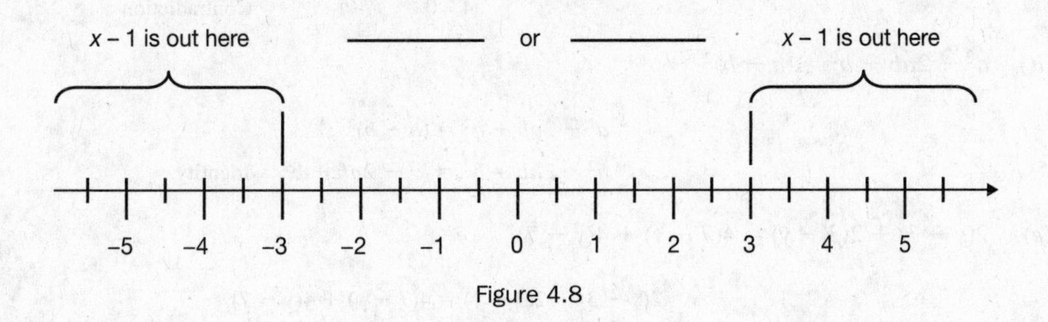

Figure 4.8

We observe that $x - 1 < -3$ or $x - 1 > 3$ must be true. We apply familiar techniques to solve the inequalities to obtain $x < -2$ or $x > 4$. The solution set is $\{x|x < -2, x \in R\} \cup \{x|x > 4, x \in R\}$ or $(-\infty, -2) \cup (4, \infty)$.

Avoid using improper compound inequality forms such as $-3 > x - 1 > 3$, since $-3 > 3$ is false!

The general relation illustrated above is now stated as property 3 below.

Property 3: If $c \geq 0$, $|q| > c$ is equivalent to $q < -c$ or $q > c$.

If the original inequality has the form $|q| \leq c$ or $|q| \geq c$ when $c \geq 0$, property 2 or property 3 is still applicable. We merely replace $<$ by $\leq$ and $>$ by $\geq$ in the associated statements. The solution sets must also include the endpoints of the relevant intervals.

See solved problem 4.31.

SOLVED PROBLEMS

4.1 Determine whether each of the following is an identity or contradiction. (Hint: Simplify each side of each equation first.)

(*a*) $x - (x - 3) = 3$

$$x - (x - 3) = 3$$
$$x - x + 3 = 3$$
$$3 = 3 \qquad \text{Identity}$$

(*b*) $2(t - 4) - 2t + 4 = 0$

$$2(t - 4) - 2t + 4 = 0$$
$$2t - 8 - 2t + 4 = 0$$
$$-4 = 0 \qquad \text{Contradiction}$$

(*c*) $\dfrac{(s+2)^2 - (s^2+4)}{s} = 0$

$$\frac{(s+2)^2 - (s^2+4)}{s} = 0$$
$$\frac{s^2 + 4s + 4 - s^2 - 4}{s} = 0$$
$$\frac{4s}{s} = 0$$
$$4 = 0 \qquad \text{Contradiction}$$

(*d*) $a^2 - 2ab + b^2 = (a - b)^2$

$$a^2 - 2ab + b^2 = (a - b)^2$$
$$a^2 - 2ab + b^2 = a^2 - 2ab + b^2 \qquad \text{Identity}$$

(*e*) $2(y - 3) + 2(3 - y) = 4(7 - y) + 4(y - 7)$

$$2(y - 3) + 2(3 - y) = 4(7 - y) + 4(y - 7)$$
$$2y - 6 + 6 - 2y = 28 - 4y + 4y - 28$$
$$0 = 0 \qquad \text{Identity}$$

4.2 Determine if the given values of the variable satisfy the equation $5x - 9 + 2(x + 1) = 0$.

(*a*) $x = 1$

Replace x by 1 and simplify. Determine if the equation is true.

$$5x - 9 + 2(x + 1) = 0$$
$$5(1) - 9 + 2(1 + 1) = 0$$
$$5 - 9 + 4 = 0$$
$$9 - 9 = 0$$
$$0 = 0 \quad \text{True}$$

Thus, $x = 1$ satisfies the equation.

(*b*) $x = 7$

Replace x by 7 and simplify.

$$5x - 9 + 2(x + 1) = 0$$
$$5(7) - 9 + 2(7 + 1) = 0$$
$$35 - 9 + 16 = 0$$
$$42 = 0 \quad \text{False}$$

Thus, $x = 7$ does not satisfy the equation.

See supplementary problem 4.1 for similar exercises.

4.3 Solve $2x + 5 = 11$ for x.

$$2x + 5 = 11$$
$$2x + 55 - 5 = 11 - 5 \quad \text{Addition (subtraction) property of equality}$$
$$2x = 6 \quad \text{Combine like terms}$$
$$\frac{2x}{2} = \frac{6}{2} \quad \text{Multiplication (division) property of equality}$$
$$x = 3 \quad \text{Simplify}$$

Three is the *solution* or *root* of the given equation. The *solution set* is {3}. It is customary to state the result simply in the form $x = 3$.

4.4 Solve.

(*a*) $3x - 7 = x - 11$

$$3x - 7 = x - 11$$
$$3x - x - 7 = x - x - 11 \quad \text{Subtract } x$$
$$2x - 7 = -11 \quad \text{Combine}$$
$$2x - 7 + 7 = -11 + 7 \quad \text{Add 7}$$
$$2x = -4 \quad \text{Combine}$$
$$\frac{2x}{2} = -\frac{4}{2} \quad \text{Divide by 2}$$
$$x = -2 \quad \text{Simplify}$$

Check:

$$3x - 7 = x - 11$$
$$3(-2) - 7 \stackrel{?}{=} -2 - 11$$
$$-6 - 7 \stackrel{?}{=} -2 - 11$$

$-13 = -13$ is true so $x = -2$ is the solution.

(*b*) $4(t+1)-3t=5t+8$

$4(t+1)-3t=5t+8$	
$4t+4-3t=5t+8$	Distribute
$t+4=5t+8$	Combine
$t-t+4=5t-t+8$	Subtract t
$4=4t+8$	Combine
$4-8=4t+8-8$	Subtract 8
$-4=4t$	Combine
$\frac{-4}{4}=\frac{4t}{4}$	Divide by 4
$-1=t$	Simplify

Check:

$$4(t+1)-3t=5t+8$$
$$4(-1+1)-3(-1)\stackrel{?}{=}5(-1)+8$$
$$4(0)+3\stackrel{?}{=}-5+8$$

$3=3$ is true so $t=-1$ is the solution.

In solved problem 4.4(*b*) above, operations were chosen so that the variable t ultimately had a positive coefficient. You should attempt to choose operations which result in positive coefficients of variables in order to reduce errors in signs when solving equations. Concentrate on the variable terms first, followed by the constant terms.

It is also desirable to perform some of the simpler operations mentally. We included more details in the above problems in order to clarify the process. We will include fewer details subsequently.

(*c*) $3y+1-(y-5)=2y-4$

$3y+1-(y-5)=2y-4$	
$3y+1-y+5=2y-4$	Distribute
$2y+6=2y-4$	Combine
$6=-4$	Subtract $2y$

The last equation is false. It is a contradiction. Therefore, there is no value of y that satisfies the original equation. There is no solution to the equation.

(*d*) $4s-2(1-s)=3(2s+1)-5$

$4s-2(1-s)=3(2s+1)-5$	
$4s-2+2s=6s+3-5$	Distribute
$6s-2=6s-2$	Combine
$-2=-2$	Subtract $6s$

The last equation is always true. It is an identity. We conclude that every value of s satisfies the original equation. The solution is the set of all real numbers R.

If some of the terms contain fractions, begin by using the multiplication property in order to "clear" fractions. Multiply both sides of the equation by the LCD of the fractions which appear in the expression. Then distribute and simplify.

(*e*) $\frac{n}{3}+4=\frac{n-1}{4}$

$$\frac{n}{3}+4=\frac{n-1}{4}$$

$$12\left(\frac{n}{3}+4\right)=12\left(\frac{n-1}{4}\right) \quad \text{Multiply by } LCD=12$$

$$4n+48=3n-3 \quad \text{Distribute}$$

$$n=-51 \quad \text{Subtract } 3n \text{ and } 48$$

Check: $\frac{n}{3}+4=\frac{n-1}{4}$

$$\frac{-51}{3}+4 \stackrel{?}{=} \frac{-51-1}{4}$$

$$-17+4 \stackrel{?}{=} \frac{-52}{4}$$

$-13=-13$ is true so $n=-51$ is the solution.

As indicated back in Section 1.6, some calculators store values by using a keyboard entry using a letter of the alphabet. If you have one of those calculators (your calculator manual will tell you whether or not you do), you can use it to check your solution of an equation rather easily. Though we prefer the process of checking your solutions as shown above for the added benefit of strengthening your arithmetic skills, we realize there are occasions when a calculator check could be beneficial. To use your calculator for a "quick" check, you could first store the solution in the storage location of the letter of your variable. You can then evaluate each side of the original equation to see if their results are the same. In solved problem 4.4(*e*) above, that would mean you would put -51 in storage location "*n*" and proceed. The process (with calculator keystrokes) is shown below:

Calculator check: $\frac{n}{3}+4=\frac{n-1}{4}$

[(−)] [5] [1] [sto►] [Alpha] [*n*]

Left side: [*n*] [÷] [3] [+] [4] [=] ⇒ Result: −13

Right side: [(] [*n*] [−] [1] [)] [÷] [4] [=] ⇒ Result: −13

Since both sides of the equation give the same result, the value, -51, is a solution. The results of the two sides being exactly the same are just as you would expect for a solution. In some cases, however, even though the value stored is a solution, the results will differ by a relatively small amount. That amount should be very small compared with the values of the results themselves and would be due to calculator round-off error. In general, you can ignore such small discrepancies.

See supplementary problem 4.2 to practice the above concepts.

4.5 Plot—that is, graph—the points with coordinates $(4, 2)$, $(-3, 3)$, $(0, 5)$, $(2, 0)$, and $(-4, -3)$ on a rectangular coordinate system.

Draw the axes and label them. Then locate the points. To graph $(4, 2)$, start at the origin and move 4 units to the right, then 2 units up. Plot the remaining points in a similar manner. See Figure 4.9.

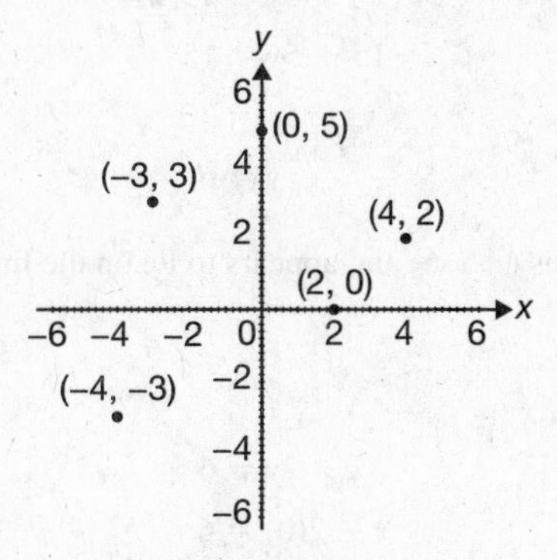

Figure 4.9

4.6 Graph the following equations. Find the intercepts and locate a third point as a check. Label the intercept points.

(*a*) $3x + 5y = 15$

Let $y = 0$ and solve for x to find the x-intercept.

$$3x + 5y = 15$$
$$3x + 5(0) = 15$$
$$3x = 15$$
$$x = 5 \text{ is the } x\text{-intercept.}$$

Let $x = 0$ and solve for y to find the y-intercept.

$$3x + 5y = 15$$
$$3(0) + 5y = 15$$
$$5y = 15$$
$$y = 3 \text{ is the } y\text{-intercept.}$$

Plot the points and draw the line. See Figure 4.10.

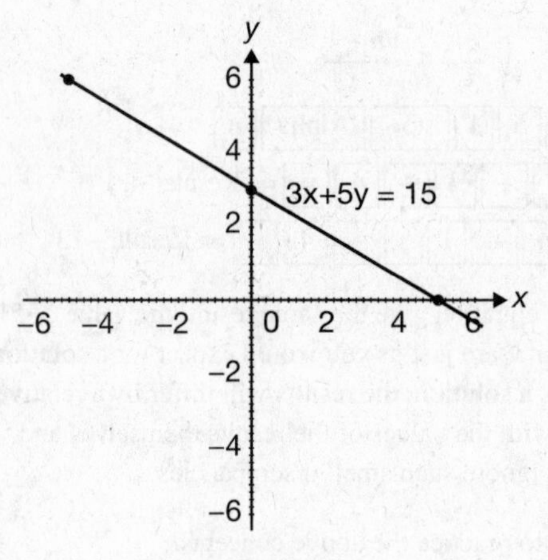

Figure 4.10

We may choose any value for either variable and then solve for the other variable to find a point. Choose $x = -5$ and solve for y to locate a third point.

$$3x + 5y = 15$$
$$3(-5) + 5y = 15$$
$$-15 + 5y = 15$$
$$5y = 30$$
$$y = 6$$

The third point has coordinates $(-5, 6)$ and appears to lie on the line. See Figure 4.10.

(*b*) $x - 2y = 6$

Choose $y = 0$ and solve for x.

$$x - 2y = 6$$
$$x - 2(0) = 6$$
$$x = 6 \text{ is the } x\text{-intercept.}$$

Choose $x = 0$ and solve for y.

$$x - 2y = 6$$
$$0 - 2y = 6$$
$$y = -3 \text{ is the } y\text{-intercept.}$$

Plot the points and draw the line. See Figure 4.11.

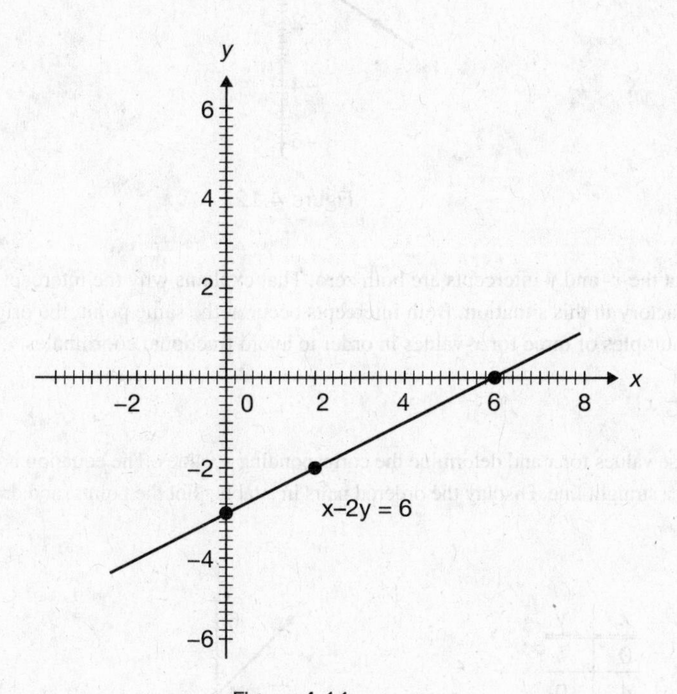

Figure 4.11

Choose $x = 2$ and solve for y to check.

$$x - 2y = 6$$
$$2 - 2y = 6$$
$$-2y = 4$$
$$y = -2$$

The point $(2, -2)$ appears to lie on the line. See Figure 4.11.

The intercept method of graphing works well on most occasions, although there are circumstances when other techniques should be employed. Solved problem 4.7 (*a*) below is such a case.

4.7 Graph the following equations.

(*a*) $y = \dfrac{2}{3}x$

The equation $y = \frac{2}{3}x$ is equivalent to $2x - 3y = 0$ so the graph is a straight line. Choose arbitrary values for x in the given equation and determine the corresponding y values. It is convenient to display the values in a table as shown below. Plot the points and draw the line. See Figure 4.12.

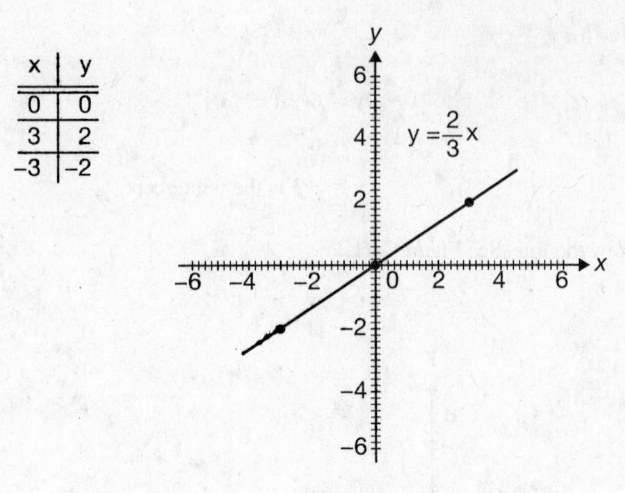

x	y
0	0
3	2
-3	-2

Figure 4.12

Note that the x- and y-intercepts are both zero. That explains why the intercept method of graphing is unsatisfactory in this situation. Both intercepts occur at the same point, the origin. Also observe that we chose multiples of three for x-values in order to avoid fractional coordinates.

(*b*) $y = 4 - x$

Choose values for x and determine the corresponding y values. The equation is equivalent to $x + y = 4$, so the graph is a straight line. Display the ordered pairs in a table, plot the points, and draw the line. See Figure 4.13.

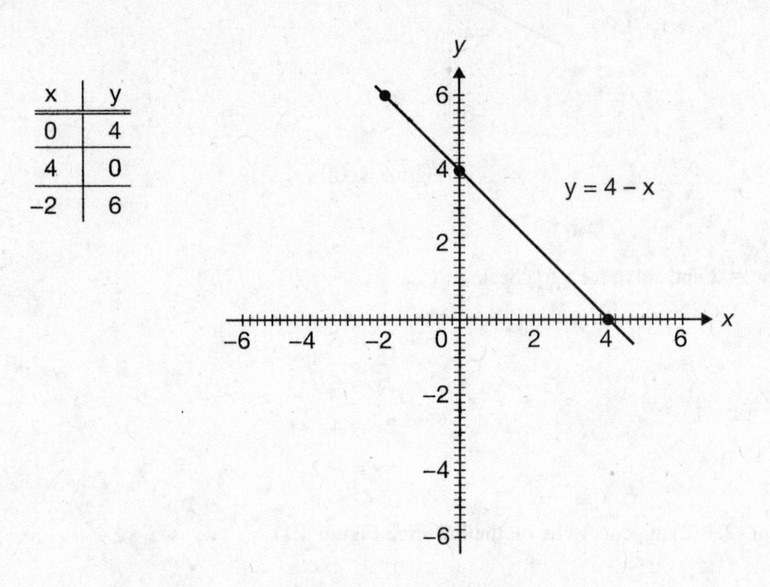

x	y
0	4
4	0
-2	6

Figure 4.13

(*c*) $x = -2$

The equation $x = -2$ can be written in the standard form $ax + by = c$ as $1x + 0y = -2$. Therefore the graph is a straight line. If we replace y in the equation $1x + 0y = -2$ by various values, the result is unchanged. That is, x must always be -2, but y can be anything. In other words, the equation $x = -2$ requires the x-coordinate of every point on the line to be -2, while the y-coordinates can be any real value. Some representative ordered pairs are shown in the table below. Plot the points (Figure 4.14 (*a*)), and draw the line (Figure 4.14 (*b*)).

x	y
-2	-3
-2	0
-2	3

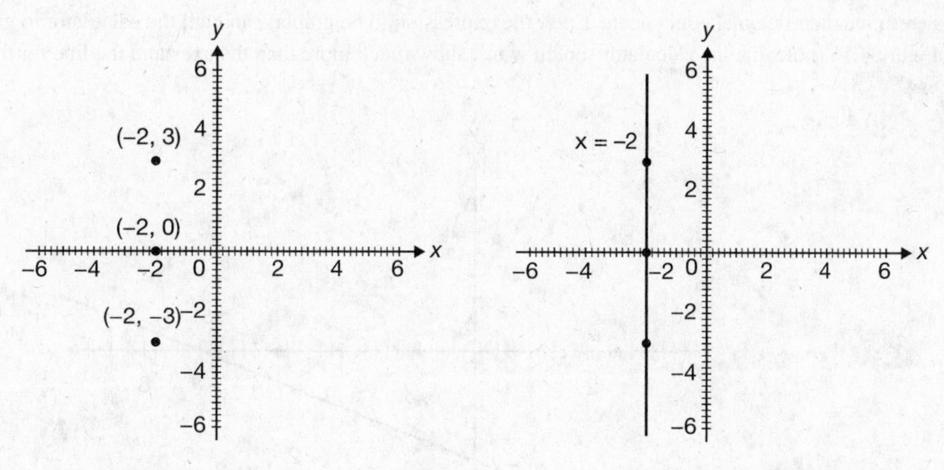

Figure 4.14 (*a*) Figure 4.14 (*b*)

It is a vertical line two units to the left of the y-axis.

(*d*) $y = 3$

The equation $y = 3$ can be written in the standard form $ax + by = c$ as $0x + 1y = 3$. Therefore, the graph is again a straight line. If x is replaced by various values in $0x + 1y = 3$, the result is unaffected. Observe that y is always 3. The equation $y = 3$ requires the y-coordinate of every point on the line to be 3, while there are no restrictions on the x-coordinate. Some typical ordered pairs are shown in the accompanying table. Plot the points and draw the line. See Figure 4.15.

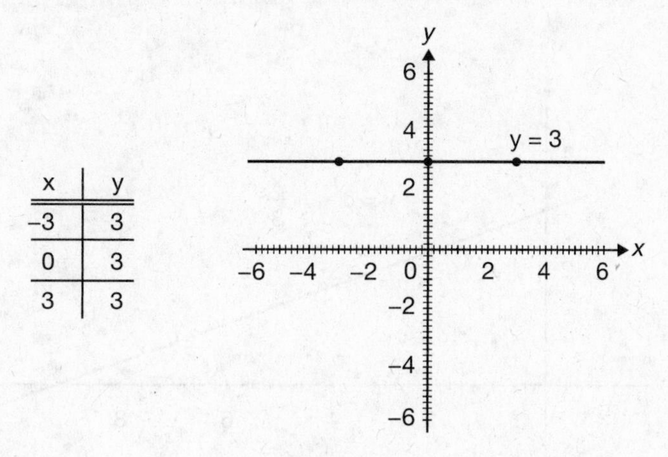

x	y
–3	3
0	3
3	3

Figure 4.15

It is a horizontal line three units above the x-axis.

4.8 Use a graphing calculator to graph the following.

(*a*) $y = \frac{2}{5}x - 2$

Determine the appropriate window. If $x = 0, y = -2$, so the y-intercept is -2. If $y = 0, x = 5$, so the x-intercept is 5. Set the range of values for x in the interval $[-2, 7]$ and the y interval to $[-3, 2]$. Note that the window extends beyond the values of the intercepts. The window may vary, but it should normally include the intercepts and the coordinate axes. Now that the graphing window is determined, we must enter the equation. Common fractions can not be entered directly on some calculators, so enter them as divisions or mentally determine their decimal equivalents. Enter the expression to be graphed and tell the calculator to graph it. See Figure 4.16 (note that the calculator screen won't show much more than the axes and the line for the graph).

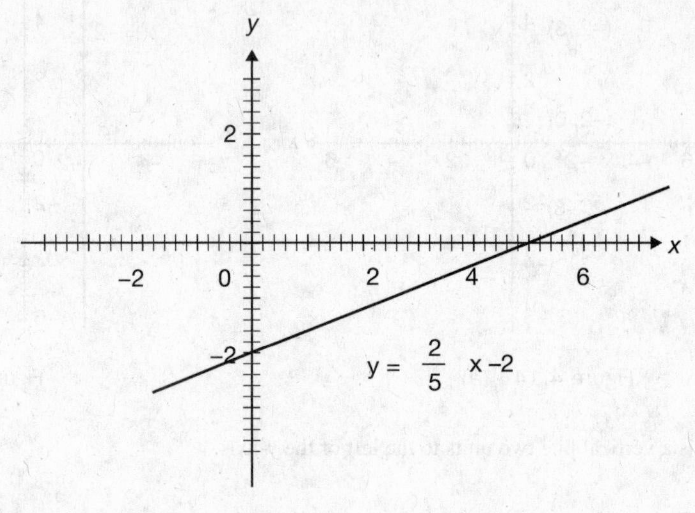

Figure 4.16

(*b*) $y = 3 - \frac{x}{3}$

Recall that intercepts occur at points which have zero as one of the coordinates. The y-intercept is 3 and the x-intercept is 9. Set the x-interval to $[-1, 10]$ and the y-interval to $[-1, 4]$. Enter the expression and graph it. See Figure 4.17.

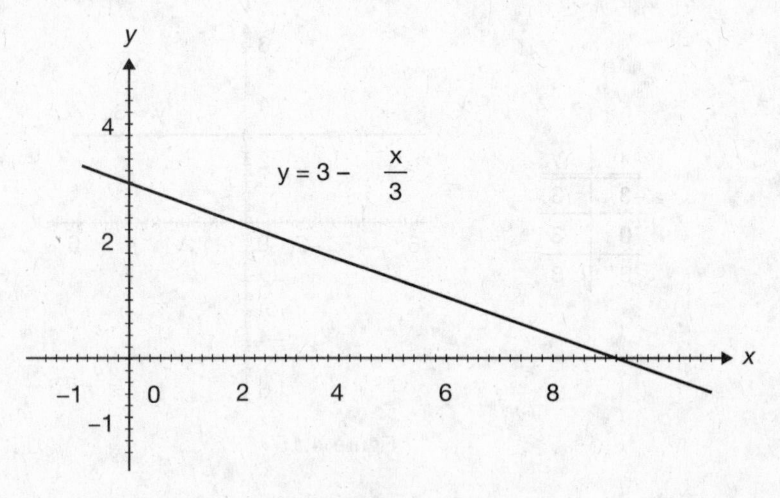

Figure 4.17

(*c*) $2x - 7y = 8$

The x-intercept is 4 and the y-intercept is $-8/7$. Set the x interval to $[-2, 7]$ and the y interval to $[-2, 2]$. We must solve for y to enter the expression to be graphed. The result is $y = \frac{2}{7}x - \frac{8}{7}$. See Figure 4.18.

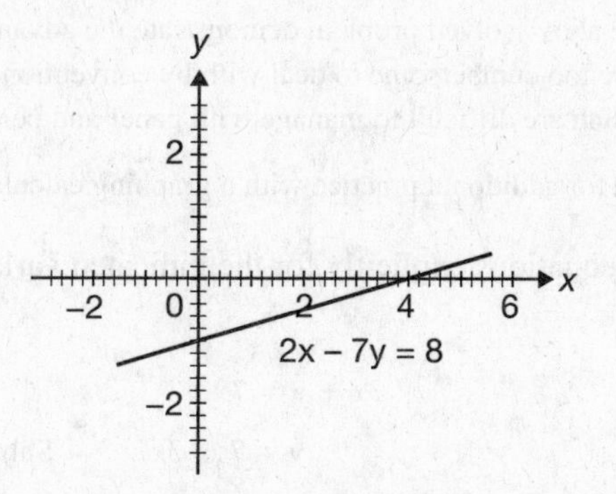

Figure 4.18

(*d*) $2.36x + 1.17y = 4.8$

The *x*-intercept is $4.8/2.36 \approx 2$, and the y-intercept is $4.8/1.17 \approx 4$. The symbol $\approx$ means "is approximately equal to." Set the *x* interval to $[-1, 3]$ and the *y* interval to $[-2, 5]$. Solve for *y* next. We find $y = (-2.36/1.17)\,x + 4.8/1.17$. Enter the expression and graph it. See Figure 4.19.

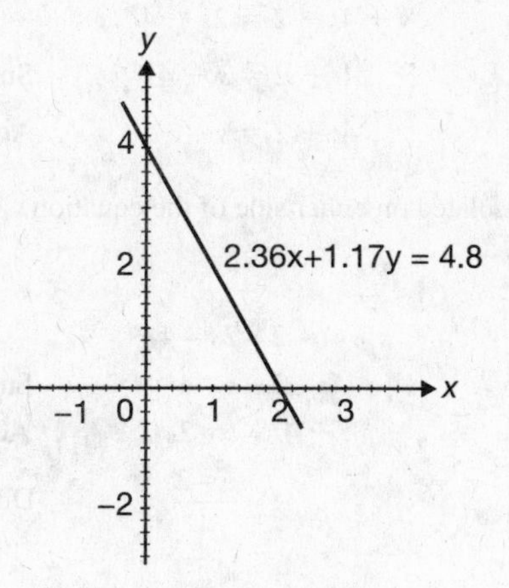

Figure 4.19

(*e*) $325x + 710y = -913$

The *x*-intercept is $-913/325 \approx -2.8$ and the *y*-intercept is $-913/710 \approx -1.3$. Set the *x* interval to $[-4, 1]$ and the *y* interval to $[-2, 1]$. We obtain $y = (-325/710)x - 913/710$ when *y* is isolated. See Figure 4.20.

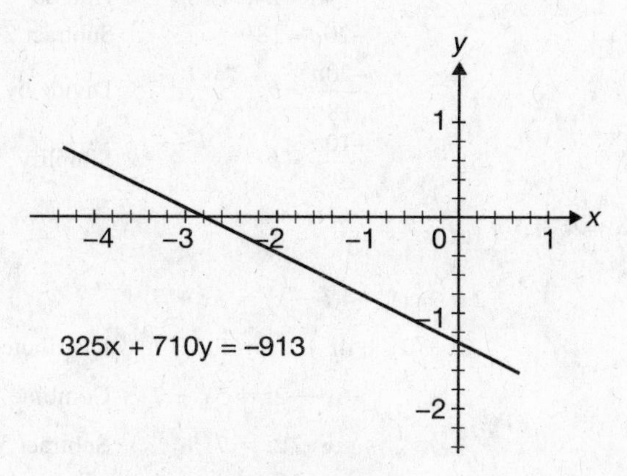

Figure 4.20

Parts (*d*) and (*e*) of the above solved problem demonstrate the advantage of using a graphing calculator. The numbers involved are too cumbersome to deal with by conventional methods. Most applications involve awkward expressions which are difficult to manage with paper and pencil.

Try supplementary problem 4.5 for additional practice with a graphing calculator.

4.9 Solve each of the following equations explicitly for the indicated variable.

(*a*) $3x + y = 7$ for y.

$$3x + y = 7$$

$$y = 7 - 3x \qquad \text{Subtract } 3x \text{ from both sides}$$

(*b*) $3x + y = 7$ for x.

$$3x + y = 7$$

$$3x = 7 - y \qquad \text{Subtract } y \text{ from both sides}$$

$$x = \frac{7-y}{3} \qquad \text{Divide both sides by 3}$$

(*c*) $s + 4t - 2 = 2s - 4$ for s.

$$s + 4t - 2 = 2s - 4$$

$$4t - 2 = s - 4 \qquad \text{Subtract } s \text{ from both sides}$$

$$4t + 2 = s \qquad \text{Add 4 to both sides}$$

Note that the variable can be isolated on either side of the equation.

(*d*) $s + 4t - 2 = 2s - 4$ for t.

$$s + 4t - 2 = 2s - 4$$

$$4t - 2 = s - 4 \qquad \text{Subtract } s$$

$$4t = s - 2 \qquad \text{Add 2}$$

$$t = \frac{s-2}{4} \qquad \text{Divide by 4}$$

(*e*) $\frac{a}{12} - \frac{b}{8} = \frac{a}{2} + \frac{b}{4}$ for b.

$$\frac{a}{12} - \frac{b}{8} = \frac{a}{2} + \frac{b}{4}$$

$$48\left(\frac{a}{12} - \frac{b}{8}\right) = 48\left(\frac{a}{2} + \frac{b}{4}\right) \qquad \text{Multiply by LCD} = 48$$

$$4a - 6b = 24a + 12b \qquad \text{Simplify}$$

$$4a = 24a + 18b \qquad \text{Add } 6b$$

$$-20a = 18b \qquad \text{Subtract } 24a$$

$$\frac{-20a}{18} = b \qquad \text{Divide by 18}$$

$$\frac{-10a}{9} = b \qquad \text{Simplify}$$

(*f*) $2(t + x) - 4(t - x) = 5x + 7$ for x.

$$2(t + x) - 4(t - x) = 5x + 7$$

$$2t + 2x - 4t + 4x = 5x + 7 \qquad \text{Distribute}$$

$$6x - 2t = 5x + 7 \qquad \text{Combine}$$

$$x - 2t = 7 \qquad \text{Subtract } 5x$$

$$x = 2t + 7 \qquad \text{Add } 2t$$

(g) $(s + 5)(t + 7) = u$ for s.

$$(s+5)(t+7) = u$$

$st + 7s + 5t + 35 = u$	Distribute
$st + 7s = u - 5t - 35$	Subtract $5t$ and 35
$s(t+7) = u - 5t - 35$	Factor
$s = \dfrac{u - 5t - 35}{t+7}$	Divide by $t + 7$

(h) $\dfrac{1}{f} = \dfrac{1}{f_1} + \dfrac{1}{f_2}$ for f_2.

$$\frac{1}{f} = \frac{1}{f_1} + \frac{1}{f_2}$$

$f f_1 f_2 \left(\dfrac{1}{f}\right) = f f_1 f_2 \left(\dfrac{1}{f_1} + \dfrac{1}{f_2}\right)$	Multiply by the LCD $= f f_1 f_2$
$f_1 f_2 = f f_2 + f f_1$	Distribute and simplify
$f_1 f_2 - f f_2 = f f_1$	Subtract $f f_2$
$f_2(f_1 - f) = f f_1$	Factor f_2 out
$f_2 = \dfrac{f f_1}{f_1 - f}$	Divide by $f_1 - f$

(i) $x = \dfrac{y-5}{y+2}$ for y.

Sol. 1 $\quad x = \dfrac{y-5}{y+2}$

$x(y+2) = y - 5$	Multiply by $y + 2$
$xy + 2x = y - 5$	Distribute
$2x + 5 = y - xy$	Subtract xy and add 5
$2x + 5 = y(1 - x)$	Factor
$\dfrac{2x+5}{1-x} = y$	Divide by $1 - x$

Sol. 2 $\quad x = \dfrac{y-5}{y+2}$

$x(y+2) = y - 5$	Multiply by $y + 2$
$xy + 2x = y - 5$	Distribute
$xy - y = -2x - 5$	Subtract $2x$ and y
$(x-1)y = -(2x+5)$	Factor
$y = \dfrac{-(2x+5)}{x-1}$	Divide by $x - 1$

Note: The reader should observe that the results in solution 1 and solution 2 above are equivalent since the numerators and denominators are opposites of each other.

See supplementary problems 4.6 and 4.7 for similar problems.

4.10 An automobile uses 7 gallons of gasoline to travel 154 miles. How many gallons are required to make a trip of 869 miles?

(Proportion Problem) Let g be the number of gallons required to make the trip. The ratio of the gallons equals the ratio of the miles. Therefore,

$$\frac{g}{7} = \frac{869}{154}$$
$$g = \frac{7(869)}{154} = 39.5$$

We conclude that 39.5 gallons will be required to make a trip of 869 miles.

Check: We can easily verify that $\frac{39.5}{7} = \frac{869}{154} = 5.6429$

Note: There are other proportions we could have used, such as $\frac{g}{869} = \frac{7}{154}$, which produce the same result.

4.11 The enrollment at a particular college decreased from 2900 to 2813 in one year. What was the percent decrease?

(Proportion Problem) Let p be the percent decrease. We need to know the amount of the decrease in order to write the proportion. The enrollment decreased by $2{,}900 - 2{,}813 = 87$ students. We can now write the appropriate proportion and solve. The ratio of the percents is equal to the ratio of the number of students.

$$\frac{p}{100} = \frac{87}{2900}$$
$$p = \frac{100(87)}{2900} = 3$$

The enrollment decreased three percent.

Check: $(3\%)(2900) = 0.03(2900) = 87$.

Note: The proportion, $p/87 = 100/2900$, produces the same result.

4.12 Use a proportion to convert 19,800 feet to miles. Recall that 5,280 feet are equivalent to 1 mile.

(Proportion Problem) Let x be the number of miles that are equivalent to 19,800 feet. The ratio of the miles equals the ratio of the feet. Therefore,

$$\frac{x}{1} = \frac{19{,}800}{5{,}280}$$
$$x = \frac{1(19{,}800)}{5{,}280} = 3.75.$$

We conclude that 19,800 feet are equivalent to 3.75 miles.

Check: $\frac{3.75}{1} = \frac{19{,}800}{5{,}280} = 3.75$

4.13 A rectangle is three feet longer than twice its width. If the perimeter is ninety feet, what are the dimensions of the rectangle?

(Geometry Problem) Refer to Steps 1 to 8 in section 4.4. We are asked to find the dimensions' that is, the length and width of the rectangle. Now draw a rectangle and label the dimensions. Refer to Figure 4.21 below.

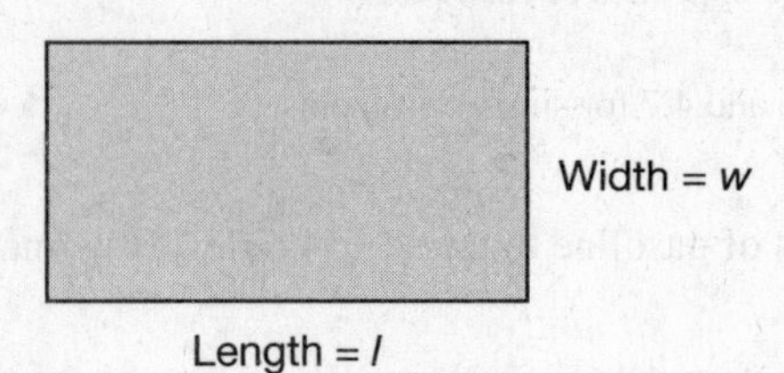

Figure 4.21

Let the length of the rectangle be l and the width be w. The problem tells us that the length is three feet longer than twice the width. Hence, $l = 2w + 3$. Additionally, the perimeter of the rectangle is ninety feet. Recall the formula $P = 2l + 2w$ for the perimeter P of a rectangle with length l and width w. Now write the equation that is applicable for the stated problem and solve it.

$$P = 2l + 2w$$
$$90 = 2(2w + 3) + 2w$$
$$90 = 4w + 6 + 2w$$
$$90 = 6w + 6$$
$$84 = 6w$$
$$14 = w$$

The solution tells us that the width of the rectangle is 14 feet. We were also asked to find the length l. We know that the length $= 2w + 3$, so $l = 2(14) + 3 = 28 + 3 = 31$. Hence, the length of the rectangle is 31 feet.

To check the problem, substitute the numerical values into the known relationships. First, 31 is 3 more than twice 14. Next, $2(31) + 2(14) = 62 + 28 = 90$. The perimeter agrees with the given value.

4.14 (*a*) A worker takes home 70% of his salary after various deductions are subtracted. What is his weekly salary if he takes home \$476 per week?

(Salary Problem) We need to know how much his salary is each week. Let s represent his weekly salary. Recall that $70\% = 0.70$. Now write the appropriate equation and solve it.

$$0.70s = 476$$
$$s = \frac{476}{0.70} = 680$$

His weekly salary is \$680. Observe that $0.70(680) = 476$, so the result checks.

(*b*) What is his hourly salary if he works forty hours per week?

Let h represent his hourly salary. Now write the equation and solve it.

$$40h = 680$$
$$h = \frac{680}{40} = 17.00$$

His hourly salary is \$17.00. Observe that $40(17.00) = 680$, so the result checks.

4.15 The total cost of manufacturing a number of items is determined by adding the fixed costs and the variable costs. If the fixed costs in a particular case are \$1150 and the variable costs are \$7.50 per item, how many items were made if the total costs are \$1465?

(Cost Problem) We are asked to find the number of items made. Let n be the number of items made. Now write the equation and solve.

$$1150 + 7.50n = 1465$$
$$7.50n = 1465 - 1150 = 315$$
$$n = \frac{315}{7.50} = 42$$

We determined that 42 items were made. Note that $1150 + 42(7.50) = 1150 + 315 = 1465$. The solution checks.

4.16 A man has an investment which earns 3.75% simple interest. The interest earned amounts to \$843.75 annually. Find the amount invested.

(Investment Problem) Recall that the simple interest formula is $I = Prt$, where I is the interest earned, P is the principal or amount invested, r is the annual interest rate, and t is the number of years. Since we need to find the amount invested, let P be that amount. We also are given that $r = 3.75\% = 0.0375$ and $t = 1$ year. We can now write the appropriate equation and solve.

$$I = Prt$$
$$843.75 = P(0.0375)\,1 = (0.0375)\,P$$
$$P = \frac{843.75}{0.0375} = 22{,}500$$

Since P represents the quantity we are seeking, the amount invested is \$22,500. To check our answer, we calculate 22,500(0.0375)(1) = 843.75

4.17 If \$8000 is invested in bonds at 4.5%, how much additional money must be invested in stocks that yield 6% to result in a total annual return of \$1044?

(Investment Problem) We again apply the simple interest formula $I = Prt$. The total return is the sum of the return on the bonds and the return on the stocks. Hence, let s be the amount invested in stocks.

$$8000\,(0.045)\,(1) + s(0.06)\,(1) = 1044$$
$$360 + 0.06\,s = 1044$$
$$0.06\,s = 684$$
$$s = \frac{684}{0.06} = 11400$$

Our results indicate that \$11,400 should be invested in stocks. We check our conclusion.
8000 (0.045) (1) + 11,400(0.06)(1) = 360 + 684 = 1044.

4.18 Person A can paint a house in 40 hours, while person B can paint the same house in 50 hours. How long will it take them to paint the house if they work together?

(Work Problem) Let t represent the number of hours needed to paint the house if they work together. We shall assume both individuals work at a constant rate for simplicity. Our basic strategy is to analyze the part of the task done by each person. If A requires 40 hours to complete the task, A must complete $\frac{1}{40}$*th* of the task each hour.

In t hours, A will complete $t\left(\frac{1}{40}\right) = t/40$ of the task. Similarly, if B requires 50 hours to complete the task,

B must complete $\frac{1}{50}$*th* of the task each hour. In t hours, B will complete $t\left(\frac{1}{50}\right) = t/50$ of the task.

{The portion done by A} + {the portion done by B} = 1 (One whole task)

$$\frac{t}{40} + \frac{t}{50} = 1$$
$$200\left(\frac{t}{40} + \frac{t}{50}\right) = 200\,(1)$$
$$5t + 4t = 200$$
$$9t = 200$$
$$t = \frac{200}{9} = 22\tfrac{2}{9}\text{ hours}$$

We have determined that $22\frac{2}{9}$ hours will be required to paint the house if they work together.

Check:

$$\frac{t}{40} + \frac{t}{50} = 1$$

$$t\left(\frac{t}{40}\right)+t\left(\frac{t}{50}\right)=1$$

$$\left(\frac{200}{9}\right)\left(\frac{1}{40}\right)+\left(\frac{200}{9}\right)\left(\frac{1}{50}\right)\stackrel{?}{=}1$$

$$\frac{5}{9}+\frac{4}{9}=\frac{9}{9}=1$$

4.19 Two planes flying in opposite directions left Denver at the same time. Plane A flew 540 mph and plane B flew 490 mph. How long will it be for them to be 1800 miles apart?

(Motion Problem) This is a distance, rate, time question. Recall that $d = rt$. We are asked to find a time quantity. Let t be the number of hours it will take for the planes to be 1800 miles apart. Make a table in order to analyze the situation and organize the knowns and unknowns. A sketch helps to visualize the problem (as shown in Figure 4.22).

	RATE IN MPH	TIME IN HOURS	DISTANCE IN MILES
Plane A	540	t	$540t$
Plane B	490	t	$490t$

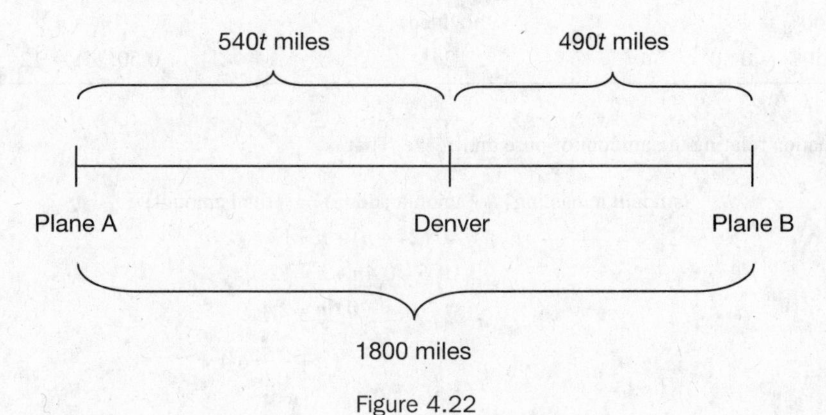

Figure 4.22

The sum of the distances traveled by the planes must equal 1800 miles. Hence,

$$490t + 540t = 1800$$
$$1030t = 1800$$
$$t = \frac{1800}{1030} \approx 1.75 \text{ hours}$$

The planes will be 1800 miles apart in approximately 1.75 hours, or 1 hour 45 minutes.

Check:

$$490t + 540t = 490(1.75) + 540(1.75)$$
$$= 1802.5 \approx 1800$$

4.20 An automobile radiator contains 24 quarts of a 40% antifreeze solution. How many quarts should be drained and replaced with pure antifreeze if the final mixture is 50% antifreeze?

(Mixture Problem) We are asked to determine how many quarts of pure (100%) antifreeze will be needed in order to increase the final concentration to 50% antifreeze. Let n be the number of quarts of antifreeze needed, and therefore also the amount drained from the radiator. See Figure 4.23.

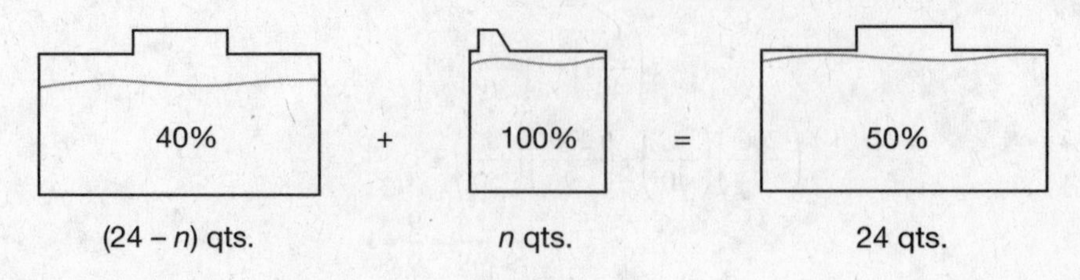

Figure 4.23

The amount of a particular substance in a mixture is obtained by multiplying the concentration of the substance by the amount of the mixture. Construct a table to help organize the important information in mixture problems. Or, in the diagram, simply multiply the number in the container by the number underneath it and use the symbols between the containers to create the equation to be solved.

% CONCENTRATION ANTIFREEZE	AMOUNT OF MIXTURE IN QUARTS	AMOUNT OF PURE ANTIFREEZE IN QUARTS
$40\% = 0.40$	24	$0.40(24) = 9.6$
$40\% = 04.40$	n (drained)	$0.40n$
$40\% = 0.40$	$24 - n$ (remains)	$0.40(24 - n)$
$100\% = 1$	n(added)	$1n = n$
$50\% = 0.50$	24	$0.50(24) = 12$

Write an equation relating the amount of pure antifreeze. That is,

$$\{\text{amount remaining}\} + \{\text{amount added}\} = \{\text{final amount}\}$$
$$0.40\,(24 - n) + n = 12$$
$$9.6 - 0.4n + n = 12$$
$$0.6n = 2.4$$
$$n = \frac{2.4}{0.6} = 4$$

We conclude that 4 quarts of pure antifreeze should be added in order to obtain a 50% solution after draining 4 quarts of the 40% solution.

Check:

$$0.40(24 - n) + n = 12$$
$$0.40(24 - 4) + 4 \stackrel{?}{=} 12$$
$$0.40(20) + 4 = 8 + 4 = 12$$

Refer to supplementary problems 4.8 through 4.16 to practice the above concepts.

4.21 Solve the following inequalities, and graph the solution set on a number line. Express the solution set using interval notation.

(*a*) $x - 4 > 2 - x$

$$x - 4 > 2 - x$$
$$2x > 6 \qquad \text{Add } x \text{ and } 4$$
$$x > 3 \qquad \text{Divide by } 2$$

The solution set in interval notation is $(3, \infty)$. The relevant graph is shown in Figure 4.24.

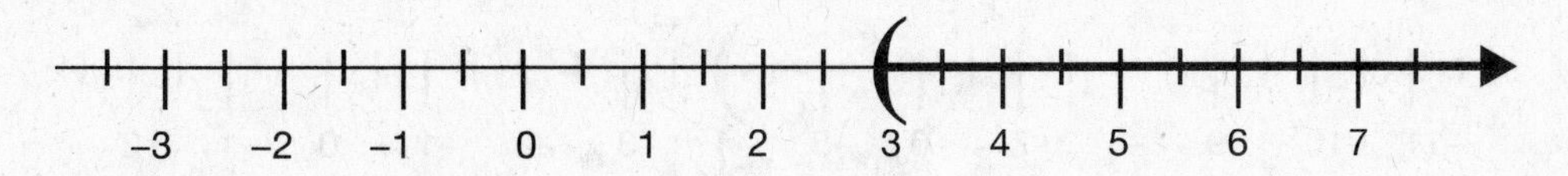

Figure 4.24

(*b*) $3(t + 4) \le t + 10$

$3(t + 4) \le t + 10$	
$3t + 12 \le t + 10$	Distribute
$2t \le -2$	Subtract t and 12
$t \le -1$	Divide by 2

The solution set in interval notation is $(-\infty, -1]$. The relevant graph is shown in Figure 4.25.

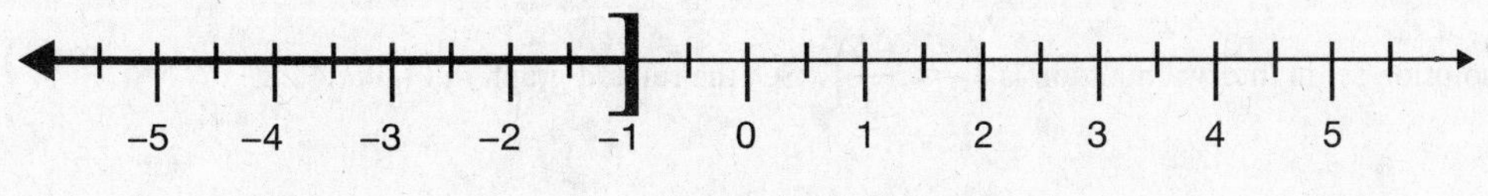

Figure 4.25

(*c*) $4 - 3s \ge 12 - s$

$4 - 3s \ge 12 - s$	
$4 \ge 12 + 2s$	Add $3s$
$-8 \ge 2s$	Subtract 12
$-4 \ge s$	Divide by 2
or $s \le -4$	Rewrite

The solution set in interval notation is $(-\infty, -4]$. The relevant graph is shown in Figure 4.26.

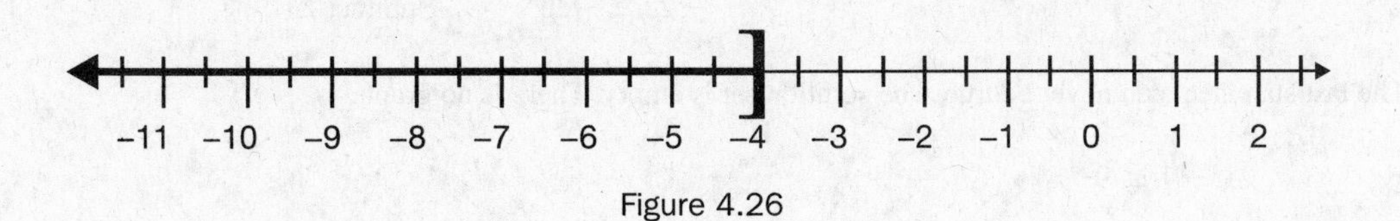

Figure 4.26

(*d*) $4(x - 3) - 5(x + 2) > -18$

$4(x - 3) - 5(x + 2) > -18$	
$4x - 12 - 5x - 10 > -18$	Distribute
$-x - 22 > -18$	Combine
$-4 > x$	Add x and 18
or $x < -4$	Rewrite

The solution set in interval notation is $(-\infty, -4)$. The relevant graph is shown in Figure 4.27.

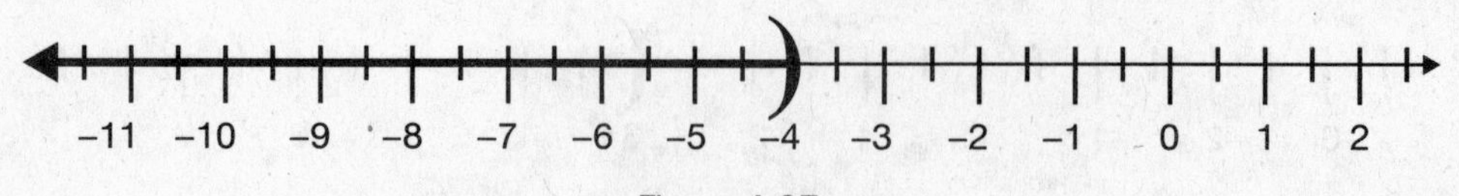

Figure 4.27

(*e*) $\frac{y+3}{4} < \frac{1}{3}$

$\frac{y+3}{4} < \frac{1}{3}$	
$3(y+3) < 4$	Multiply by 12 and simplify
$3y + 9 < 4$	Distribute
$3y < -5$	Subtract 9
$y < \frac{-5}{3}$	Divide by 3

The solution set in interval notation is $\left(-\infty, \frac{-5}{3}\right)$. See the related graph in Figure 4.28.

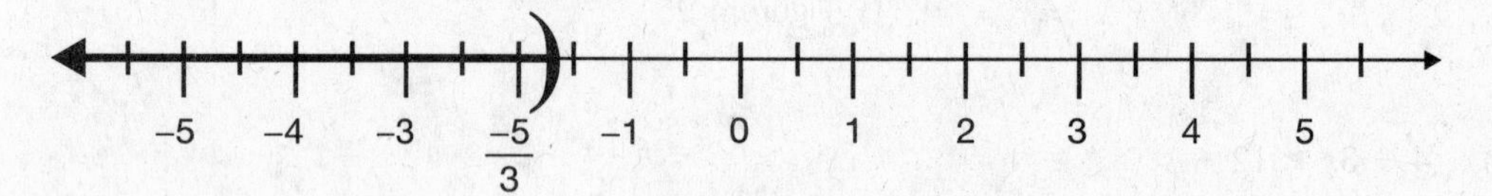

Figure 4.28

(*f*) $5(n-2) - 3(n+4) \geq 2n - 20$

$5(n-2) - 3(n+4) \geq 2n - 20$	
$5n - 10 - 3n - 12 \geq 2n - 20$	Distribute
$2n - 22 \geq 2n - 20$	Combine
$-22 \geq -20$	Subtract $2n$

The last statement can never be true. The solution set is empty. There is no graph.

(*g*) $5 < 2x - 1 \leq 9$

$5 < 2x - 1 \leq 9$	
$6 < 2x \leq 10$	Add 1 to each member
$3 < x \leq 5$	Divide by 2

The solution set in interval notation is (3, 5]. The relevant graph is shown in Figure 4.29.

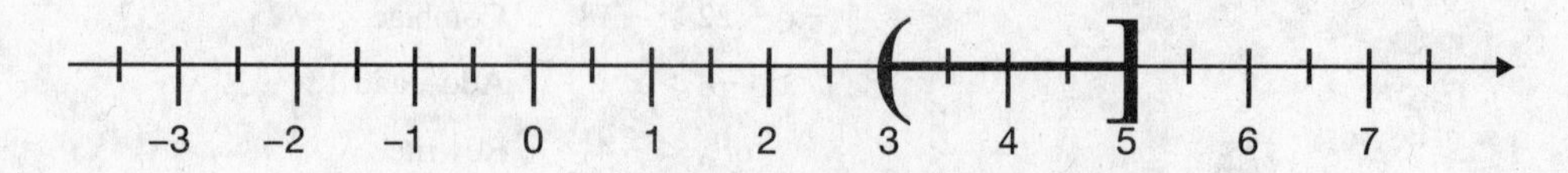

Figure 4.29

(*h*) $-2 \le 5 - s \le 0$

$$
\begin{array}{rll}
 & -2 \le 5 - s \le 0 & \\
 & -7 \le \ -s \ \le -5 & \text{Subtract 5} \\
 & 7 \ge \ s \ \ge 5 & \text{Multiply or divide by } -1 \\
\text{or} & 5 \le \ s \ \le 7 & \text{Rewrite}
\end{array}
$$

The solution set in interval notation is [5, 7]. The relevant graph is shown in Figure 4.30.

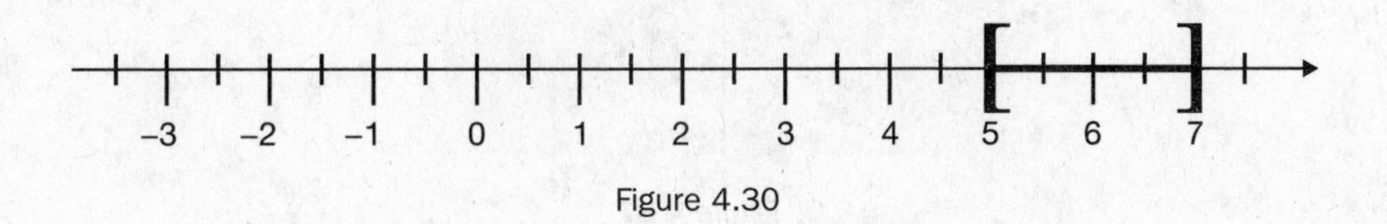

Figure 4.30

Refer to supplementary problem 4.17 for similar problems.

4.22 Graph the following inequalities.

(*a*) $x + y \le 3$

Graph the equation $x + y = 3$ first using the intercept method. The intercepts occur at (3, 0) and (0, 3). The graph is a line passing through those points. It is the boundary of the half-plane which represents the solution set of the given inequality. Draw the line on the coordinate system. We must now determine the ordered pairs which satisfy the inequality $x + y < 3$. Those ordered pairs are associated with points on one side of the line. We must determine which side is appropriate. Simply identify the coordinates of a point not on the line. Choose the origin if it is not on the boundary. The origin has coordinates (0, 0), and $0 + 0 < 3$ is true. We conclude that points on the origin side of the line satisfy the stated inequality. Shade the origin side of the boundary to represent the solution set of the inequality $x + y \le 3$. See Figure 4.31.

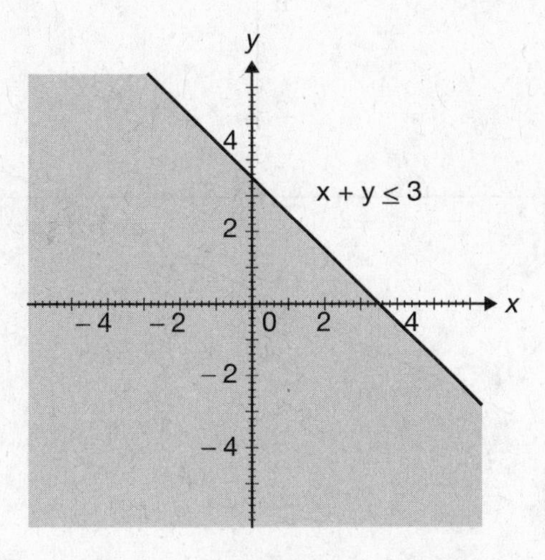

Figure 4.31

(*b*) $x + y < 3$

The graph is similar to the graph in part (*a*). The points on the boundary $x + y = 3$ are excluded this time since we are given a strict inequality. We draw the boundary as a dashed line to indicate that the points on the line are not included. The points on the origin side of the line are included in the graph. See Figure 4.32.

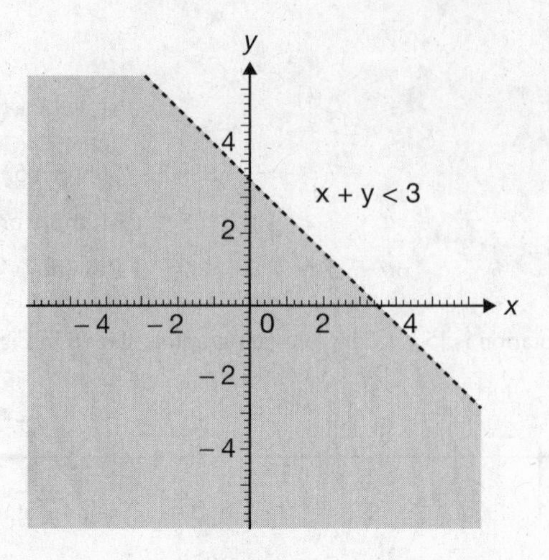

Figure 4.32

(c) $x - 2y \geq 4$

The intercepts of the graph of $x - 2y = 4$ are points with coordinates $(4, 0)$ and $(0, -2)$. Graph the boundary as a solid line, since the coordinates of points on the line satisfy the given inequality. The origin is not on the boundary, and $0 - 2(0) \geq 4$ is false. The half-plane not containing the origin is shaded. We can verify this result by choosing a point in the shaded region. Substitute the coordinates of the point into the inequality, and determine if the statement is true. We illustrate by choosing the point with coordinates $(5, 0)$. $5 - 2(0) \geq 4$ is true. Our results are confirmed. See Figure 4.33.

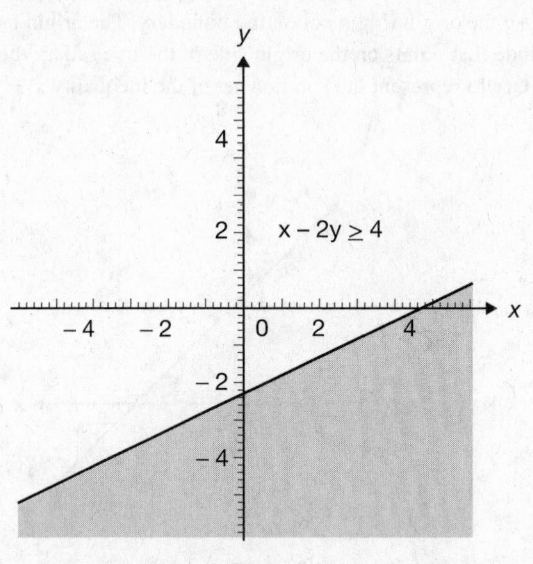

Figure 4.33

(d) $y > \dfrac{-1}{4}x$

Graph the equation $y = (-1/4)\,x$ as a dashed line, since the points on the boundary do not satisfy the given inequality. Since the origin lies on the boundary, another point must be chosen to determine which side of the line to shade. We arbitrarily choose $(4, 0)$. Since $0 > (-1/4)(4) = -1$ is true, the half-plane containing $(4, 0)$ is shaded. See Figure 4.34.

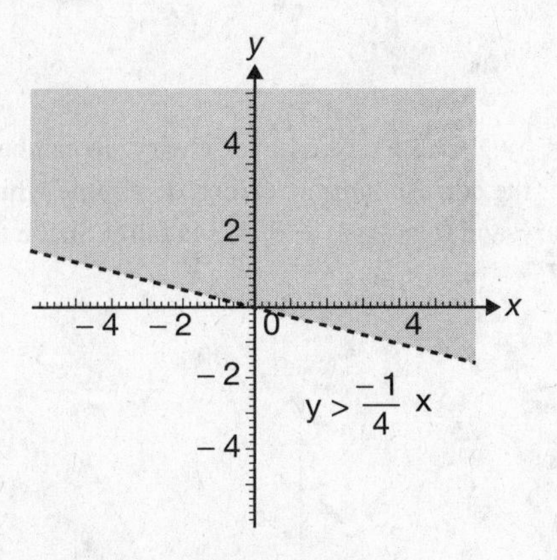

Figure 4.34

(*e*) $x \geq -3$

The graph of $x = -3$ is a vertical line three units to the left of the *y*-axis. The boundary points are included in a ≥ relation; therefore, a solid line is drawn. The origin may be employed to determine appropriate shading. $0 \geq -3$ is true, so the origin side of the line is shaded. See Figure 4.35.

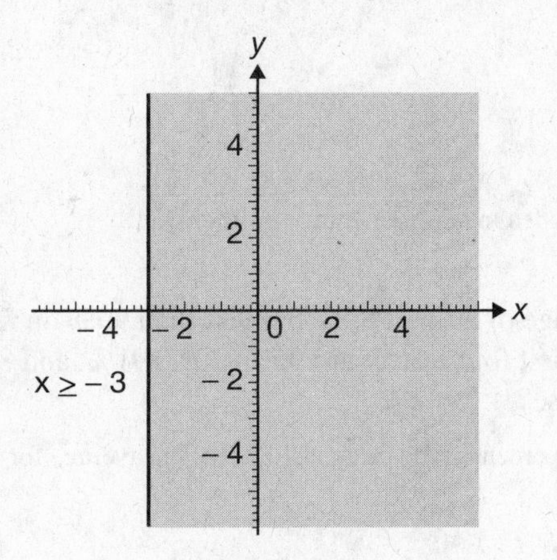

Figure 4.35

(*f*) $y < 5$

The graph of $y = 5$ is a horizontal line five units above the *x*-axis. Draw a dashed line, since the points on the boundary are excluded. We use the origin as our test point and note that $0 < 5$ is true. Shade the region below the line. See Figure 4.36.

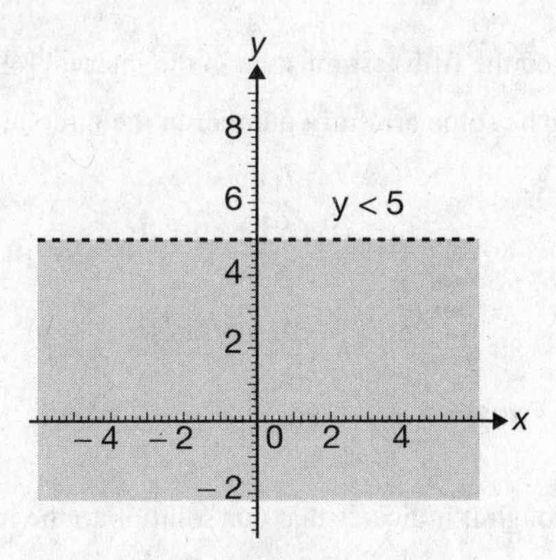

Figure 4.36

(g) $x > \frac{1}{2}y + 3$

Graph the boundary $x = \frac{1}{2}y + 3$ as a dashed line. The graph can be determined by choosing arbitrary values for y and solving for the corresponding x. Next, determine which half-plane should be shaded. The origin is not on the boundary, and $0 > \frac{1}{2}(0) + 3 = 3$ is false. Shade the half-plane which does not contain the origin. See Figure 4.37.

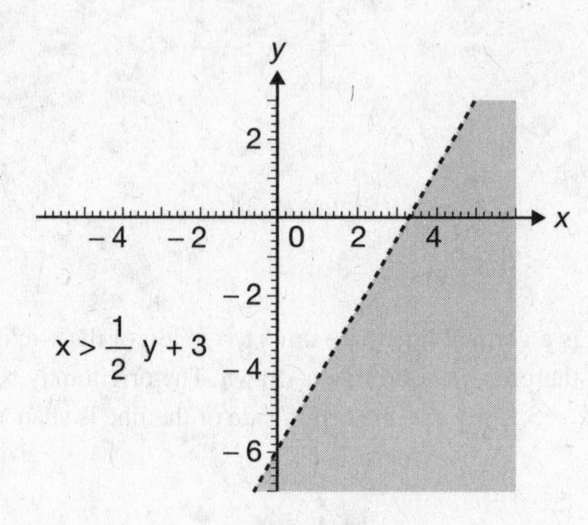

Figure 4.37

Practice graphing linear inequalities in supplementary problem 4.18

4.23 A student must have an average of at least 80% but less than 90% on five tests to receive a B in a course. The student's grades on the first four tests were 95%, 79%, 91%, and 86%. What grade on the fifth test will result in a B for the course?

(Grade Problem) Let p be the percent grade on the fifth test. The average for the 5 tests is $\frac{95 + 79 + 91 + 86 + p}{5}$.

Then $80 \le \frac{95 + 79 + 91 + 86 + p}{5} < 90$ must be satisfied. Apply the techniques of Chapter 4 to solve the inequality.

$$80 \le \frac{95 + 79 + 91 + 86 + p}{5} < 90$$
$$400 \le 351 + p < 450$$
$$49 \le p < 99$$

The student's percentage grade on the fifth test must be in the interval [49, 99) to receive a B in the course.

Check: To check the result, let p be some arbitrary number in the interval [49, 99), say 50. Then

$$80 \le \frac{95 + 79 + 91 + 86 + p}{5} < 90$$
$$80 \overset{?}{\le} \frac{95 + 79 + 91 + 86 + 50}{5} \overset{?}{<} 90$$
$$80 \le 80.2 < 90 \text{ is true.}$$

Our check is not foolproof, although it indicates that our solution to the inequality is probably correct.

4.24 Car rental company X rents a particular car for \$54 per day. Rental company Y rents a similar car for \$49 per day plus an initial charge of \$75. For what rental period is it cheaper to rent from company Y?

(Consumer Problem) Let d be the number of days the car is rented. Then it costs \$54d to rent from company X and it costs \$(49d + 75) to rent from company Y. Now solve $49d + 75 < 54d$.

$$49d + 75 < 54d$$
$$75 < 5d$$
$$15 < d \text{ or } d > 15$$

Company Y is cheaper if the rental period is more than 15 days.

Check: Choose an arbitrary rental period of 18 days to check our conclusion. The costs for companies X and Y respectively are

$$54d = 54(18) = 972 \text{ for company X,}$$
$$49d + 75 = 49(18) + 75 = 957$$

The results substantiate our conclusion.

4.25 A retiree requires an annual income of at least \$2000 from an investment that earns interest at 4% per year. What is the smallest amount the retiree must invest in order to achieve the desired return?

(Interest Problem) Recall that $I = Prt$. Let P be the amount the retiree invests. Then

$$I = Prt \geq 2000$$
$$P(0.04)1 \geq 2000$$
$$P \geq \frac{2000}{0.04} = 50{,}000$$

The smallest amount the retiree can invest is \$50,000.

Check: To check the result, try $P = \$55{,}000$. Then $I = 55{,}000\,(0.04)(1) = 2200 > 2000$. Apparently our conclusion is correct.

4.26 An organization plans to raise money by sponsoring a concert featuring local talent. The organization will sell 1000 reserved-seat tickets and 500 general-admission tickets at the door. Reserved-seat tickets cost \$3 more than general-admission tickets. What is the minimum price of a reserved-seat seat ticket if the organization must gross at least \$25,500 in order to make a profit?

(Profit Problem) Let p be the price of a reserved-seat ticket in dollars. Then general-admission tickets will be $(p - 3)$ dollars. The gross receipts are then $1000p + 500(p - 3)$. Now write the appropriate inequality and solve.

$$1000p + 500(p - 3) \geq 25{,}500$$
$$1500p - 1500 \geq 25{,}500$$
$$1500p \geq 27{,}000$$
$$p \geq 18$$

The minimum price of a reserved-seat ticket is \$18.

Check: To check, suppose the price p is \$18. The gross receipts are then

$$1000p + 500(p - 3) = 1000(18) + 500(18 - 3)$$
$$= 18{,}000 + 7500 = 25{,}500 \geq 25{,}500.$$

4.27 A store has space for 200 items of a seasonal type of apparel it plans to sell. The items are available in small, medium, and large sizes. Past records indicate that demand for the medium-size items is the greatest, and that 1/3 as many small-size items and 1/2 as many large-size items as medium-size items will be needed. What is the largest number of medium-size items for which the store has space?

(Retail Problem) Let m be the number of medium-size items. Then the number of small-size and large-size items is $\frac{1}{3}m$ and $\frac{1}{2}m$, respectively. Now solve,

$$\frac{1}{3}m + m + \frac{1}{2}m \leq 200$$
$$2m + 6m + 3m \leq 1200$$
$$11m \leq 1200$$
$$m \leq \frac{1200}{11} \approx 109.09 \rightarrow 109$$

The largest number of medium-size items for which the store has space is 109. In order to have a whole number of items, we must drop back to the largest integer less than or equal to 109.09.

Check: Let's use $m = 108$ to check. Then

$$\frac{1}{3}m + m + \frac{1}{2}m = \frac{1}{3}(108) + 108 + \frac{1}{2}(108)$$
$$= 36 + 108 + 54 = 198 \leq 200$$

4.28 A supplier received an order, to be shipped by truck, for 100 boxes of merchandise that weigh 39 pounds each. In addition, the order must include the largest number of 65-pound boxes of merchandise that the truck can carry. The truck can carry 5 tons safely. How many 65-pound boxes of merchandise can be included in the order by the supplier? (1 ton = 2000 lb.)

(Supplier Problem) Let n be the number of 65 pound boxes of merchandise that can be included. Then

$$100(39) + 65n \leq 5(2000) = 10{,}000$$
$$3900 + 65n \leq 10{,}000$$
$$65n \leq 6100$$
$$n \leq \frac{6100}{65} \approx 93.846154 \rightarrow 93$$

The supplier can include no more than 93 of the 65-pound boxes of merchandise in the order.

Check: $100\,(39) + 65\,(93) = 3900 + 6045 = 9945 \leq 10{,}000$

4.29 The length of a rectangle is three times its width. If the width varies from 9 to 12 feet, what is the range of values for the perimeter?

(Geometry Problem) Let P be the perimeter of the rectangle. Recall that $P = 2l + 2w$, where l and w are the length and width of the rectangle, respectively. For this rectangle, $l = 3w$. See Figure 4.38.

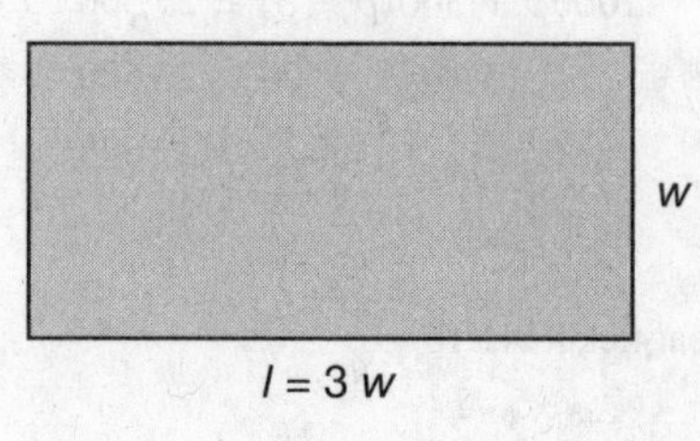

Figure 4.38

The perimeter of the rectangle is $P = 2l + 2w = 2(3w) + 2w = 8w$. Therefore, since the width is restricted in the given information,

$$9 \le w \le 12$$
$$8(9) \le 8w \le 8(12)$$
$$72 \le P \le 96$$

The perimeter lies in the interval $[72, 96]$.

Check: To check, suppose $w = 10$ feet. (w varies from 9 to 12 feet.) Then the perimeter is $8w = 8(10) = 80$. Note that 80 lies in the indicated interval.

Refer to supplementary problems 4.19 through 4.23 for similar problems.

4.30 Use property 1 of section 4.8 to solve the following. State the solution set for each.

(*a*) $|t| = 6$

$$t = -6 \quad \text{or} \quad t = 6$$
$$\{-6, 6\}$$

Check:

$$t = -6 \qquad t = 6$$
$$|-6| = 6 \qquad |6| = 6$$

(*b*) $|s + 5| = 4$

$$s + 5 = -4 \quad \text{or} \quad s + 5 = 4$$
$$s = -9 \qquad s = -1$$
$$\{-9, -1\}$$

Check:

$$s = -9 \qquad s = -1$$
$$|-9 + 5| = |-4| = 4 \qquad |-1 + 5| = |4| = 4$$

(*c*) $|4y - 7| = 5$

$$4y - 7 = -5 \quad \text{or} \quad 4y - 7 = 5$$
$$4y = 2 \qquad 4y = 12$$
$$y = \frac{1}{2} \qquad y = 3$$
$$\left\{\frac{1}{2}, 3\right\}$$

Check:

$$y = \frac{1}{2} \qquad y = 3$$
$$\left|4\left(\frac{1}{2}\right) - 7\right| = |2 - 7| \qquad |4(3) - 7| = |12 - 7|$$
$$= |-5| = 5 \qquad = |5| = 5$$

(*d*) $|2n - 7| = -3$

The equation states that the distance between $2n - 7$ and zero is minus three. Distance is not negative. There is no solution to the equation. The solution set is the empty set Ø. Property 1 is not applicable if the constant $c < 0$.

(*e*) $|3 - x| - 8 = -5$

$|3 - x| - 8 = -5$ is equivalent to $|3 - x| = 3$. Now apply property 1.

$$3 - x = -3 \quad \text{or} \quad 3 - x = 3$$
$$-x = -6 \qquad\qquad -x = 0$$
$$x = 6 \qquad\qquad x = 0$$
$$\{6, 0\}$$

Check:

$$x = 6 \qquad\qquad x = 0$$
$$|3 - 6| - 8 = |-3| - 8 \qquad |3 - 0| - 8 = |3| - 8$$
$$= 3 - 8 = -5 \qquad\qquad = 3 - 8 = -5$$

In solved problem 4.30 (*e*), we rewrote the original equation in an equivalent form prior to the application of property 1. Be sure to isolate the absolute value expression first or property 1 is not applicable.

(*f*) $|7x - 14| = 0$

$$7x - 14 = -0 \quad \text{or} \quad 7x - 14 = 0$$

Since $-0 = 0$, there is only one equation to solve. We shall choose $7x - 14 = 0$

$$7x - 14 = 0$$
$$7x = 14$$
$$x = 2 \quad \text{so } \{2\}$$

Check: $x = 2$

$$|7\,(2) - 14| = |14 - 14| = |0| = 0$$

(*g*) $|4s - 6| = s$

If $s < 0$, there is no solution. If $s > 0$, we proceed in the usual manner.

$$4s - 6 = -s \quad \text{or} \quad 4s - 6 = s$$
$$5s - 6 = 0 \qquad\qquad 3s - 6 = 0$$
$$5s = 6 \qquad\qquad 3s = 6$$
$$s = \frac{6}{5} \qquad\qquad s = 2$$
$$\left\{\frac{6}{5}, 2\right\}$$

Check:

$$s = \frac{6}{5} \qquad\qquad s = 2$$
$$\left|4\left(\frac{6}{5}\right) - 6\right| \stackrel{?}{=} \frac{6}{5} \qquad\qquad |4\,(2) - 6| \stackrel{?}{=} 2$$
$$\left|\frac{24}{5} - \frac{30}{5}\right| \stackrel{?}{=} \frac{6}{5} \qquad\qquad |8 - 6| \stackrel{?}{=} 2$$
$$\left|\frac{-6}{5}\right| \stackrel{?}{=} \frac{6}{5} \qquad\qquad |2| \stackrel{?}{=} 2$$
$$\frac{6}{5} = \frac{6}{5} \qquad\qquad 2 = 2$$

(*h*) $|2w + 6| = |w - 8|$

In order for the absolute value of two quantities to be equal, they must be the same distance from zero. This can be true if the two quantities are equal or if they are opposites of one another. Therefore, our absolute value equation is true if

$$2w + 6 = w - 8 \qquad \text{or} \qquad 2w + 6 = -(w - 8)$$
$$w + 6 = -8 \qquad\qquad 2w + 6 = -w + 8$$
$$w = -14 \qquad\qquad 3w + 6 = 8$$
$$3w = 2$$
$$w = \frac{2}{3}$$

$$\left\{-14, \frac{2}{3}\right\}$$

Check:

$$w = -14 \qquad\qquad w = \frac{2}{3}$$
$$|2(-14) + 6| \stackrel{?}{=} |-14 - 8| \qquad \left|2\left(\frac{2}{3}\right) + 6\right| \stackrel{?}{=} \left|\frac{2}{3} - 8\right|$$
$$|-28 + 6| \stackrel{?}{=} |-22| \qquad \left|\frac{4}{3} + \frac{18}{3}\right| \stackrel{?}{=} \left|\frac{2}{3} - \frac{24}{3}\right|$$
$$|-22| \stackrel{?}{=} |-22| \qquad \left|\frac{22}{3}\right| \stackrel{?}{=} \left|\frac{-22}{3}\right|$$
$$22 = 22 \qquad \frac{22}{3} = \frac{22}{3}$$

4.31 Use either property 2 or property 3 of section 4.8 to solve the following inequalities. Express the solution set in interval notation.

(*a*) $|t| < 7$

$|t| < 7$ is equivalent to $-7 < t < 7$ by property 2. The solution set is $(-7, 7)$.

(*b*) $|s| \leq 4$

$|s| \leq 4$ is equivalent to $-4 \leq s \leq 4$ by property 2. The solution set is $[-4, 4]$.

(*c*) $|x| \geq 5$

$|x| \geq 5$ is equivalent to $x \leq -5$ or $x \geq 5$ by property 3. The solution set is $(-\infty, -5] \cup [5, \infty)$.

(*d*) $|w + 3| > 4$

$|w + 3| > 4$ is equivalent to $w + 3 < -4$ or $w + 3 > 4$ by property 3. Therefore, $w < -7$ or $w > 1$. The solution set is $(-\infty, -7) \cup (1, \infty)$.

(*e*) $|y - 6| < 4$

$|y - 6| < 4$ is equivalent to $-4 < y - 6 < 4$ by property 2. $-4 < y - 6 < 4$ implies $2 < y < 10$. The solution set is $(2, 10)$.

(*f*) $|p + 2| + 3 \leq 5$

Subtract 3 from both members to obtain $|p + 2| \leq 2$. Now apply property 2 and solve. $-2 \leq p + 2 \leq 2$ implies $-4 \leq p \leq 0$. The solution set is $[-4, 0]$.

(*g*) $3\,|x - 8| > 12$

Divide both members by 3 to obtain $|x - 8| > 4$. Now apply property 3 and solve. $|x - 8| > 4$ is equivalent to $x - 8 < -4$ or $x - 8 > 4$. Hence, $x < 4$ or $x > 12$. The solution set is $(-\infty, 4) \cup (12, \infty)$.

(*h*) $|4y + 3| - 5 < 10$

$|4y + 3| - 5 < 10$ is equivalent to $|4y + 3| < 15$. Therefore,

$$-15 < 4y + 3 < 15$$
$$-15 < 4y < 12$$
$$\frac{-9}{2} < y < 3$$

The solution set is $\left(\frac{-9}{2}, 3\right)$.

(*i*) $\left|3 - \frac{x}{4}\right| \geq 1$

$\left|3 - \frac{x}{4}\right| \geq 1$ is equivalent to

$$3 - \frac{x}{4} \leq -1 \quad \text{or} \quad 3 - \frac{x}{4} \geq 1$$
$$12 - x \leq -4 \qquad 12 - x \geq 4$$
$$16 \leq x \qquad 8 \geq x$$

The solution set is $(-\infty, 8] \cup [16, \infty)$.

(*j*) $|5 - 2y| + 4 < 3$

$|5 - 2y| + 4 < 3$ is equivalent to $|5 - 2y| < -1$. $|5 - 2y| < -1$ means that the distance between $5 - 2y$ and 0 must be less than -1. That is not possible. The statement is a contradiction. There is no solution. The solution set is empty and is represented by Ø.

(*k*) $|5 - 2y| + 4 > 3$

$|5 - 2y| + 4 > 3$ is equivalent to $|5 - 2y| > -1$. $|5 - 2y| > -1$ means that the distance between $5 - 2y$ and 0 must be more than -1. That is always true. The statement is an identity. All real numbers satisfy the inequality. The solution set is $(-\infty, \infty)$.

Refer to supplementary problems 4.24 and 4.25 to practice solving similar equations and inequalities.

SUPPLEMENTARY PROBLEMS

4.1 Determine if the given values of the variable satisfy the equation.

(*a*) $5(t - 2) - (t + 2) = 3 - t$ for $t = 0$ and $t = 3$.

(*b*) $s^2 - 4s = 4 - s$ for $s = -1$ and $s = 4$

4.2 Solve.

(*a*) $3x + 5 = 20$

(*b*) $2t - 7 = 13$

(*c*) $4(s - 2) = s + 10$

(*d*) $5n - 10 = 7n + 3$

(*e*) $\frac{a}{2} + \frac{a}{3} + \frac{a}{4} = 13$

(*f*) $t = \frac{t}{2} + 11$

(*g*) $3(n - 14) - 7(n - 18) = 2$

(*h*) $4n - 6(1 - n) = 4(2n + 3)$

(*i*) $3(x + 2) - 5x = 2(3 - x)$

(*j*) $3(t + 1) - t = 2(t + 5)$

(*k*) $7 - 5(x + 4) = -4(x + 1) + (x - 7)$

(*l*) $s - \frac{2}{3}(4 + s) = \frac{-2}{9}$

4.3 Use the intercept method to graph the following.

(*a*) $x - y = 4$

(*b*) $y - x = 4$

(*c*) $3x + 4y = 12$

(*d*) $2x + 4y = 12$

(*e*) $2y - 3x = 6$

(*f*) $2x + 3y = 8$

4.4 Graph the following.

(*a*) $y = -x$ (*b*) $y = \frac{x}{2} - 3$ (*c*) $y = \frac{5}{3x} - 2$

(*d*) $y = -3$ (*e*) $x = 4$ (*f*) $x = 2y + 4$

4.5 Use a graphing calculator to graph the following. Your graphing window may vary slightly from those shown.

(*a*) $y = \frac{3}{4}x - 3$ (*b*) $y = 2 - \frac{x}{5}$ (*c*) $3.28y - 5.62x = 11.8$

(*d*) $4x + 7y = 4$ (*e*) $413x - 248y = -1{,}279$ (*f*) $y = 0.46x + \sqrt{11}$

4.6 Solve each of the following equations explicitly for the indicated variable.

(*a*) $2x + y - 3 = 2$ for y (*b*) $2x + y - 3 = 2$ for x

(*c*) $2a - 3b + 5 = a + 8$ for a (*d*) $2a - 3b + 5 = a + 8$ for b

(*e*) $\frac{x}{3} - \frac{y}{6} = \frac{x}{2} + \frac{y}{4}$ for y (*f*) $3(s + t) - 2(3s - 4t) = t - 4$ for t

(*g*) $y = \frac{x - 2}{x + 2}$ for x (*h*) $(3a - 5)(2b + 7) = -20$ for b

4.7 Each of the following is a formula employed in mathematics, the social sciences, or the physical sciences. Solve each formula for the indicated variable.

(*a*) $P = 2l + 2w$ for w (*b*) $C = 2\pi r$ for r

(*c*) $\frac{P_1}{V_1} = \frac{P_2}{V_2}$ for V_1 (*d*) $A = P + Prt$ for t

(*e*) $\frac{1}{R} = \frac{1}{R_1} + \frac{1}{R_2}$ for R (*f*) $C = \frac{5}{9}(F - 32)$ for F

(*g*) $S = s_0 + v_0 t + \frac{1}{2}gt^2$ for v_0 (*h*) $a_n = a_1 + (n - 1)d$ for n

4.8 Two cars leave an intersection traveling in opposite directions. Their average speeds are 54 and 58 mph. How long will it be until they are 308 miles apart?

4.9 Convert 200 grams to ounces, if 1 gram = 0.0353 ounces.

4.10 A baby weighed 6 pounds 4 ounces at birth. Six months later the baby weighed 10 pounds 10 ounces. What was the percent increase in the baby's weight? (Hint: Use 1 lb= 16 oz to express the weights in pounds.)

4.11 A store sells its merchandise for cost plus 15%. (*a*) What is the selling price of an item which costs the store \$67.60? (*b*) A buyer of the item must pay 6.5% sales tax. What is the total paid by the buyer for the item?

4.12 Factory A can produce 5,200 blobs in 20 days while factory B can produce 3,390 blobs in 30 days. How many days will be required for the factories combined production to fill an order for 5,595 blobs?

4.13 A man has an annual income of \$5,550 from two investments. He has \$10,000 more invested at 5% than he has in a riskier investment at 7.5%. How much does he have invested at each rate?

4.14 How many square yards of carpet will be needed to carpet a room with the indicated dimensions in feet in Figure 4.39 below?

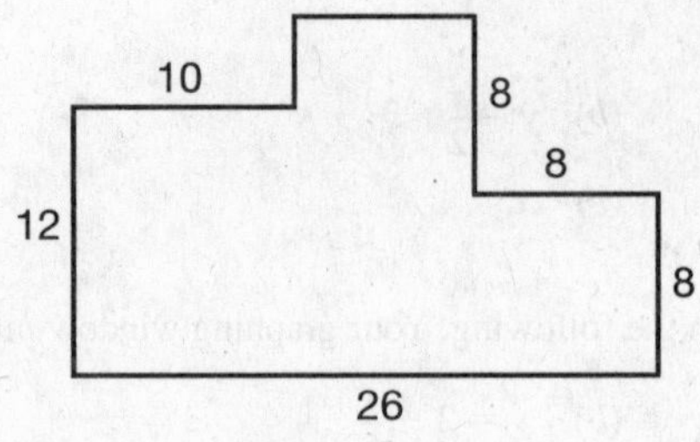

Figure 4.39

4.15 How many liters of a 16% solution of acid should be mixed with 20 liters of 50% solution of acid in order to obtain a 36% solution?

4.16 It takes an airplane 15 minutes longer to complete a flight between two cities when it has a 50 mph headwind than it takes when the wind isn't blowing. The plane normally cruises at 500 mph when the wind is not blowing. (*a*) How long does the trip take when the plane has the headwind? (*b*) How far apart are the cities?

4.17 Solve the following inequalities and graph the solution set on a number line. Express the solution set using interval notation.

(*a*) $2s - 1 < 5 - s$ (*b*) $2x + 3 \geq 3(x + 2)$ (*c*) $2 - 4t \leq 5 - t$

(*d*) $4y - 14 - 6(y + 1) > 3y$ (*e*) $4(t - 1) + 3t + 10 \leq 7(t + 2) - 3$ (*f*) $\frac{x+6}{3} \geq \frac{5}{2}$

(*g*) $s + 1 \leq 2s + 1 < 4$ (*h*) $\frac{3}{2} \leq y + \frac{5}{2} \leq 3$

4.18 Graph the following linear inequalities.

(*a*) $x - y \geq -3$ (*b*) $2x + y < 8$ (*c*) $x - 3y > 9$ (*d*) $y \geq -4$

(*e*) $x < 0$ (*f*) $y > \frac{2}{3}x$ (*g*) $x \leq \frac{4}{3}y + 4$

4.19 A student must have an average of at least 80% but less than 90% on four hour tests and a comprehensive final exam to receive a B in a course. The student's grades on the first four tests were 94%, 79%, 91%, and 86%. What is the minimum grade the student needs on the final exam in order to receive a B for the course if the final exam counts twice as much as an hour test?

4.20 Rental agency R charges $12 plus $8 per hour to rent a rototiller. Rental agency S charges $10 plus $8.50 per hour to rent the rototiller. For which rental period (in hours) is it cheaper to rent from agency R?

4.21 Ray needs a combined annual return of at least $5200 from two investments. A reasonably. safe investment earns 3.75% per year, while a more risky investment earns 5.5% per year. He has half as much invested in the risky investment as he has invested in the more safe investment. Find the minimum amount that can be applied to the more risky investment.

4.22 Kerry makes $8.00 per hour working in a fast-food restaurant. He also does lawn jobs that net him $150 per week. Find the minimum number of hours he must work in the restaurant in order to earn enough to pay his car insurance premium of $734.00. The premium is due in three weeks and he has saved nothing to date.

4.23 A manufacturer anticipates it will need a new two-level rectangular storage facility for its inventory. Available ground space for the facility consists of 18,000 square feet. The normal inventory of materials for medium-size metal sheds is 200 items that require 50 square feet each. An unknown number of large-size metal sheds that require 80 square feet each for materials is to be determined. Find the maximum number of large-size metal sheds for which the manufacturer will have space in the new facility.

4.24 Use property 1 to solve the following. State the solution set for each.

(*a*) $|2x| = 8$ (*b*) $|3t - 8| = 17$ (*c*) $|2 - x| = 10$

(*d*) $|5s + 6| - 8 = -3$ (*e*) $|3n + 5| = -2$ (*f*) $|6 + 4a| = 2$

(*g*) $|4y - 1| = 3 - y$ (*h*) $|x + 5| = |5x - 8|$ (*i*) $6 = |4k + 10|$

(*j*) $2t + 5 = |t + 4|$ (*k*) $\left|\frac{2}{3}w + 1\right| = \frac{1}{4}$ (*l*) $\frac{1}{2}|s - 3| = \frac{2}{3}$

4.25 Use either property 2 or property 3 to solve the following inequalities. Express the solution set in interval notation.

(*a*) $|x| \geq 4$ (*b*) $|t| < 8$ (*c*) $|s + 1| \leq 7$

(*d*) $|w - 5| > 6$ (*e*) $|2n - 1| - 4 \leq -1$ (*f*) $|4 - 3x| \geq 8$

(*g*) $|3s + 1| \leq -5$ (*h*) $|3s + 1| \geq -5$ (*i*) $\left|\frac{2}{3}p + \frac{1}{4}\right| < 1$

(*j*) $4\left|\frac{3}{5} - \frac{n}{2}\right| > 8$ (*k*) $7 - \left|\frac{r}{2} + 2\right| \leq 4$ (*l*) $\left|\frac{x}{4} - 3\right| - 4 \geq \frac{-1}{3}$

ANSWERS TO SUPPLEMENTARY PROBLEMS

4.1 (*a*) $t = 0$ does not satisfy the equation; $t = 3$ satisfies the equation.

(*b*) $s = -1$ and $s = 4$ both satisfy the equation.

4.2 (*a*) $x = 5$ (*b*) $t = 10$ (*c*) $s = 6$

(*d*) $n = \frac{-13}{2}$ (*e*) $a = 12$ (*f*) $t = 22$

(*g*) $n = \frac{41}{2}$ (*h*) $n = 9$ (*i*) All real numbers

(*j*) No solution (*k*) $x = -1$ (*l*) $s = \frac{22}{3}$

4.3 (*a*) See Figure 4.40. (*b*) See Figure 4.41. (*c*) See Figure 4.42.

(*d*) See Figure 4.43. (*e*) See Figure 4.44. (*f*) See Figure 4.45.

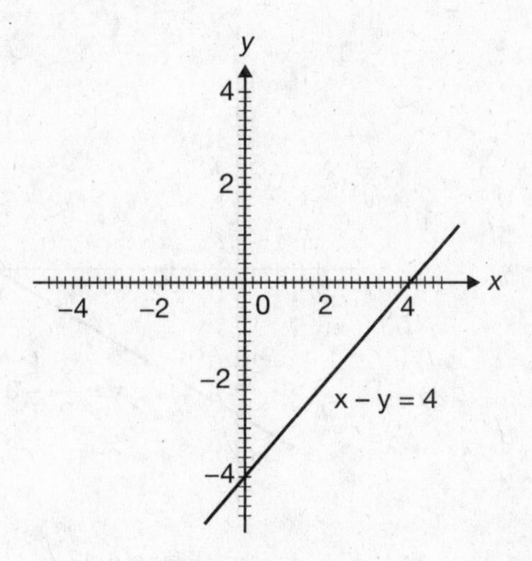

Figure 4.40

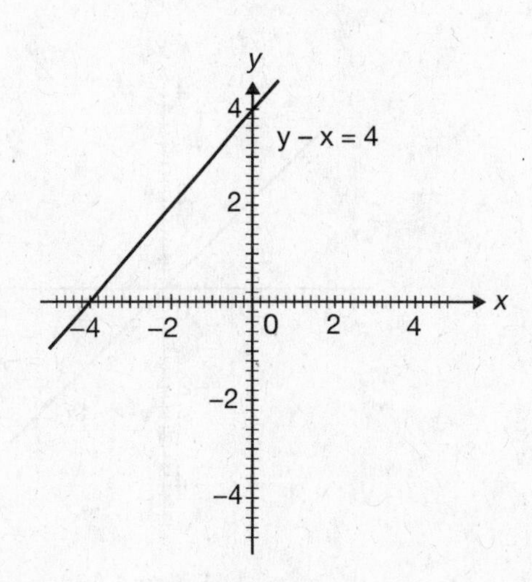

Figure 4.41

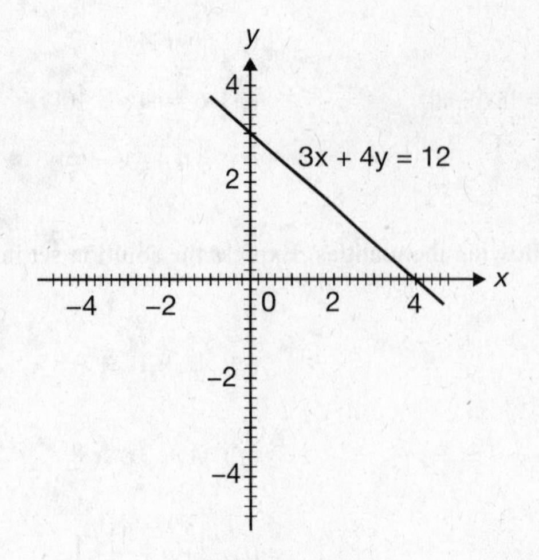

Figure 4.42

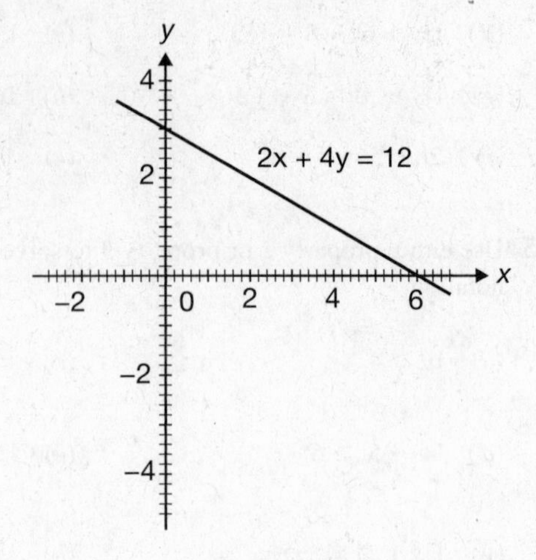

Figure 4.43

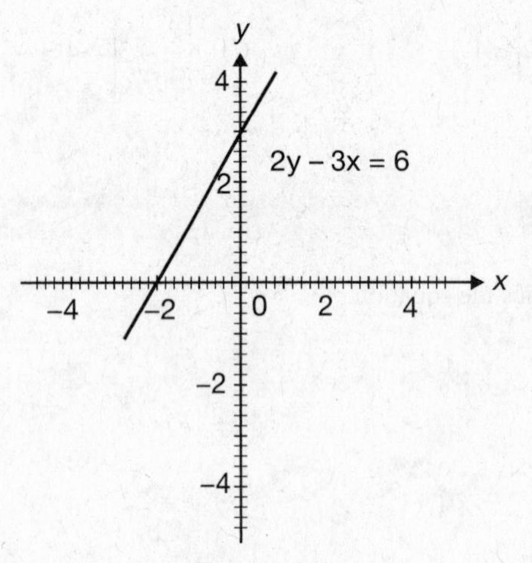

Figure 4.44

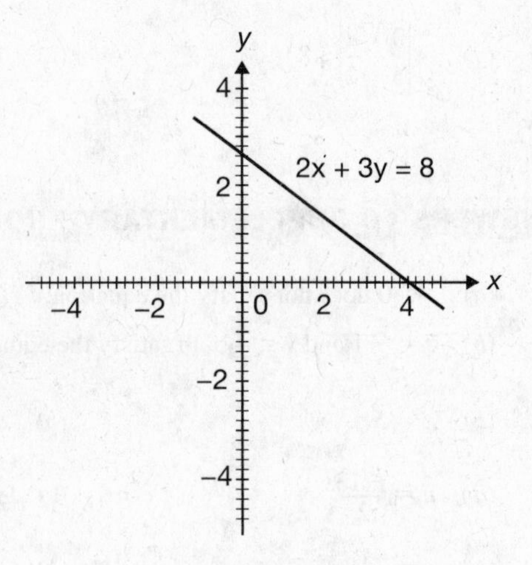

Figure 4.45

4.4 (*a*) See Figure 4.46. (*b*) See Figure 4.47. (*c*) See Figure 4.48.
(*d*) See Figure 4.49. (*e*) See Figure 4.50. (*f*) See Figure 4.51.

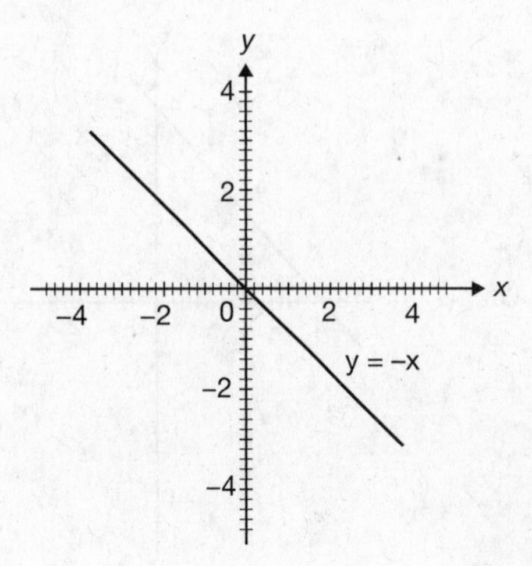

Figure 4.46

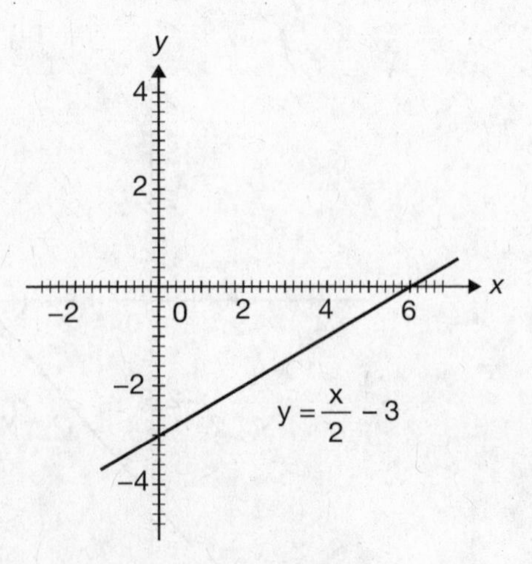

Figure 4.47

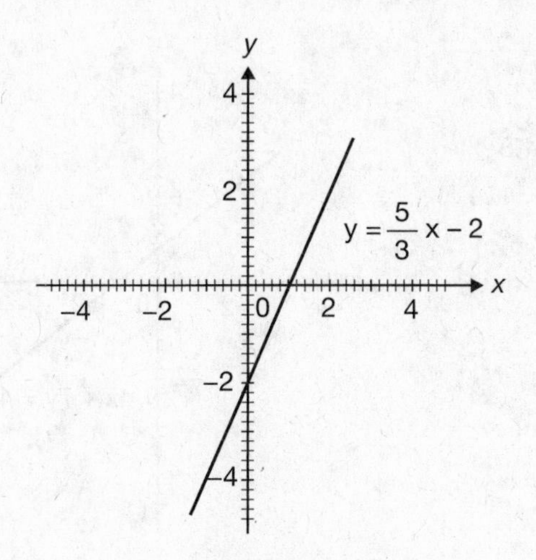

Figure 4.48

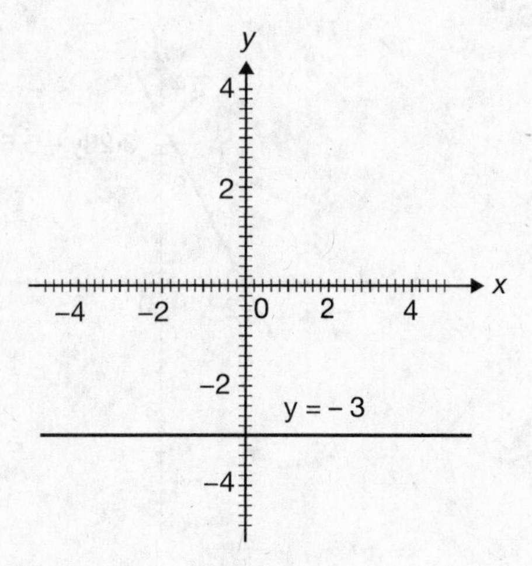

Figure 4.49

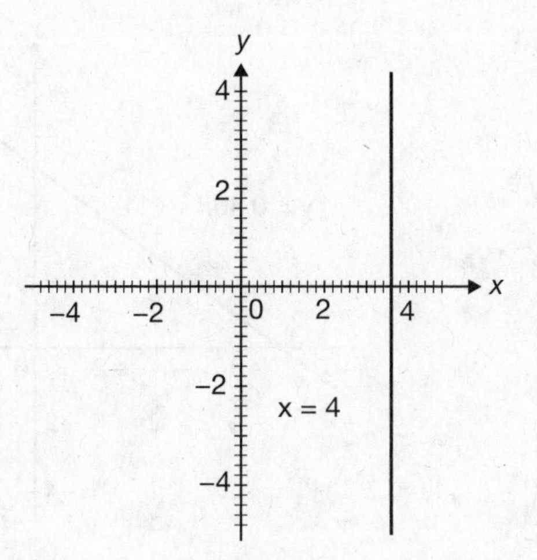

Figure 4.50

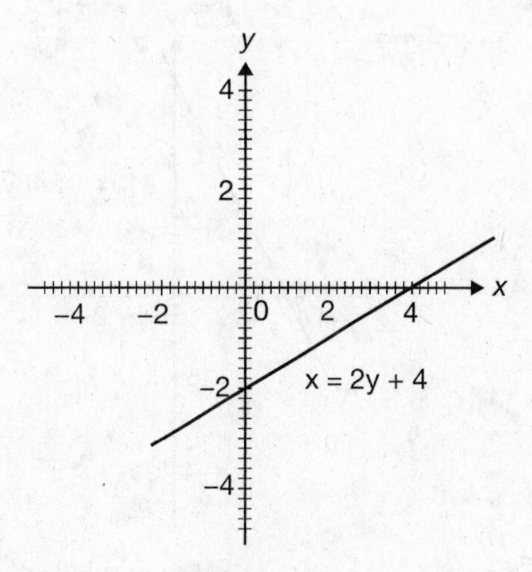

Figure 4.51

4.5 (*a*) See Figure 4.52. (*b*) See Figure 4.53. (*c*) See Figure 4.54.
(*d*) See Figure 4.55. (*e*) See Figure 4.56. (*f*) See Figure 4.57.

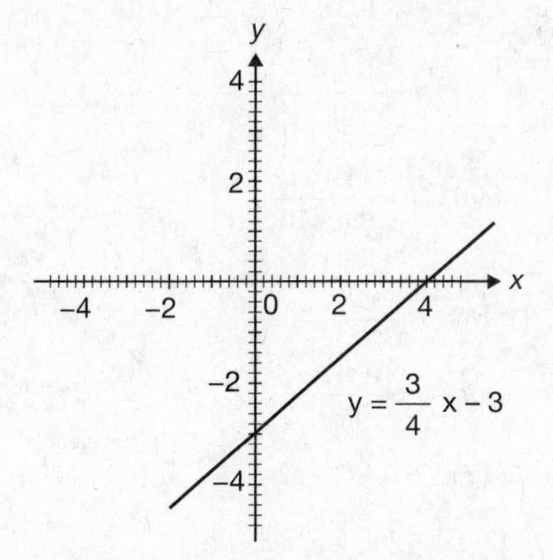

Figure 4.52

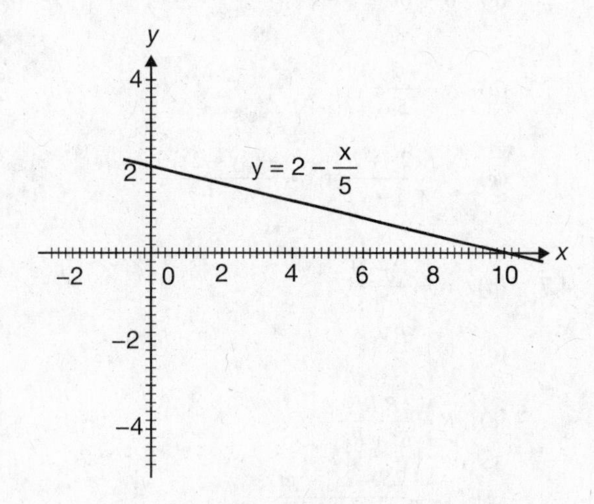

Figure 4.53

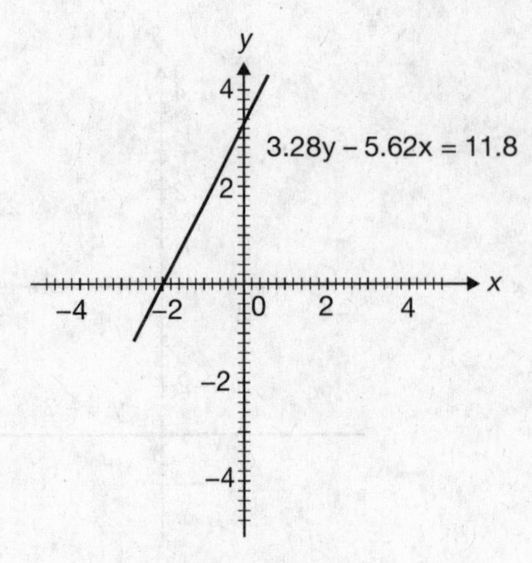

Figure 4.54

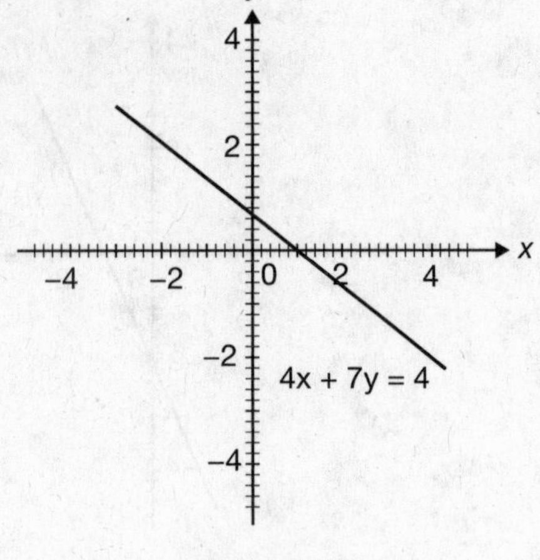

Figure 4.55

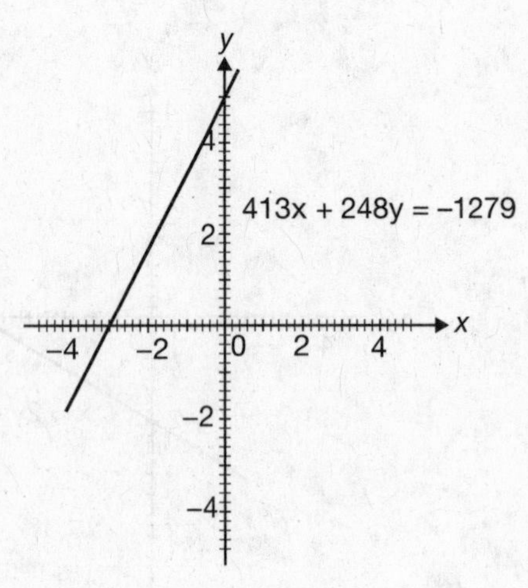

Figure 4.56

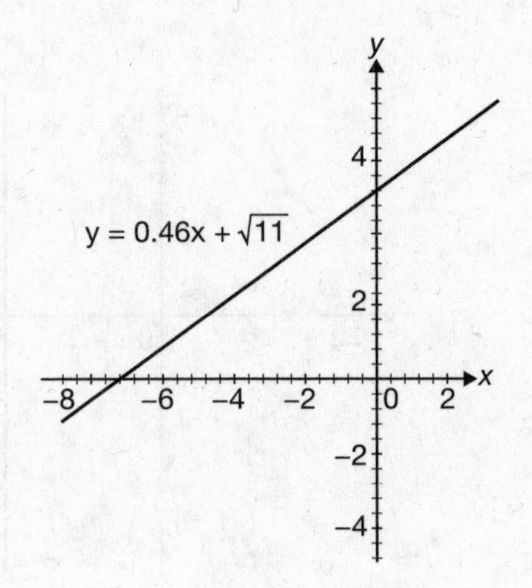

Figure 4.57

4.6 (*a*) $y = 5 - 2x$ (*b*) $x = \dfrac{5-y}{2}$

(*c*) $a = 3b + 3$ (*d*) $b = \dfrac{a-3}{3}$

(*e*) $y = \dfrac{-2}{5}x$ (*f*) $t = \dfrac{3s-4}{10}$

(*g*) $x = \dfrac{2y+2}{1-y}$ or $x = \dfrac{-(2y+2)}{y-1}$ (*h*) $b = \dfrac{15-21a}{6a-10}$

4.7 (*a*) $w = \dfrac{P-2l}{2}$ (*b*) $r = \dfrac{C}{2\pi}$

(*c*) $V_r = \dfrac{P_1V_2}{P_2}$ (*d*) $t = \dfrac{A-P}{Pr}$

(*e*) $R = \dfrac{R_1R_2}{R_1+R_2}$ (*f*) $F = \dfrac{9}{5}C + 32$

(*g*) $v_0 = \dfrac{2S - 2s_0 - gt^2}{2t}$ (*h*) $n = \dfrac{a_n - a_1 + d}{d}$

4.8 (Motion Problem) 2.75 hours.

4.9 (Proportion Problem) 7.06 oz.

4.10 (Proportion Problem) 70%.

4.11 (Consumer Problem) (*a*) \$77.74; (*b*) \$82.79.

4.12 (Work Problem) 15 days.

4.13 (Investment Problem) \$50,400 at 5% and \$40,400 at 7.5%.

4.14 (Geometry Problem) Approximately 35 square yards.

4.15 (Mixture Problem) 14 liters.

4.16 (Motion Problem) (*a*) 2.5 hours or 2 hours 30 minutes; (*b*) 1,125 miles.

4.17 (*a*) $(-\infty, 2)$. See Figure 4.58. (*b*) $(-\infty, -3]$. See Figure 4.59.

(*c*) $[-1, \infty)$. See Figure 4.60. (*d*) $(-\infty, -4)$. See Figure 4.61.

(*e*) $(-\infty. \infty)$. See Figure 4.62. (*f*) $\left[\frac{3}{2}, \infty\right)$. See Figure 4.63.

(*g*) $\left[0, \frac{3}{2}\right)$. See Figure 4.64. (*h*) $\left[-1, \frac{1}{2}\right)$. See Figure 4.65.

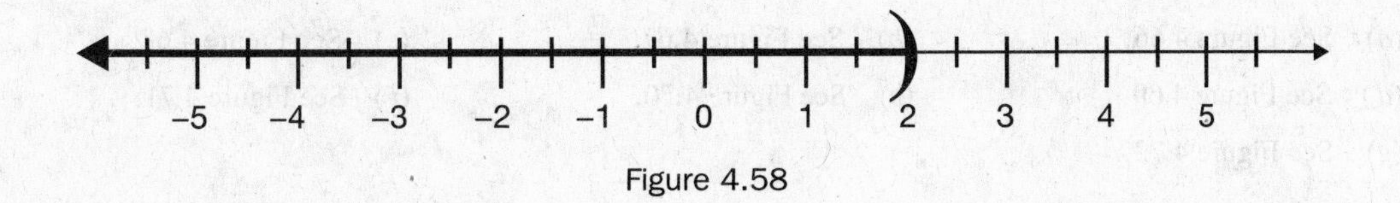

Figure 4.58

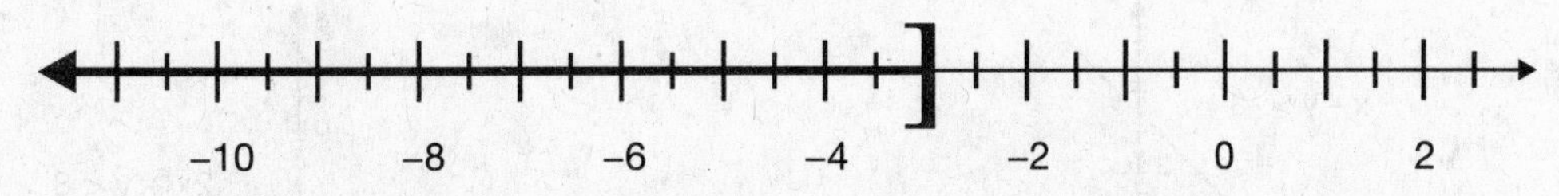

Figure 4.59

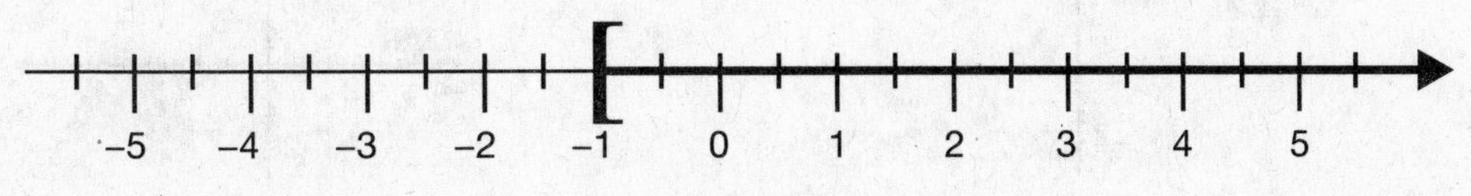

Figure 4.60

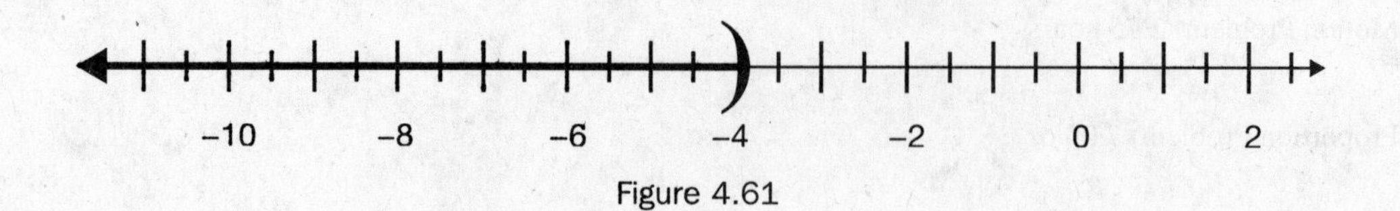

Figure 4.61

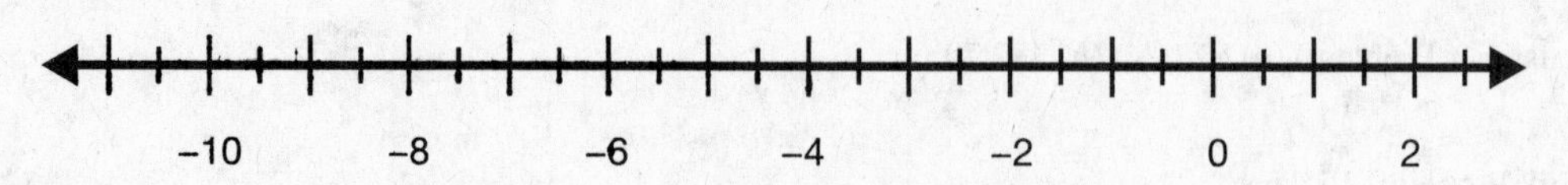

Figure 4.62

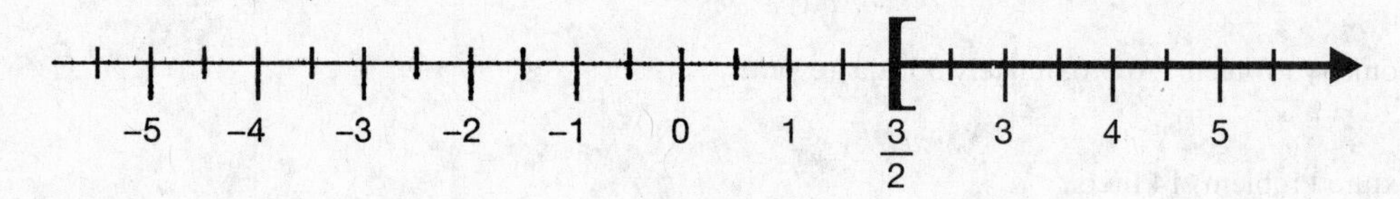

Figure 4.63

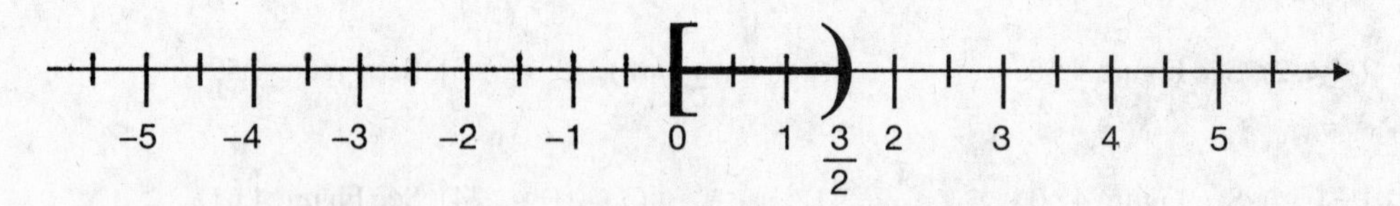

Figure 4.64

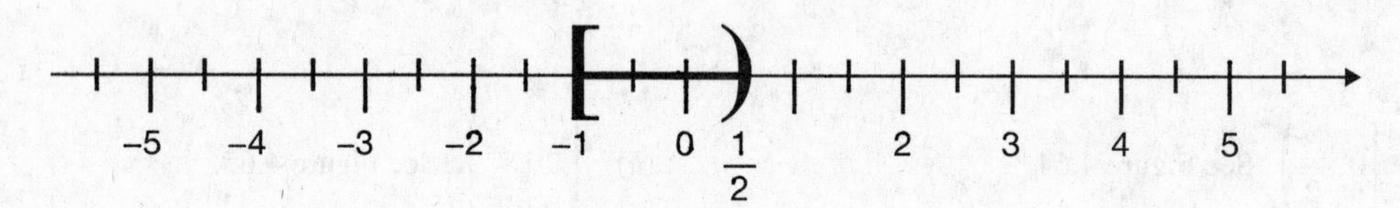

Figure 4.65

4.18 (*a*) See Figure 4.66. (*b*) See Figure 4.67. (*c*) See Figure 4.68.
(*d*) See Figure 4.69. (*e*) See Figure 4.70. (*f*) See Figure 4.71.
(*g*) See Figure 4.72.

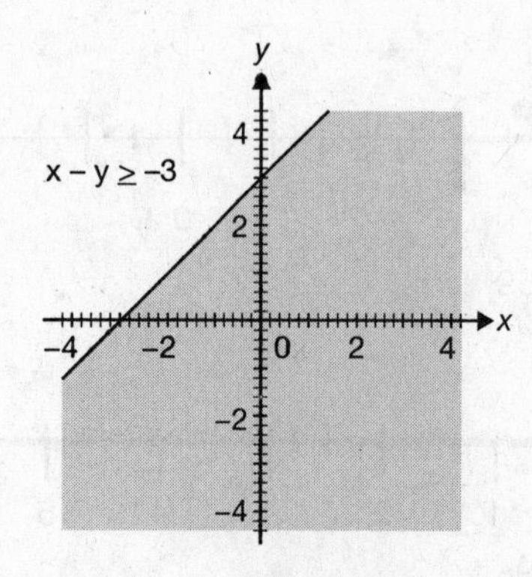

Figure 4.66

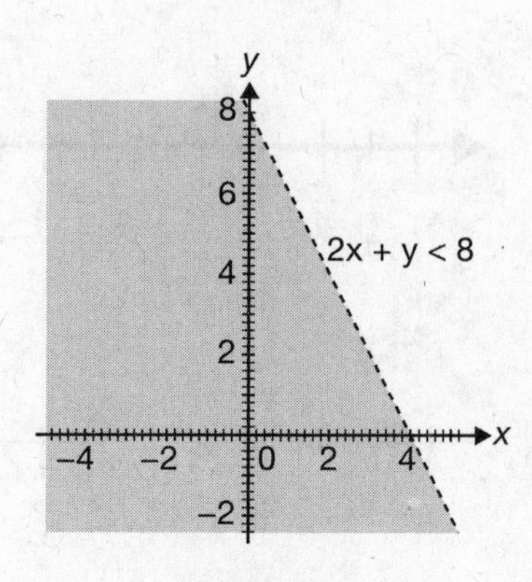

Figure 4.67

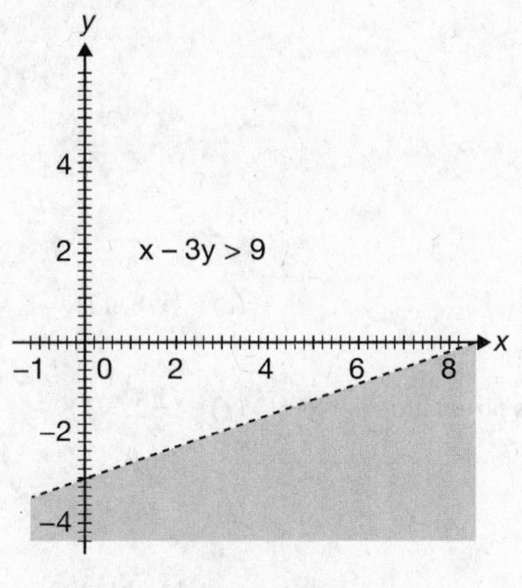

Figure 4.68

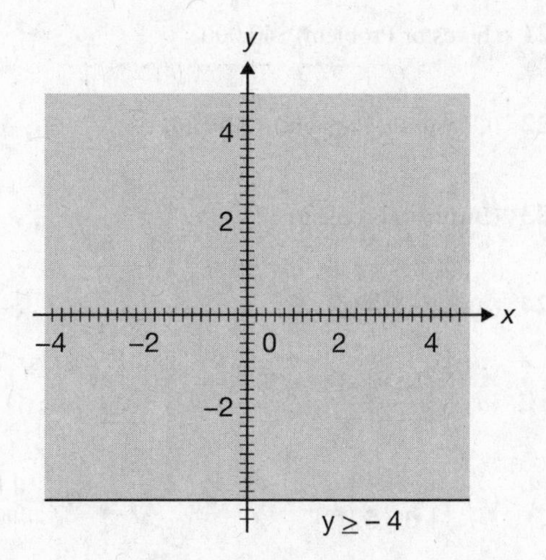

Figure 4.69

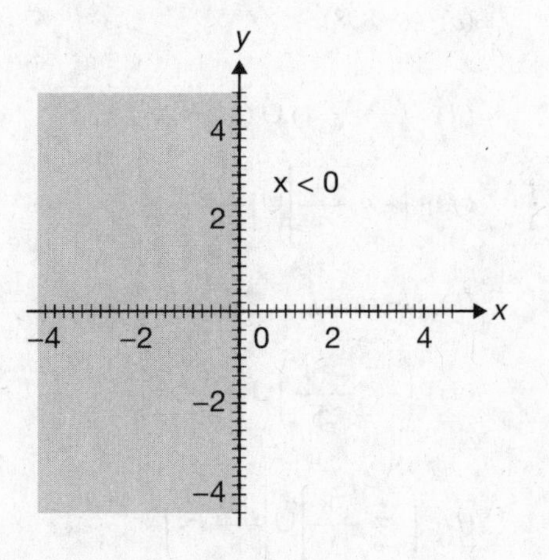

Figure 4.70

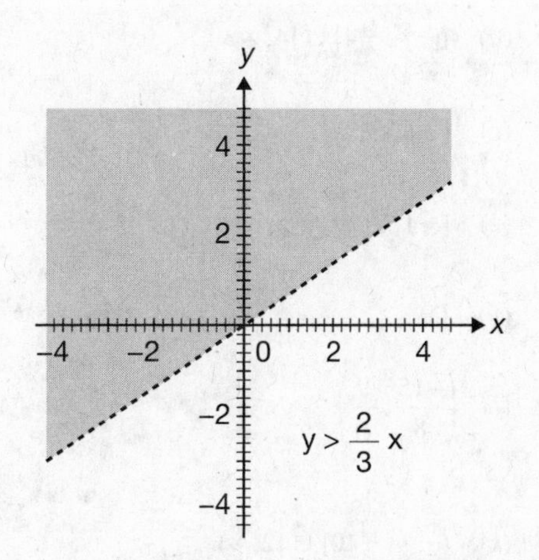

Figure 4.71

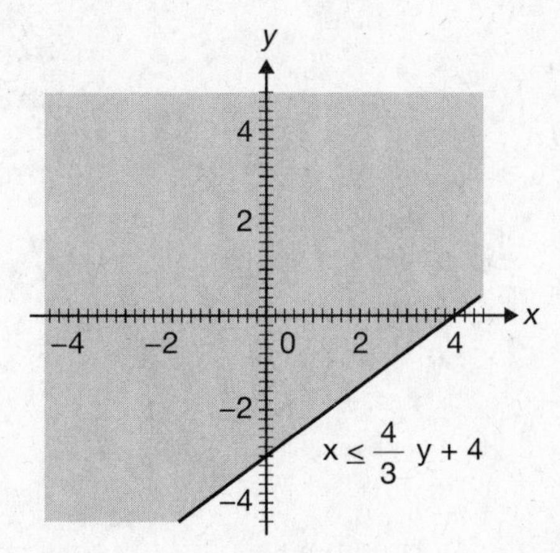

Figure 4.72

4.19 (Grade Problem) 65%.

4.20 (Consumer Problem) More than 4 hours.

4.21 (Investor Problem) \$40,000.

4.22 (Consumer Problem) 73 hours.

4.23 (Business Problem) 325.

4.24 (*a*) $\{-4, 4\}$ (*b*) $\left\{-3, \frac{25}{3}\right\}$ (*c*) $\{-8, 12\}$

(*d*) $\left\{\frac{-11}{5}, \frac{-1}{5}\right\}$ (*e*) Ø There is no solution. (*f*) $\{-2, -1\}$

(*g*) $\left\{\frac{-2}{3}, \frac{4}{5}\right\}$ (*h*) $\left\{\frac{1}{2}, \frac{13}{4}\right\}$ (*i*) $\{-4, -1\}$

(*j*) $\{-1\}$ (*k*) $\left\{\frac{-15}{8}, \frac{-9}{8}\right\}$ (*l*) $\left\{\frac{5}{3}, \frac{13}{3}\right\}$

4.25 (*a*) $(-\infty, -4] \cup [4, \infty)$ (*b*) $(-8, 8)$

(*c*) $[-8, 6]$ (*d*) $(-\infty, -1) \cup (11, \infty)$

(*e*) $[-1, 2]$ (*f*) $\left(-\infty, \frac{-4}{3}\right] \cup [4, \infty)$

(*g*) Ø (*h*) $(-\infty, \infty)$

(*i*) $\left(\frac{-15}{8}, \frac{9}{8}\right)$ (*j*) $\left(-\infty, \frac{-14}{5}\right) \cup \left(\frac{26}{5}, \infty\right)$

(*k*) $(-\infty, -10] \cup [2, \infty)$ (*l*) $\left(-\infty, \frac{-8}{3}\right] \cup \left[\frac{80}{3}, \infty\right)$

Exponents, Roots, and Radicals

5.1 Zero and Negative-Integer Exponents

Zero Exponents

We know that a nonzero quantity divided by itself equals one. In particular, if n is a positive integer and $b \neq 0$, $b^n/b^n = 1$. Also, if the fourth law of exponents is to hold when $m = n$ and $b \neq 0$, $b^m/b^n = b^n/b^n = b^{n-n} = b^0$. It follows that b^0 must be equivalent to 1.

Definition 1. If $b \neq 0, b^0 = 1$.

A nonzero quantity raised to the zero power is equal to 1.

EXAMPLE 1. (*a*) $5^0 = 1$ (*b*) $(-8)^0 = 1$ (*c*) $(x - 8)^0 = 1, x \neq 8$

Note: The expression 0^0 is indeterminate. We are unable to give the expression meaning in this text.

Negative-integer Exponents

Let us now consider expressions which contain negative-integer exponents. We would like the laws of exponents introduced in Chapter 2 to hold for negative exponents also. In particular, if $b \neq 0, b^n \cdot b^{-n} = b^{n+(-n)} = b^0 = 1$.

The multiplicative inverse property states that if $b \neq 0, b^n \cdot 1/b^n = 1$. Consistency compels us to state the following definition.

Definition 2. If $b \neq 0, b^{-n} = 1/b^n$ where n is a positive integer.

In other words, an expression with a negative exponent is equivalent to the reciprocal of that expression with a positive exponent. Assume no denominator is zero in examples 2 and 3.

EXAMPLE 2. (*a*) $7^{-2} = \dfrac{1}{7^2} = \dfrac{1}{49}$ (*b*) $x^{-4} = \dfrac{1}{x^4}$ (*c*) $(s+t)^{-8} = \dfrac{1}{(s+t)^8}$

Now consider an expression with a negative exponent in the denominator. If n is a natural number (i.e., a positive integer) and $b \neq 0$,

$$\frac{1}{b^{-n}} = \frac{1}{\frac{1}{b^n}} \qquad \text{Definition of a negative exponent}$$

$$= 1 \cdot \frac{b^n}{1} \qquad \text{Definition of division}$$

$$= b^n \qquad \text{Multiplicative identity}$$

In other words, a nonzero expression with a negative exponent in the denominator is equivalent to the same expression with a positive exponent in the numerator.

EXAMPLE 3. (*a*) $\frac{1}{4^{-1}} = 4^1 = 4$ (*b*) $\frac{1}{t^{-3}} = t^3$ (*c*) $\frac{1}{(a+b)^{-2}} = (a+b)^2$

Be aware that if $b \neq 0$ and n is a positive integer, the important ideas stated above are

$$(i) \quad b^0 = 1$$

$$(ii) \quad b^{-n} = \frac{1}{b^n}$$

$$(iii) \quad \frac{1}{b^{-n}} = b^n$$

It can be shown that the laws of exponents stated for positive integers in Chapter 2 are valid for zero and negative-integer exponents as well. This is an extremely useful consequence of our definitions. We may apply the laws of exponents in any order. In general we attempt to choose the order which is most efficient, although the results will be the same regardless of the order chosen.

Since we now understand how to interpret expressions with any integer exponent, laws (4*a*) and (4*b*) stated back in Section 2.5 may be combined into one statement. The result for our exponent laws is shown in the summary below.

Laws and Definitions for Integer Exponents

Law 1	$b^m \cdot b^n = b^{m+n}$
Law 2	$(b^m)^n = b^{mn}$
Law 3	$(ab)^n = a^n \cdot b^n$
Law 4	$\frac{b^m}{b^n} = b^{m-n}, b \neq 0$
Law 5	$\left(\frac{a}{b}\right)^n = \frac{a^n}{b^n}, b \neq 0$
Definition 1	$b^0 = 1, b \neq 0$
Definition 2	$b^{-n} = \frac{1}{b^n}, b \neq 0$

The simplest form of an expression with exponents is the form which contains only positive exponents and no base is repeated in a product or quotient. With law 4, it is sometimes necessary to apply the definition of negative exponents in order to obtain the result in simplest form. We now illustrate a variety of problems. Remember to apply the appropriate order of operations when applicable.

See solved problems 5.1–5.2.

5.2 Scientific Notation

Scientific notation is particularly useful in applications which involve very large or very small mathematical quantities. This notation allows us to compress the representation of very large or very small numbers. You may have seen numbers like 8.36*E*11 and 4.82*E* − 8 when using a scientific calculator or computer. The expression 8.36*E*11 means 8.36×10^{11} or 836,000,000,000 and, similarly, 4.82*E* − 8 means 4.82×10^{-8} or 0.0000000482.

Now let us compare some integer-exponent powers of ten and their decimal representations.

$$10^0 = 1 \qquad 10^{-1} = \frac{1}{10} = 0.1$$

$$10^1 = 10 \qquad 10^{-2} = \frac{1}{10^2} = \frac{1}{100} = 0.01$$

$$10^2 = 100 \qquad 10^{-3} = \frac{1}{10^3} = \frac{1}{1{,}000} = 0.001$$

$$10^3 = 1{,}000 \qquad 10^{-4} = \frac{1}{10^4} = \frac{1}{10{,}000} = 0.0001$$

$$10^4 = 10{,}000 \qquad 10^{-5} = \frac{1}{10^5} = \frac{1}{100{,}000} = 0.00001$$

Standard notation or *decimal notation* is the form we normally use to express numerical values.

Note in the above display that the exponent on ten tells us the number of places and the direction to move the decimal point from one in order to express the number in decimal notation. If the exponent is positive, the movement is to the right, and if the exponent is negative, the movement is to the left. Stated another way, if the exponent on ten is positive, we are multiplying by that number of factors of ten. Similarly, if the exponent on ten is negative, we are dividing by that number of factors of ten.

A number is written in *scientific notation* if it has the form $q \times 10^i$ where q is in the interval $[1, 10)$ and i is an integer (the "×" here is the multiplication operation, not the variable).

The procedure used to write a number in scientific notation follows.

1. Count the number of places from the decimal point that are required to obtain a q in $[1,10)$. Position the decimal point in q. The number of places counted is the magnitude of the exponent on ten. If movement of the decimal point was to the left, the exponent on ten is positive. If movement of the decimal point was to the right, the exponent on ten is negative.
2. Write the number in the form $q \times 10^i$.

Observe that the exponent on ten is positive if the original number is greater than or equal to ten. Similarly, the exponent on ten is negative if the original number is less than one.

See solved problem 5.3.

Think about the meaning of 7.89×10^3, for example. The symbols indicate that we are to multiply 7.89 times three factors of ten. The result is 7,890. In other words, $7.89 \times 10^3 = 7{,}890$. We have illustrated how to convert from scientific notation to standard form if the exponent on ten is positive.

Now consider a number of the form 4.56×10^{-2}. The symbols indicate that we are to multiply 4.56 times $10^{-2} = 1/10^2$. Hence, $4.56 \times 10^{-2} = 4.56 \times 1/10^2 = 4.56/10^2 = 456/100$. In other words, divide 4.56 by two factors of ten (or 100). The result is 0.0456. This illustrates how to convert scientific notation to standard form if ten has a negative exponent.

The procedure used to convert a number in scientific notation to standard or decimal form follows.

If the exponent on ten is positive, move the decimal point to the right the number of places equal to the exponent on ten. If the exponent on ten is negative, determine the absolute value of the exponent and move the decimal point that number of places to the left.

See solved problem 5.4.

There are circumstances in which it is more appropriate to express a number as a product of a number that is *not* in the interval [1,10) and a power of ten. Any of the following forms of 4,350 may be useful in certain situations, but only one of them is in scientific notation. Which one?

$0.435 \times 10^4 \qquad 4.35 \times 10^3 \qquad 43.5 \times 10^2 \qquad 435 \times 10$

The factored form which contains no decimals is frequently the easiest to use.

See solved problem 5.5.

Significant Digits

Numbers obtained by measurement are approximate. Knowing the accuracy of measurements obtained in applications is extremely important. The *accuracy* of a measurement indicates the preciseness of that measurement. Scientific notation is used to indicate the accuracy of measurements. Stated another way, scientific notation indicates which digits are *significant* in a number. The *significant digits* are simply the digits that have meaning in a numerical representation. They are the digits used to specify q in scientific notation.

All nonzero digits are significant. Zeros between nonzero digits are significant. Zeros to the right of the decimal point are significant. Zeros used as placeholders to position the decimal point are *not* significant.

If a number such as 7,900 is encountered in an application, the final zeros are not assumed to be significant unless further information about the accuracy of the number is available. For example in the statement, "The jets are 7,900 feet apart," we assume that the number has two significant digits and that the measurement is accurate to the nearest hundred feet.

The table below illustrates the number of significant digits and the accuracy of several forms of the number 7,900.

NUMBER	NUMBER OF SIGNIFICANT DIGITS	ACCURACY TO THE NEAREST
$7.9 \times 10^3 = 7{,}900$	2	Hundred
$7.90 \times 10^3 = 7{,}900$	3	Ten
$7.900 \times 10^3 = 7{,}900$	4	One or Unit
$7.9000 \times 10^3 = 7{,}900.0$	5	One-tenth
$7.90000 \times 10^3 = 7{,}900.00$	6	One-hundredth

As another example, $2.59 \times 10^{-4} = 0.000259$ has 3 significant digits. (Leading zeros in standard form are not significant.)

Work supplementary problem 5.6 to check your understanding.

5.3 Rational Exponents and Roots

Our original discussion of exponents required the exponents to be natural numbers. In Section 5.1 integer exponents were defined. We give meaning to rational exponents in this section. Recall that a rational number is a number of the form a/b where a and b are integers and $b \neq 0$.

Consider an expression such as $b^{\frac{1}{2}}$. What should $b^{\frac{1}{2}}$ mean? If Law 2 of exponents holds, $(b^{\frac{1}{2}})^2 = b^{\frac{1}{2}\cdot 2} = b^1 = b$. In particular, $(25^{\frac{1}{2}})^2 = 25^{\frac{1}{2}\cdot 2} = 25^1 = 25$. Therefore, $25^{\frac{1}{2}}$ is the number whose square is 25. We also know that $5^2 = 25$ and $(-5)^2 = 25$. Hence, $25^{\frac{1}{2}}$ must be another name for 5 or (-5). To avoid ambiguity, the nonnegative value, $25^{\frac{1}{2}} = 5$, is chosen.

Definition 3. $b^{\frac{1}{2}}$ is the nonnegative quantity which, when squared, is equal to b. It is called the *principal square root* of b.

See solved problem 5.6.

We develop a definition of $b^{\frac{1}{3}}$ in a similar manner. Apply Law 2 of exponents again to obtain $(b^{\frac{1}{3}})^3 = b^{\frac{1}{3}\cdot 3} = b^1 = b$. In particular, $(8^{\frac{1}{3}})^3 = 8^{\frac{1}{3}\cdot 3} = 8^1 = 8$. Hence, $8^{\frac{1}{3}}$ is the number whose cube is 8. Note that $2^3 = 8$ also. Therefore, $8^{\frac{1}{3}}$ and 2 must be different names for the same number.

Definition 4. $b^{\frac{1}{3}}$ is the quantity which, when cubed, is equal to b. It is called the *cube root* of b.

See solved problem 5.7.

Consider $b^{\frac{1}{4}}$ next. Apply Law 2 of exponents once more to obtain $(b^{\frac{1}{4}})^4 = b^{\frac{1}{4}\cdot 4} = b^1 = b$. Specifically $(16^{\frac{1}{4}})^4 = 16^{\frac{1}{4}\cdot 4} = 16^1 = 16$. We know that $2^4 = 16$ and $(-2)^4 = 16$. So $16^{\frac{1}{4}}$ is the same as 2 or (-2). We have encountered a situation like that encountered with the square root. We choose the positive value again for $16^{\frac{1}{4}}$. That is, $16^{\frac{1}{4}} = 2$.

Definition 5. $b^{\frac{1}{4}}$ is the nonnegative quantity which, when raised to the fourth power, is equal to b. It is called the *principal fourth root* of b.

See solved problem 5.8.

The discussion above illustrates that the possibility of ambiguity occurs when the root is even. The inclusion of the term *nonnegative* eliminates the ambiguity for square roots and fourth roots. There is only one real value for cube roots, so no ambiguity is involved. We now state a general definition.

Definition 6. If n is a natural number, $b^{\frac{1}{n}}$ is the real number which when raised to the *nth* power is b. It is positive when n is even. It is called the *principal nth root* of b.

The preceding definition is stated another way below.

Definition 7. $b^{\frac{1}{n}} = \begin{cases} |q| \text{ if } n \text{ is even, } b \geq 0 \text{ and } q^n = b. \\ q \text{ if } n \text{ is odd and } q^n = b. \end{cases}$

If $b < 0$ and n is even, there is no real-number root. We will address this situation in Section 5.6.

See solved problem 5.9.

Roots are often represented symbolically in a notation called *radical notation*. The symbols are shown below.

EXAMPLE 4. $b^{\frac{1}{n}}$ is written as $\sqrt[n]{b}$, if n is a natural number greater than one.

EXAMPLE 5. (*a*) $8^{\frac{1}{3}}$ is written as $\sqrt[3]{8}$.

(*b*) $16^{\frac{1}{4}}$ is written as $\sqrt[4]{16}$

(*c*) $9^{\frac{1}{2}}$ is written as $\sqrt[2]{9} = \sqrt{9}$.

We stated that $b^{\frac{1}{n}}$ and $\sqrt[n]{b}$ are symbols which represent the *nth* root of b. $\sqrt[n]{b}$ is the quantity which, when raised to the *nth* power, yields b. $\sqrt[n]{b}$ is nonnegative if n is even. In $\sqrt[n]{b}$, b is called the *radicand*, $\sqrt{\ }$ is called the *radical sign*, and n is called the *index* of the radical. The following example illustrates the terminology.

EXAMPLE 6.

EXPRESSION	RADICAND	INDEX	ROOT
$\sqrt[3]{5t}$	$5t$	3	Cube
$\sqrt[5]{x+y}$	$x + y$	5	Fifth
$\sqrt[8]{w^{10}}$	w^{10}	8	Eighth
$\sqrt{81a^3b}$	$81a^3b$	2	Square

Note: The radical symbol $\sqrt{\ }$ without an index showing always means square root. The index is understood to be 2 when not showing.

We have explained the meaning of rational exponents whose numerators were one. We shall now employ Law 2 of exponents to define rational exponents in general.

Definition 8. If $b^{\frac{1}{n}}$ is a real number, then $b^{\frac{m}{n}} = (b^{\frac{1}{n}})^m$.

In other words, $b^{\frac{m}{n}}$ is the *nth* root of b raised to the *mth* power.

The statement, "$b^{\frac{1}{n}}$ is a real number," merely excludes even roots of negative numbers.

Fortunately, the laws of exponents stated previously for integers apply to rational exponents also, provided the root is a real number. Therefore, it follows that

$$b^{\frac{m}{n}} = (b^{\frac{1}{n}})^m = (b^m)^{\frac{1}{n}}.$$

In other words, as long as all of the roots are real, the results are the same if we first determine the *nth* root of b and raise the result to the *mth* power, or if we first raise b to the *mth* power followed by finding the *nth* root of the result.

See solved problem 5.10.

We previously stated that $b^{\frac{1}{n}}$ and $\sqrt[n]{b}$ are equivalent expressions. Therefore, if $b^{\frac{1}{n}}$ is a real number and m/n is reduced to lowest terms, the following are equivalent expressions also.

$$b^{\frac{m}{n}} = \left(b^{\frac{1}{n}}\right)^m = \left(\sqrt[n]{b}\right)^m$$

$$= (b^m)^{\frac{1}{n}} = \sqrt[n]{b^m}$$

See solved problem 5.11.

Some of the results in the solved problem 5.11 are not simplified. We will explain simplification of radicals in the next section.

Simplification of expressions which contain rational exponents is analogous to expressions which have integer exponents. The sequence of operations can vary without changing the result. Normally operations within parentheses should be performed first. In other instances one should do exponentiation first. Analyze the expression to be simplified to determine the most efficient method to employ. In any case, the order of operations stated previously must be adhered to. A particular base should appear as few times as is possible and all exponents should be positive.

See solved problems 5.12–5.14.

5.4 Simplifying Radicals

Now that we know how to interpret radicals, we must learn how to simplify them. The definition of a radical and the following three properties provide almost all of the necessary tools.

Properties of Radicals

If $\sqrt[n]{a}$ and $\sqrt[n]{b}$ are real numbers, then:

1. $\sqrt[n]{ab} = \sqrt[n]{a}\sqrt[n]{b}$ (The *nth* root of a product is the product of the *nth* roots of the factors.)
2. $\sqrt[n]{\dfrac{a}{b}} = \dfrac{\sqrt[n]{a}}{\sqrt[n]{b}}, b \neq 0$ (The *nth* root of a quotient is the quotient of the *nth* roots of the numerator and denominator provided the denominator is not zero.)
3. $\sqrt[kn]{b^{km}} = \sqrt[n]{b^m}$ or $b \geq 0$ (If the index and the exponent of the radicand contain a common factor, the common factor may be divided out.)

Properties 1 and 2 are actually Laws 3 and 5 of exponents. Property 3 is obtained if we rewrite the radical in exponential form and reduce the exponent. The steps involved follow.

$$\sqrt[kn]{b^{km}} = \left(b^{km}\right)^{\frac{1}{kn}} = b^{\frac{km}{kn}} = b^{\frac{m}{n}} = \sqrt[n]{b^m}$$

In order to simplify radical expressions, we must know what is meant by the simplest radical form. The requirements are given below.

Simplest Radical Form

A radical expression is in simplest form if:

1. All factors of the radicand have exponents less than the index.
2. The radicand contains no fractions.
3. No denominator contains a radical.
4. The index and the exponents of the factors in the radicand have no common factor other than one.

A perfect *nth* power of a factor must possess an exponent that is a multiple of n. Thus, perfect square factors must have exponents that are multiples of two, perfect cube factors must have exponents that are multiples of three, etc. Keep this idea in mind as you simplify radicals.

See solved problem 5.15.

Rationalizing the Denominator

There are circumstances which result in radicals in the denominator such that none of the factors of the radicand have exponents which are multiples of the index. This situation violates criterion 3 for simplifying radicals. We apply the fundamental principle of fractions to obtain a rational expression in the denominator. This process is called *rationalizing the denominator*. The technique employed is to multiply the numerator and denominator by the expression which changes the denominator into the *nth* root of a perfect *nth* power. That is, the denominator is a rational expression if it is the square root of a perfect square, or the cube root of a perfect cube, or, in general, the *nth* root of a perfect *nth* power.

See solved problem 5.16.

We next illustrate how property 3 of radicals is used to reduce the index or order of a radical. Recall that property 3 allows us to divide out common factors of the index and exponents in the radicand.

See solved problem 5.17.

Finally we illustrate how to find products and quotients of radicals with *different indices*. Property 3 of radicals may be used, although it is easier to convert radicals to their equivalent exponential form, then simplify.

See solved problem 5.18.

5.5 Operations on Radical Expressions

Adding and Subtracting

The distributive properties, $a\,(b + c) = ab + ac$ and $(b + c)\,a = ba + ca$, were used previously to combine like terms. These same properties can be employed to combine like radicals. *Like radicals* are radicals with the same indices and radicands. Like radicals can be manipulated just as like terms can be manipulated.

See solved problem 5.19.

Multiplying and Dividing

We employ the distributive properties to multiply expressions which contain radicals. The process is analogous to multiplying polynomials. Radicals multiplied by factors should be written with the factors preceding the radical in order to avoid ambiguity. That is, write $y\sqrt{x}$ rather than $\sqrt{x}y$. The latter may be construed as $\sqrt{xy}$ incorrectly.

See solved problem 5.20.

We now turn our attention to division. If the denominator contains irrational terms, division is accomplished by rationalizing the denominator. Refer to Section 5.4 if necessary.

See solved problem 5.21.

5.6 Complex Numbers

Consider the equation $x^2 = -9$. There is no real number solution, since there is no real number whose square is negative. We define a number system in this section which will satisfy our needs. Our first step is to define a quantity i.

Definition 9. The *imaginary unit*, i, is defined by the equation $i^2 = -1$.

The imaginary unit is the number whose square is -1. Therefore, $\sqrt{i^2} = i = \sqrt{-1}$. We list some powers of i below.

$$i = \sqrt{-1}$$
$$i^2 = \left(\sqrt{-1}\right)^2 = -1$$
$$i^3 = i^2 \cdot i = (-1)\,i = -i$$
$$i^4 = i^2 \cdot i^2 = (-1)\,(-1) = 1$$
$$i^5 = i^4 \cdot i = 1 \cdot i = i$$
$$i^6 = i^4 \cdot i^2 = (1)\,(-1) = -1$$

Note that the cycle repeats beginning with i^5. In other words, any power of i is i, $-i$, 1, or -1.

We employ the number i to represent the square root of a negative number. The process is illustrated below.

EXAMPLE 7. (*a*) $\sqrt{-9} = \sqrt{-1 \cdot 9} = \sqrt{-1}\sqrt{9} = i \cdot 3 = 3i$

or $\sqrt{-9} = \sqrt{-1 \cdot 9} = \sqrt{i^2 \cdot 9} = \sqrt{i^2}\,\sqrt{9} = i \cdot 3 = 3i$

(*b*) $\sqrt{-5} = \sqrt{-1 \cdot 5} = \sqrt{-1}\sqrt{5} = i\sqrt{5}$

or $\sqrt{-5} = \sqrt{-1 \cdot 5} = \sqrt{i^2}\,\sqrt{5} = i\sqrt{5}$

We now employ the imaginary unit, i, to define a new type of number. The new number is called a complex number.

Definition 10. A *complex number* is a number of the form $a + bi$, where a and b are real numbers, and i is the imaginary unit. a is called the *real part* and b is called the *imaginary part* of the complex number. We use C to represent the set of complex numbers.

In the complex number $4 + 5i$, 4 is the real part and 5 is the imaginary part.

All real numbers may be written in complex number form. We assume $0i = 0$. See the equalities below.

$$8 = 8 + 0i$$
$$\frac{-1}{2} = \frac{-1}{2} + 0i$$
$$\sqrt{3} = \sqrt{3} + 0i$$

Since every real number is also a complex number, $R \subseteq C$. Does C, the set of complex numbers, contain numbers which are not real numbers? The answer is yes! The numbers given in example 7, $\sqrt{-9} = 3i$ and $\sqrt{-5} = i\sqrt{5}$, are not real numbers. These numbers are called *pure imaginary numbers*. Pure imaginary numbers may be written in complex number or $a + bi$ form also. Refer to the equalities below.

$$3i = 0 + 3i$$
$$i\sqrt{5} = 0 + i\sqrt{5}$$
$$-\frac{3}{4}i = 0 - \frac{3}{4}i$$

Pure imaginary numbers are merely complex numbers whose real part is zero. No pure imaginary number is a member of the set of real numbers. Therefore, the set of real numbers, R, is a proper subset of the set of complex numbers C. That is, $R \subset C$.

If b is an irrational number in the $a + bi$ expression, we write the i factor first. We write $3 + i\sqrt{2}$ rather than $3 + \sqrt{2}i$ in order to avoid ambiguity. Hence, the complex number form is sometimes $a + ib$.

See solved problem 5.22.

Definition 11. $a + bi = c + di$ if and only it $a = c$ and $b = d$.

The above definition states that two complex numbers are equal if and only if their real parts are the same and their imaginary parts are the same also.

We shall now consider operations on complex numbers.

Addition: $(a + bi) + (c + di) = (a + c) + (b + d)i$

Subtraction: $(a + bi) - (c + di) = (a - c) + (b - d)i$

To add or subtract complex numbers, simply add or subtract their real parts and their imaginary parts, respectively.

See solved problem 5.23.

We multiply complex numbers as if they were two binomials. That is, distribute. You may use the FOIL method. Read the following sequence of statements.

$$\begin{aligned}(a + bi)(c + di) &= ac + adi + bci + bdi^2 \\ &= ac + adi + bci + bd(-1) \\ &= [ac + bd(-1)] + [adi + bci] \\ &= (ac - bd) + (ad + bc)i\end{aligned}$$

Multiplication: $(a + bi)(c + di) = (ac - bd) + (ad + bc)i$.

It is not necessary to memorize the above equation. It is preferable to treat the complex numbers as binomials, multiply, then combine like terms.

Take another look at the right side of the multiplication equation.

ac is the product of the first terms—F

$-bd$ is the product of the last terms—L

adi is the product of the outside terms—O

bci is the product of the inside terms—I

It may be helpful to think of multiplication of complex numbers in terms of the FLOI rule. This sequence places like terms in juxtaposition in the product. Like terms are therefore more readily combined.

There is one more item we must clarify prior to proceeding. Compare the statements shown below.

$$\sqrt{-3}\sqrt{-5} = (i\sqrt{3})(i\sqrt{5}) = i^2\sqrt{15} = -\sqrt{15}$$
$$\sqrt{-3}\sqrt{-5} \neq \sqrt{(-3)(-5)} = \sqrt{15}$$

The product of the square roots of negative quantities must be determined as in the first sequence. If $b > 0$, rewrite $\sqrt{-b}$ in the form $i\sqrt{b}$ before performing any computations. *This is important!*

See solved problem 5.24.

Definition 12. $a + bi$ and $a - bi$ are *complex conjugates*.

In general,

$$\begin{aligned}(a + bi)(a - bi) &= a^2 - (bi)^2\\ &= a^2 - b^2i^2\\ &= a^2 - b^2(-1)\\ &= a^2 + b^2\end{aligned}$$

Since $a, b \in \mathbf{R}$, $(a^2 + b^2) \in \mathbf{R}$. In other words, the product of two complex conjugates is a real number. This result is useful when we divide certain complex numbers.

The quotient of a complex number and a real number is found in a manner similar to dividing a polynomial by a monomial. In other words, if $a, b, c \in \mathbf{R}$, and $c \neq 0$, then $\frac{a+bi}{c} = \frac{a}{c} + \frac{b}{c}i$.

If the divisor is a complex number that is not real, the division is accomplished by multiplying the numerator and denominator by a factor which results in a real number in the denominator. The appropriate factor is usually the conjugate of the divisor. We express these ideas symbolically below.

Division: If $a, b, c \in \mathrm{R}$, and $c \neq 0$, then $\frac{a+bi}{c} = \frac{a}{c} + \frac{b}{c}i$ and

if $a, b, c, d \in \mathrm{R}$ and $c, d \neq 0$, then $\frac{a+bi}{c+di} = \frac{a+bi}{c+di} \cdot \frac{c-di}{c-di} = \left(\frac{ac+bd}{c^2+d^2}\right) + \left(\frac{bc-ad}{c^2+d^2}\right)i.$

If the divisor has the form ci and $c \neq 0$, the division can be accomplished by multiplying the numerator and denominator by i. Observe that $ci \cdot i = ci^2 = c(-1) = -c$ is a real number.

See solved problem 5.25.

We conclude the chapter by summarizing some important ideas. Complex numbers and their operations have been defined. The numbers and the operations satisfy the same properties as the real numbers; associative, commutative, inverse, etc. The numbers together with their properties form a system called the *complex number system.* We use C to represent the set of complex numbers.

The Venn diagram shown in Figure 5.1 below illustrates the relationships between the various sets of numbers we have introduced. The letters used to represent the various sets of numbers shown in the diagram are restated for you also.

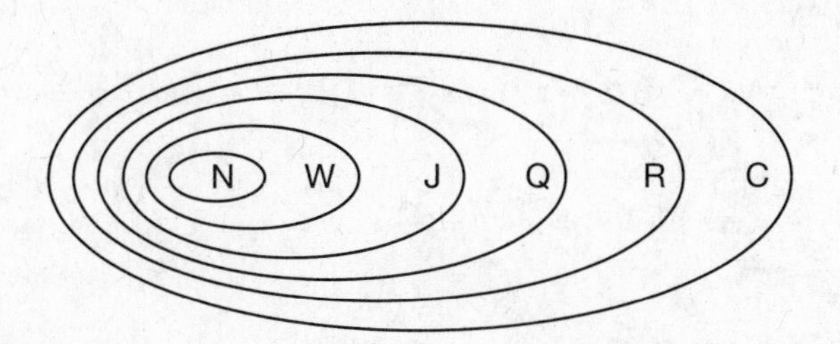

N – Natural numbers
W – Whole numbers
J – Integers
Q – Rational numbers
R – Real numbers
C – Complex numbers

Figure 5.1

SOLVED PROBLEMS

5.1 Perform the indicated operations and simplify.

(*a*) 3^{-3}

$$3^{-3} = \frac{1}{3^3} = \frac{1}{27}$$

(*b*) $9 \cdot 3^{-3}$

$$9 \cdot 3^{-3} = 9 \cdot \frac{1}{3^3} = 9 \cdot \frac{1}{27} = \frac{9}{27} = \frac{1}{3}$$

(*c*) $\frac{2}{4^{-2}}$

$$\frac{2}{4^{-2}} = 2 \cdot 4^2 = 2 \cdot 16 = 32$$

(*d*) $\left(\frac{3}{2}\right)^{-1}$

$$\left(\frac{3}{2}\right)^{-1} = \frac{3^{-1}}{2^{-1}} = \frac{2^1}{3^1} = \frac{2}{3}.$$ As you may note, this indicates that $\left(\frac{a}{b}\right)^{-n} = \left(\frac{b}{a}\right)^{n}$.

(*e*) $(-3)^{-2}$

$$(-3)^{-2} = \frac{1}{(-3)^{2}} = \frac{1}{9}$$

(*f*) $\frac{2 \cdot 3^{-3}}{6^{-2}}$

$$\frac{2 \cdot 3^{-3}}{6^{-2}} = \frac{2 \cdot 6^2}{3^3} = \frac{2 \cdot 36}{27} = \frac{2 \cdot 4}{3} = \frac{8}{3}$$

(*g*) $\left(\frac{3}{4}\right)^{0} - \frac{1}{7^0}$

$$\left(\frac{3}{4}\right)^{0} - \frac{1}{7^0} = 1 - \frac{1}{1} = 1 - 1 = 0$$

(*h*) $5 - 5^{-1}$

$$5 - 5^{-1} = 5 - \frac{1}{5} = \frac{25 - 1}{5} = \frac{24}{5}$$

Refer to supplementary problem 5.1 for practice problems.

5.2 Perform the indicated operations and simplify. Assume that no variable is zero.

(a) $x^6 \cdot x^{-1} \cdot x^{-3}$

$$x^6 \cdot x^{-1} \cdot x^{-3} = x^{6+(-1)+(-3)} = x^2$$

(b) $(s^{-2} \cdot t^{-3})(s^4 \cdot t^{-2})$

$$(s^{-2} \cdot t^{-3})(s^4 \cdot t^{-2}) = (s^{-2} \cdot s^4)(t^{-3} \cdot t^{-2}) = s^{-2+4} \cdot t^{-3+(-2)} = s^2 \cdot t^{-5} = \frac{s^2}{t^5}$$

(c) $(x^{-2} \cdot y^3)^{-2}$

$$(x^{-2} \cdot y^3)^{-2} = x^{-2(-2)}y^{3(-2)} = x^4y^{-6} = \frac{x^4}{y^6}$$

(d) $(s^{-2}t)^{-3}(st)^{-2}$

$$(s^{-2}t)^{-3}(st)^{-2} = s^{-2(-3)}t^{-3}s^{-2}t^{-2} = s^{6+(-2)}t^{-3+(-2)} = s^4t^{-5} = \frac{s^4}{t^5}$$

(e) $(-3ab^2)^{-3}(2a^2b^{-3})^{-2}$

$$(-3ab^2)^{-3}(2a^2b^{-3})^{-2} = (-3)^{-3}a^{-3}(b^2)^{-3}2^{-2}(a^2)^{-2}(b^{-3})^{-2} = \frac{a^{-3}b^{-6}a^{-4}b^6}{(-3)^3 2^2}$$

$$= \frac{a^{-7}b^0}{-27 \cdot 4} = \frac{-1}{108a^7}$$

(f) $\dfrac{s^{-3}t^6}{s^5t^4}$

$$\frac{s^{-3}t^6}{s^5t^4} = \frac{t^{6-4}}{s^{5-(-3)}} = \frac{t^2}{s^8}$$

(g) $\left(\dfrac{d^{-2}d^0d^3}{d^5d^{-2}}\right)^{-2}$

$$\left(\frac{d^{-2}d^0d^3}{d^5d^{-2}}\right)^{-2} = \left(\frac{d^{-2+0+3}}{d^{5+(-2)}}\right)^{-2} = \left(\frac{d}{d^3}\right)^{-2} = (d^{-2})^{-2} = d^{-2(-2)} = d^4$$

(h) $\left(\dfrac{4^{-1}a^2b^{-3}}{ab^{-4}}\right)^{-2}\left(\dfrac{12^{-1}a^{-2}b}{a^2b^{-4}}\right)^2$

$$\left(\frac{4^{-1}a^2b^{-3}}{ab^{-4}}\right)^{-2}\left(\frac{12^{-1}a^{-2}b}{a^2b^{-4}}\right)^2 = (4^{-1}a^{2-1}b^{-3-(-4)})^{-2}(12^{-1}a^{-2-2}b^{1-(-4)})^2$$

$$= (4^{-1}ab)^{-2}(12^{-1}a^{-4}b^5)^2$$

$$= 4^{(-1)(-2)}a^{-2}b^{-2} \cdot 12^{-1(2)}a^{-4(2)}b^{5(2)}$$

$$= 4^2a^{-2}b^{-2} \cdot 12^{-2}a^{-8}b^{10} = 4^2 \cdot 12^{-2}a^{-2+(-8)}b^{-2+10}$$

$$= \frac{4^2b^8}{12^2a^{10}} = \frac{4^2b^8}{(3 \cdot 4)^2a^{10}} = \frac{4^2b^8}{3^2 \cdot 4^2a^{10}}$$

$$= \frac{b^8}{9a^{10}}$$

(i) $\dfrac{x^{-1} - y^{-1}}{(xy)^{-1}}$

$$\frac{x^{-1} - y^{-1}}{(xy)^{-1}} = \frac{\frac{1}{x} - \frac{1}{y}}{\frac{1}{xy}} = \frac{\left(\frac{1}{x} - \frac{1}{y}\right)(xy)}{\left(\frac{1}{xy}\right)(xy)} = \frac{\frac{1}{x}(xy) - \frac{1}{y}(xy)}{1} = \frac{y - x}{1} = y - x$$

(j) $\left(\dfrac{c^{2n}}{c^{n+1}}\right)^{-1}$

$$\left(\frac{c^{2n}}{c^{n+1}}\right)^{-1} = (c^{2n-(n+1)})^{-1} = (c^{2n-n-1})^{-1} = (c^{n-1})^{-1}c^{-1(n-1)} = c^{-n+1} = c^{1-n}$$

(*k*) $(x^{-1}+y^{-1})^{-1}$

$$(x^{-1}+y^{-1})^{-1} = \left(\frac{1}{x}+\frac{1}{y}\right)^{-1} = \left(\frac{1}{x}\cdot\frac{y}{y}+\frac{1}{y}\cdot\frac{x}{x}\right)^{-1} = \left(\frac{y}{xy}+\frac{x}{xy}\right)^{-1}$$

$$= \left(\frac{y+x}{xy}\right)^{-1} = \frac{xy}{y+x} \text{ or } \frac{xy}{x+y}$$

(*l*) $\left(\dfrac{a^3b+ab^2}{a^{-2}b^3}\right)^0$

$$\left(\frac{a^3b+ab^2}{a^{-2}b^3}\right)^0 = 1 \quad \text{(The zero exponent law)}$$

Refer to supplementary problem 5.2 for similar problems.

5.3 Write the following numbers in scientific notation.

(*a*) 30,000

The decimal point must be moved four places to the left to obtain the appropriate q.

$$30{,}000 = 3 \times 10^4$$

(*b*) 254,000

The decimal point must be moved five places to the left to obtain the appropriate q.

$$254{,}000 = 2.54 \times 10^5$$

(*c*) 5,085,273

The decimal point must be moved six places to the left to obtain the appropriate q.

$$5{,}085{,}273 = 5.085273 \times 10^6$$

(*d*) 0.09

The decimal point must be moved two places to the right to obtain the appropriate q.

$$0.09 = 9 \times 10^{-2}$$

(*e*) 0.000076

The decimal point must be moved five places to the right to obtain the appropriate q.

$$0.000076 = 7.6 \times 10^{-5}$$

(*f*) 0.0004050

The decimal point must be moved four places to the right to obtain the appropriate q.

$$0.0004050 = 4.050 \times 10^{-4}$$

(*g*) 4.78

The decimal point must be moved zero places to the right to obtain the appropriate q.

$$4.78 = 4.78 \times 10^0$$

See supplementary problem 5.3.

5.4 Write the following numbers in standard form.

(*a*) 4.69×10^3

Move the decimal point three places to the right, since we are multiplying by three factors of ten.

$$4.69 \times 10^3 = 4{,}690$$

(*b*) 4.69×10^{-4}

Move the decimal point four places to the left, since we are dividing by four factors of ten

$$4.69 \times 10^{-4} = 0.000469$$

(*c*) 1.4×10^6

Move the decimal point six places to the right, since we are multiplying by six factors of ten.

$$1.4 \times 10^6 = 1{,}400{,}000$$

(*d*) 3.215×10^4

Move the decimal point four places to the right, since we are multiplying by four factors of ten.

$$3.215 \times 10^4 = 32{,}150$$

(*e*) 7.3213×10^{-4}

Move the decimal point four places to the left, since we are dividing by four factors of ten.

$$7.3213 \times 10^{-4} = 0.00073213$$

(*f*) 9×10^5

Move the decimal point five places to the right, since we are multiplying by five factors of ten.

$$9 \times 10^5 = 900{,}000$$

(*g*) 9×10^{-5}

Move the decimal point five places to the left, since we are dividing by five factors of ten.

$$9 \times 10^{-5} = 0.00009$$

See supplementary problem 5.4 for more conversion problems.

5.5 Perform the indicated operations using the appropriate factored form of each number. Express the result in standard form.

(*a*) $\dfrac{325{,}000}{0.025}$

$$\frac{325{,}000}{0.025} = \frac{325 \times 10^3}{25 \times 10^{-3}} = 13 \times 10^6 = 13{,}000{,}000$$

(*b*) $\dfrac{(0.004)\,(0.000036)}{0.00016}$

$$\frac{(0.004)\,(0.000036)}{0.00016} = \frac{4 \times 10^{-3} \times 36 \times 10^{-6}}{16 \times 10^{-5}} = \frac{36 \times 10^{-9}}{4 \times 10^{-5}} = 9 \times 10^{-4} = 0.0009$$

See supplementary problem 5.5 for practice.

5.6 Evaluate the following.

(*a*) $9^{\frac{1}{2}}$

$$9^{\frac{1}{2}} = 3 \text{ since } 3^2 = 9$$

(*b*) $100^{\frac{1}{2}}$

$$100^{\frac{1}{2}} = 10 \text{ since } 10^2 = 100$$

(*c*) $-16^{\frac{1}{2}}$

$-16^{\frac{1}{2}} = -(16^{\frac{1}{2}}) = -(4) = -4$ since $(4^2) = 16$ and $-(4^2) = -16$ (Remember the order of operations.)

5.7 Evaluate the following.

(*a*) $64^{\frac{1}{3}}$

$64^{\frac{1}{3}} = 4$ since $4^3 = 64$. Another way of looking at these is by using the laws of exponents developed earlier. $64^{\frac{1}{3}} = (4^3)^{\frac{1}{3}} = 4^{3\cdot\frac{1}{3}} = 4$

(*b*) $1000^{\frac{1}{3}}$

$$1000^{\frac{1}{3}} = 10 \text{ since } 10^3 = 1000$$

(*c*) $(-27)^{\frac{1}{3}}$

$$(-27)^{\frac{1}{3}} = -3 \text{ since } (-3)^3 = -27$$

5.8 Evaluate the following.

(*a*) $10{,}000^{\frac{1}{4}}$

$$10{,}000^{\frac{1}{4}} = 10 \text{ since } 10^4 = 10{,}000$$

(*b*) $81^{\frac{1}{4}}$

$$81^{\frac{1}{4}} = 3 \text{ since } 3^4 = 81$$

(*c*) $-81^{\frac{1}{4}}$

$-81^{\frac{1}{4}} = -3$ since $3^4 = 81$ (Remember the order of operations.)

5.9 Evaluate the following.

(*a*) $36^{\frac{1}{2}}$

$$36^{\frac{1}{2}} = |6| = 6 \text{ since } 6^2 = 36$$

(*b*) $\left(\frac{1}{4}\right)^{\frac{1}{2}}$

$$\left(\frac{1}{4}\right)^{\frac{1}{2}} = \left|\frac{1}{2}\right| = \frac{1}{2} \text{ since } \left(\frac{1}{2}\right)^2 = \frac{1}{4}$$

(*c*) $(64)^{\frac{1}{3}}$

$$(64)^{\frac{1}{3}} = 4 \text{ since } 4^3 = 64$$

(*d*) $\left(\frac{-1}{27}\right)^{\frac{1}{3}}$

$$\left(\frac{-1}{27}\right)^{\frac{1}{3}} = \frac{-1}{3} \text{ since } \left(\frac{-1}{3}\right)^3 = \frac{-1}{27}$$

(*e*) $\left(\frac{81}{16}\right)^{\frac{1}{4}}$

$$\left(\frac{81}{16}\right)^{\frac{1}{4}} = \left|\frac{3}{2}\right| = \frac{3}{2} \text{ since } \left(\frac{3}{2}\right)^4 = \frac{81}{16}$$

(*f*) $0^{\frac{1}{2}}$

$$0^{\frac{1}{2}} = |0| = 0 \text{ since } 0^2 = 0$$

5.10 Evaluate the following.

(*a*) $64^{\frac{2}{3}}$

$$64^{\frac{2}{3}} = \left(64^{\frac{1}{3}}\right)^2 = 4^2 = 16$$

(*b*) $32^{\frac{3}{5}}$

$$32^{\frac{3}{5}} = \left(32^{\frac{1}{5}}\right)^3 = 2^3 = 8$$

(*c*) $(-27)^{\frac{4}{3}}$

$$(-27)^{\frac{4}{3}} = \left(-27^{\frac{1}{3}}\right)^4 = (-3)^4 = 81$$

(*d*) $-27^{\frac{4}{3}}$

$$-27^{\frac{4}{3}} = \left(27^{\frac{1}{3}}\right)^4 = -(3)^4 = -81 \text{ (Compare with part } c.)$$

(*e*) $(-8)^{\frac{5}{3}}$

$$(-8)^{\frac{5}{3}} = \left[(-8)^{\frac{1}{3}}\right]^5 = [-2]^5 = -32$$

(*f*) $(-125)^{\frac{-2}{3}}$

$$(-125)^{\frac{-2}{3}} = \frac{1}{(-125)^{\frac{2}{3}}} = \frac{1}{\left[(-125)^{\frac{1}{3}}\right]^2} = \frac{1}{[-5]^2} = \frac{1}{25}$$

(*g*) $\left(\frac{-1}{125}\right)^{\frac{-1}{3}}$

$$\left(\frac{-1}{125}\right)^{\frac{-1}{3}} = (-125)^{\frac{1}{3}} = -5 \text{ (We applied the definition of negative exponents.)}$$

5.11 Express the following in radical notation.

(*a*) $t^{\frac{1}{5}}$

$$t^{\frac{1}{5}} = \sqrt[5]{t}$$

(*b*) $x^{\frac{2}{3}}$

$$x^{\frac{2}{3}} = (\sqrt[3]{x})^2 \text{ or } \sqrt[3]{x^2}$$

(*c*) $7^{\frac{6}{5}}$

$$7^{\frac{6}{5}} = (\sqrt[5]{7})^6 \text{ or } \sqrt[5]{7^6}$$

(*d*) $s^{\frac{-3}{2}}$

$$s^{\frac{-3}{2}} = \frac{1}{s^{\frac{3}{2}}} = \frac{1}{(\sqrt{s})^3} \text{ or } \frac{1}{\sqrt{s^3}}$$

(*e*) $(8y)^{\frac{1}{4}}$

$$(8y)^{\frac{1}{4}} = \sqrt[4]{8y}$$

(*f*) $8y^{\frac{1}{4}}$

$$8y^{\frac{1}{4}} = 8\sqrt[4]{y} \quad \text{(The exponent applies to } y \text{ only.)}$$

See supplementary problem 5.7.

5.12 Perform the indicated operations and simplify. Assume all variables represent positive real numbers.

(*a*) $x^{\frac{2}{3}}x^{\frac{1}{4}}$

$$x^{\frac{2}{3}}x^{\frac{1}{4}} = x^{\frac{2}{3}+\frac{1}{4}} = x^{\frac{8+3}{12}} = x^{\frac{11}{12}}$$

(*b*) $s^{\frac{3}{5}}s^{\frac{-1}{3}}$

$$s^{\frac{3}{5}}s^{\frac{-1}{3}} = s^{\frac{3}{5}+\frac{-1}{3}} = s^{\frac{9-5}{15}} = s^{\frac{4}{15}}$$

(*c*) $\dfrac{t^{\frac{4}{5}}}{t^{\frac{1}{6}}}$

$$\frac{t^{\frac{4}{5}}}{t^{\frac{1}{6}}} = t^{\frac{4}{5}-\frac{1}{6}} = t^{\frac{24-5}{30}} = t^{\frac{19}{30}}$$

(*d*) $\dfrac{b^{\frac{1}{2}}b^{\frac{-1}{3}}}{b^{\frac{3}{4}}}$

$$\frac{b^{\frac{1}{2}}b^{\frac{-1}{3}}}{b^{\frac{3}{4}}} = \frac{b^{\frac{1}{2}+(\frac{-1}{3})}}{b^{\frac{3}{4}}} = \frac{b^{\frac{3-2}{6}}}{b^{\frac{3}{4}}} = \frac{b^{\frac{1}{6}}}{b^{\frac{3}{4}}} = b^{\frac{1}{6}-\frac{3}{4}} = b^{\frac{2-9}{12}} = b^{\frac{-7}{12}} = \frac{1}{b^{\frac{7}{12}}}$$

(*e*) $\left(x^{\frac{-1}{2}}x^{\frac{3}{2}}\right)^3$

$$\left(x^{\frac{-1}{2}}x^{\frac{3}{2}}\right)^3 = \left(x^{\frac{-1}{2}+\frac{3}{2}}\right)^3 = \left(x^{\frac{2}{2}}\right)^3 = (x^1)^3 = x^3$$

(*f*) $\dfrac{\left(y^{-1}z^{\frac{1}{2}}\right)^{\frac{-1}{4}}}{y^{\frac{1}{2}}z^{\frac{-1}{3}}}$.

$$\frac{\left(y^{-1}z^{\frac{1}{2}}\right)^{\frac{-1}{4}}}{y^{\frac{1}{2}}z^{\frac{-1}{3}}} = \frac{y^{(-1)\left(\frac{-1}{4}\right)}z^{\left(\frac{1}{2}\right)\left(\frac{-1}{4}\right)}}{y^{\frac{1}{2}}z^{\frac{-1}{3}}} = \frac{y^{\frac{1}{4}}z^{\frac{-1}{8}}}{y^{\frac{1}{2}}z^{\frac{-1}{3}}} = y^{\frac{1}{4}-\frac{1}{2}}z^{\frac{-1}{8}-\frac{-1}{3}} = y^{\frac{1-2}{4}}z^{\frac{-3+8}{24}}$$

$$= y^{\frac{-1}{4}}z^{\frac{5}{24}} = \frac{z^{\frac{5}{24}}}{y^{\frac{1}{4}}}$$

(*g*) $\left(\dfrac{s^{\frac{-1}{2}}t^{\frac{1}{4}}}{s^{\frac{-1}{4}}}\right)^4$

Do the external exponentiation first, since four is a multiple of each of the rational exponents.

$$\left(\frac{s^{\frac{-1}{2}}t^{\frac{1}{4}}}{s^{\frac{-1}{4}}}\right)^4 = \frac{s^{\left(\frac{-1}{2}\right)(4)}t^{\left(\frac{1}{4}\right)(4)}}{s^{\left(\frac{-1}{4}\right)(4)}} = \frac{s^{-2}t}{s^{-1}} = \frac{t}{s^{-1-(-2)}} = \frac{t}{s}$$

(*h*) $\left(x^{\frac{1}{2}}+y^{\frac{1}{2}}\right)^2$

This is a special product.

$$\left(x^{\frac{1}{2}}+y^{\frac{1}{2}}\right)^2 = \left(x^{\frac{1}{2}}\right)^2 + 2x^{\frac{1}{2}}y^{\frac{1}{2}} + \left(y^{\frac{1}{2}}\right)^2 = x + 2x^{\frac{1}{2}}y^{\frac{1}{2}} + y$$

(*i*) $\left(s^{\frac{1}{3}}-t^{\frac{1}{3}}\right)s^{\frac{2}{3}}$

$$\left(s^{\frac{1}{3}}-t^{\frac{1}{3}}\right)s^{\frac{2}{3}} = s^{\frac{1}{3}}s^{\frac{2}{3}} - t^{\frac{1}{3}}s^{\frac{2}{3}} = s^{\frac{1}{3}+\frac{2}{3}} - s^{\frac{2}{3}}t^{\frac{1}{3}} = s - s^{\frac{2}{3}}t^{\frac{1}{3}}$$

(*j*) $\left(x^{\frac{1}{2}}+y^{\frac{1}{2}}\right)\left(x^{\frac{1}{2}}-y^{\frac{1}{2}}\right)$

This is a special product.

$$\left(x^{\frac{1}{2}}+y^{\frac{1}{2}}\right)\left(x^{\frac{1}{2}}-y^{\frac{1}{2}}\right) = \left(x^{\frac{1}{2}}\right)^2 - \left(y^{\frac{1}{2}}\right)^2 = x^{\frac{1}{2}(2)} - y^{\frac{1}{2}(2)} = x - y$$

(*k*) $\left(3a^{\frac{1}{2}}+b^{\frac{1}{2}}\right)\left(a^{\frac{1}{2}}-4b^{\frac{1}{2}}\right)$

Apply the FOIL method to distribute.

$$\left(3a^{\frac{1}{2}}+b^{\frac{1}{2}}\right)\left(a^{\frac{1}{2}}-4b^{\frac{1}{2}}\right) = 3a^{\frac{1}{2}}a^{\frac{1}{2}} + 3a^{\frac{1}{2}}\left(-4b^{\frac{1}{2}}\right) + b^{\frac{1}{2}}a^{\frac{1}{2}} + b^{\frac{1}{2}}\left(-4b^{\frac{1}{2}}\right) = 3a - 11a^{\frac{1}{2}}b^{\frac{1}{2}} - 4b$$

Refer to supplementary problem 5.8 for similar problems.

5.13 Evaluate the following.

(*a*) $\sqrt[3]{27}$

$$\sqrt[3]{27} = 3 \text{ since } 3^3 = 27$$

(*b*) $\sqrt{64}$

$$\sqrt{64} = |8| = 8 \text{ since } 8^2 = 64$$

(*c*) $\sqrt[5]{-32}$

$$\sqrt[5]{-32} = -2 \text{ since } (-2)^5 = -32$$

(*d*) $\sqrt[4]{16}$

$$\sqrt[4]{16} = |2| = 2 \text{ since } 2^4 = 16$$

5.14 Express in exponential form.

(*a*) $\sqrt[7]{s-t}$

$$\sqrt[7]{s-t} = (s-t)^{\frac{1}{7}}$$

(*b*) $\sqrt[4]{x^3}$

$$\sqrt[4]{x^3} = (x^3)^{\frac{1}{4}} = x^{\frac{3}{4}}$$

(*c*) $\sqrt{100r^3s}$

$$\sqrt{100r^3s} = (100r^3s)^{\frac{1}{2}}$$

(*d*) $8\sqrt[3]{l^3 - w^3}$

$$8\sqrt[3]{l^3 - w^3} = 8(l^3 - w^3)^{\frac{1}{3}}$$

(*e*) $9\sqrt[5]{(2x-y)^3}$

$$9\sqrt[5]{(2x-y)^3} = 9\left[(2x-y)^3\right]^{\frac{1}{5}} = 9(2x-y)^{\frac{3}{5}}$$

(*f*) $\sqrt[n]{a^{n+1}b^{n-2}}$

$$\sqrt[n]{a^{n+1}b^{n-2}} = (a^{n+1}b^{n-2})^{\frac{1}{n}}$$

See supplementary problem 5.9.

5.15 Use the properties of radicals to express the following in simplest radical form. For simplicity, assume all variables represent positive numbers.

(*a*) $\sqrt{20}$

Identify perfect square factors of the radicand. Employ the appropriate properties next.

$$\sqrt{20} = \sqrt{4 \cdot 5} = \sqrt{2^2 \cdot 5} = \sqrt{2^2}\,\sqrt{5} = 2\sqrt{5}$$

(*b*) $\sqrt{500}$

Identify perfect square factors of the radicand. Employ appropriate properties next.

$$\sqrt{500} = \sqrt{100 \cdot 5} = \sqrt{10^2 \cdot 5} = \sqrt{10^2}\,\sqrt{5} = 10\sqrt{5}$$

(*c*) $\sqrt[3]{5000}$

Identify perfect cube factors of the radicand. Then employ the appropriate properties.

$$\sqrt[3]{5000} = \sqrt[3]{1000 \cdot 5} = \sqrt[3]{10^3 \cdot 5} = \sqrt[3]{10^3}\,\sqrt[3]{5} = 10\sqrt[3]{5}$$

(*d*) $\sqrt[3]{54x^5}$

Identify perfect cube factors of the radicand. Then employ the appropriate properties.

$$\sqrt[3]{54x^5} = \sqrt[3]{27 \cdot 2x^3x^2} = \sqrt[3]{3^3\, x^3 2x^2} = \sqrt[3]{3^3x^3}\,\sqrt[3]{2x^2} = 3x\sqrt[3]{2x^2}$$

(*e*) $\sqrt[4]{32a^{11}b^{17}}$

Solution 1: Identify perfect fourth powers of factors of the radicand. Apply the appropriate properties next.

$$\sqrt[4]{32a^{11}b^{17}} = \sqrt[4]{16 \cdot 2a^8a^3b^{16}b} = \sqrt[4]{2^4a^8b^{16} \cdot 2a^3b} = \sqrt[4]{2^4(a^2)^4(b^4)^4 \cdot 2a^3b}$$
$$= \sqrt[4]{2^4(a^2)^4(b^4)^4}\,\sqrt[4]{2a^3b} = 2a^2b^4\sqrt[4]{2a^3b}$$

Solution 2: The division algorithm may be employed to determine the exponents of factors that are removed from the radicand as well as the exponents of factors that remain in the radicand. Divide the exponents of factors in the radicand by the index of the radical. $\sqrt[4]{32a^{11}b^{17}} = \sqrt[4]{2^5a^{11}b^{17}}$ so $\frac{5}{4} = 1$ with rem. 1; $\frac{11}{4} = 2$ with rem. 3; and $\frac{17}{4} = 4$ with rem. 1. The quotients of the divisions are the exponents of the factors preceding the radical, while the remainders are the exponents of the factors in the radicand. Therefore, $\sqrt[4]{2^5a^{11}b^{17}} = 2^1\,a^2b^4\,\sqrt[4]{2^1a^3b^1} = 2a^2b^4\,\sqrt[4]{2a^3b}$.

(*f*) $\sqrt{3x^2y}\,\sqrt{6xy^3}$

Multiply the radicals first, then simplify.

$$\sqrt{3x^2y}\sqrt{6xy^3} = \sqrt{3x^2y \cdot 6xy^3} = \sqrt{18x^3y^4} = \sqrt{9 \cdot 2x^2x(y^2)^2} = \sqrt{9x^2(y^2)^2}\,\sqrt{2x}$$
$$= 3xy^2\sqrt{2x}$$

(*g*) $3s\sqrt{2t} \cdot 4t\sqrt{10st}$

Employ the commutative and associative laws, multiply the radicals, then simplify.

$$3s\sqrt{2t} \cdot 4t\sqrt{10st} = 3s\,(4t)\,\sqrt{2t\,(10st)} = 12st\sqrt{20st^2} = 12st\sqrt{4 \cdot 5st^2} = 12st\sqrt{4t^2}\,\sqrt{5s}$$
$$= 12st \cdot 2t\sqrt{5s} = 24st^2\sqrt{5s}$$

(*h*) $\sqrt[5]{\dfrac{s^{10}}{t^{20}}}$

Determine perfect fifth powers of factors in the radicand first, then simplify.

$$\sqrt[5]{\frac{s^{10}}{t^{20}}} = \sqrt[5]{\frac{(s^2)^5}{(t^4)^5}} = \frac{\sqrt[5]{(s^2)^5}}{\sqrt[5]{(t^4)^5}} = \frac{s^2}{t^4}$$

(*i*) $\sqrt[3]{\dfrac{32x}{y^4}}\,\sqrt[3]{\dfrac{2x^2}{y^2}}$

Multiply first. Determine perfect third powers of factors in the radicand next, then simplify.

$$\sqrt[3]{\frac{32x}{y^4}}\,\sqrt[3]{\frac{2x^2}{y^2}} = \sqrt[3]{\frac{32x}{y^4} \cdot \frac{2x^2}{y^2}} = \sqrt[3]{\frac{64x^3}{y^6}} = \frac{\sqrt[3]{4^3x^3}}{\sqrt[3]{(y^2)^3}} = \frac{4x}{y^2}$$

(*j*) $\sqrt[4]{(a+b)^5}$

Determine perfect fourth powers of factors in the radicand, then simplify.

$$\sqrt[4]{(a+b)^5} = \sqrt[4]{(a+b)^4\,(a+b)} = \sqrt[4]{(a+b)^4}\,\sqrt[4]{(a+b)} = (a+b)\,\sqrt[4]{(a+b)}$$

Try supplementary problem 5.10 to improve your skills.

5.16 Simplify. Assume all radicands are positive numbers.

(*a*) $\dfrac{4}{\sqrt{5}}$

We must obtain a perfect square in the radicand since the index is 2. Multiply the numerator and denominator by $\sqrt{5}$ to accomplish the task.

$$\frac{4}{\sqrt{5}}=\frac{4}{\sqrt{5}}\cdot\frac{\sqrt{5}}{\sqrt{5}}=\frac{4\sqrt{5}}{\sqrt{5^2}}=\frac{4\sqrt{5}}{5}$$

(*b*) $\dfrac{7}{3\sqrt{7}}$

We must obtain a perfect square in the radicand since the index is 2. Multiply the numerator and denominator by $\sqrt{7}$ to accomplish the task.

$$\frac{7}{3\sqrt{7}}=\frac{7}{3\sqrt{7}}\cdot\frac{\sqrt{7}}{\sqrt{7}}=\frac{7\sqrt{7}}{3\sqrt{7^2}}=\frac{7\sqrt{7}}{3\cdot 7}=\frac{\sqrt{7}}{3}$$

(*c*) $\dfrac{8\sqrt{2}}{\sqrt{5}}$

We must obtain a perfect square in the radicand since the index is 2. Multiply the numerator and denominator by $\sqrt{5}$ to accomplish the task.

$$\frac{8\sqrt{2}}{\sqrt{5}}=\frac{8\sqrt{2}}{\sqrt{5}}\cdot\frac{\sqrt{5}}{\sqrt{5}}=\frac{8\sqrt{2\cdot 5}}{\sqrt{5^2}}=\frac{8\sqrt{10}}{5}$$

(*d*) $\dfrac{1}{\sqrt[3]{4}}$

Solution 1: We need a perfect cube in the radicand this time. Multiply the numerator and denominator by $\sqrt[3]{2}$, since $8 = 2^3$ is a perfect cube.

$$\frac{1}{\sqrt[3]{4}}=\frac{1}{\sqrt[3]{4}}\cdot\frac{\sqrt[3]{2}}{\sqrt[3]{2}}=\frac{\sqrt[3]{2}}{\sqrt[3]{8}}=\frac{\sqrt[3]{2}}{2}$$

Solution 2: The exponential form of the expression may be employed also. The exponent of all factors of the radicand in the denominator must ultimately be multiples of the index of the radical for the denominator to be rationalized.

$$\frac{1}{\sqrt[3]{4}}=\frac{1}{4^{\frac{1}{3}}}=\frac{1}{(2^2)^{\frac{1}{3}}}=\frac{1}{2^{\frac{2}{3}}}\cdot\frac{2^{\frac{1}{3}}}{2^{\frac{1}{3}}}=\frac{2^{\frac{1}{3}}}{2}=\frac{\sqrt[3]{2}}{2}$$

(*e*) $\dfrac{9}{\sqrt[3]{x}}$

We need a perfect cube in the radicand again. Multiply the numerator and denominator by $\sqrt[3]{x^2}$.

$$\frac{9}{\sqrt[3]{x}}=\frac{9}{\sqrt[3]{x}}\cdot\frac{\sqrt[3]{x^2}}{\sqrt[3]{x^2}}=\frac{9\sqrt[3]{x^2}}{\sqrt[3]{x^3}}=\frac{9\sqrt[3]{x^2}}{x}$$

(*f*) $\sqrt{\dfrac{3}{2t}}$

Apply property 2 of radicals first. Rationalize the denominator next.

$$\sqrt{\frac{3}{2t}} = \frac{\sqrt{3}}{\sqrt{2t}} = \frac{\sqrt{3}}{\sqrt{2t}} \cdot \frac{\sqrt{2t}}{\sqrt{2t}} = \frac{\sqrt{3 \cdot 2t}}{\sqrt{(2t)^2}} = \frac{\sqrt{6t}}{2t}$$

(*g*) $\sqrt[3]{\dfrac{3}{2t}}$

$$\sqrt[3]{\frac{3}{2t}} = \frac{\sqrt[3]{3}}{\sqrt[3]{2t}} = \frac{\sqrt[3]{3}}{\sqrt[3]{2t}} \cdot \frac{\sqrt[3]{(2t)^2}}{\sqrt[3]{(2t)^2}} = \frac{\sqrt[3]{3(2t)^2}}{\sqrt[3]{(2t)^3}} = \frac{\sqrt[3]{12t^2}}{2t}$$

(*h*) $\sqrt[4]{\dfrac{5}{8y^2}}$

$$\sqrt[4]{\frac{5}{8y^2}} = \frac{\sqrt[4]{5}}{\sqrt[4]{2^3 y^2}} = \frac{\sqrt[4]{5}}{\sqrt[4]{2^3 y^2}} \cdot \frac{\sqrt[4]{2y^2}}{\sqrt[4]{2y^2}} = \frac{\sqrt[4]{5 \cdot 2y^2}}{\sqrt[4]{2^4 y^4}} = \frac{\sqrt[4]{10y^2}}{\sqrt[4]{(2y)^4}} = \frac{\sqrt[4]{10y^2}}{2y}$$

(*i*) $\sqrt{\dfrac{1}{s-t}}$

$$\sqrt{\frac{1}{s-t}} = \frac{\sqrt{1}}{\sqrt{s-t}} = \frac{\sqrt{1}}{\sqrt{s-t}} \cdot \frac{\sqrt{s-t}}{\sqrt{s-t}} = \frac{\sqrt{s-t}}{\sqrt{(s-t)^2}} = \frac{\sqrt{s-t}}{s-t}$$

(*j*) $\dfrac{4}{2+\sqrt{6}}$

Recall that $(a + b)(a - b) = a^2 - b^2$. Multiply the numerator and denominator by $2 - \sqrt{6}$.

$$\frac{4}{2+\sqrt{6}} = \frac{4}{2+\sqrt{6}} \cdot \frac{2-\sqrt{6}}{2-\sqrt{6}} = \frac{4(2-\sqrt{6})}{2^2-(\sqrt{6})^2} = \frac{4(2-\sqrt{6})}{4-6} = \frac{4(2-\sqrt{6})}{-2}$$
$$= -2(2-\sqrt{6}) = 2\sqrt{6} - 4$$

(*k*) $\dfrac{7}{\sqrt{x}-5}$

Multiply the numerator and denominator by $\sqrt{x} + 5$.

$$\frac{7}{\sqrt{x}-5} = \frac{7}{\sqrt{x}-5} \cdot \frac{\sqrt{x}+5}{\sqrt{x}+5} = \frac{7(\sqrt{x}+5)}{(\sqrt{x}-5)(\sqrt{x}+5)} = \frac{7(\sqrt{x}+5)}{(\sqrt{x})^2-5^2} = \frac{7(\sqrt{x}+5)}{x-25}$$

Try supplementary problem 5.11 for drill.

5.17 Simplify. Assume radicands are positive numbers.

(*a*) $\sqrt[6]{b^3}$

$$\sqrt[6]{b^3} = \sqrt[2\cdot 3]{b^3} = \sqrt[2]{b} = \sqrt{b}$$

(*b*) $\sqrt[4]{9t^2}$

$$\sqrt[4]{9t^2} = \sqrt[4]{3^2 t^2} = \sqrt[2\cdot 2]{(3t)^2} = \sqrt{3t}$$

(*c*) $\sqrt[9]{64x^6}$

$$\sqrt[9]{64x^6} = \sqrt[9]{4^3 x^6} = \sqrt[3\cdot 3]{(4x^2)} = \sqrt[3]{4x^2}$$

(*d*) $\sqrt[10]{32(x-y)^5}$

$$\sqrt[10]{32(x-y)^5} = \sqrt[10]{2^5(x-y)^5} = \sqrt[2\cdot 5]{[2(x-y)]^5} = \sqrt{2(x-y)}$$

5.18 Simplify. Assume radicands are positive numbers.

(a) $\sqrt{x}\sqrt[3]{x}$

$$\sqrt{x}\sqrt[3]{x} = x^{\frac{1}{2}}x^{\frac{1}{3}} = x^{\frac{1}{2}+\frac{1}{3}} = x^{\frac{3+2}{6}} = x^{\frac{5}{6}} = \sqrt[6]{x^5}$$

(b) $\dfrac{\sqrt{x}}{\sqrt[3]{x}}$

$$\frac{\sqrt{x}}{\sqrt[3]{x}} = \frac{x^{\frac{1}{2}}}{x^{\frac{1}{3}}} = x^{\frac{1}{2}-\frac{1}{3}} = x^{\frac{3-2}{6}} = x^{\frac{1}{6}} = \sqrt[6]{x}$$

(c) $\sqrt[3]{t^2}\sqrt[5]{t^4}$

$$\sqrt[3]{t^2}\sqrt[5]{t^4} = (t^2)^{\frac{1}{3}}(t^4)^{\frac{1}{5}} = t^{\frac{2}{3}}t^{\frac{4}{5}} = t^{\frac{2}{3}+\frac{4}{5}} = t^{\frac{10-12}{15}} = t^{\frac{-2}{15}} = \frac{1}{t^{\frac{2}{15}}} = \frac{1}{\sqrt[15]{t^2}}$$

$$= \frac{1}{\sqrt[15]{t^2}} \cdot \frac{\sqrt[15]{t^{13}}}{\sqrt[15]{t^{13}}} = \frac{\sqrt[15]{t^{13}}}{\sqrt[15]{t^{15}}} = \frac{\sqrt[15]{t^{13}}}{t}$$

(d) $\dfrac{\sqrt{s}\sqrt[3]{s}}{\sqrt[4]{s}}$

$$\frac{\sqrt{s}\sqrt[3]{s}}{\sqrt[4]{s}} = \frac{s^{\frac{1}{2}}s^{\frac{1}{3}}}{s^{\frac{1}{4}}} = s^{\frac{1}{2}+\frac{1}{3}-\frac{1}{4}} = s^{\frac{6+4-3}{12}} = s^{\frac{7}{12}} = \sqrt[12]{s^7}$$

See supplementary problem number 5.12.

5.19 Simplify. Assume variables represent positive numbers.

(a) $3\sqrt{5} + 8\sqrt{5}$

$$3\sqrt{5} + 8\sqrt{5} = (3+8)\sqrt{5} = 11\sqrt{5}$$

(b) $7\sqrt[3]{5x} + 8\sqrt[3]{5x} - 3\sqrt[3]{5x}$

$$7\sqrt[3]{5x} + 8\sqrt[3]{5x} - 3\sqrt[3]{5x} = (7+8-3)\sqrt[3]{5x} = 12\sqrt[3]{5x}$$

(c) $4\sqrt{t} - 3\sqrt{t} - \left(2\sqrt{u} - 6\sqrt{u}\right)$

$$4\sqrt{t} - 3\sqrt{t} - \left(2\sqrt{u} - 6\sqrt{u}\right) = 4\sqrt{t} - 3\sqrt{t} - 2\sqrt{u} + 6\sqrt{u} = (4-3)\sqrt{t} + (-2+6)\sqrt{u} = \sqrt{t} + 4\sqrt{u}$$

(d) $4\sqrt{50y} - 5\sqrt{72y}$

Simplify the radicals first.

$$4\sqrt{50y} - 5\sqrt{72y} = 4\sqrt{25\cdot 2y} - 5\sqrt{36\cdot 2y} = 4\sqrt{25}\sqrt{2y} - 5\sqrt{36}\sqrt{2y} = 4\cdot 5\sqrt{2y} - 5\cdot 6\sqrt{2y}$$

$$= 20\sqrt{2y} - 30\sqrt{2y} = (20-30)\sqrt{2y} = -10\sqrt{2y}$$

(e) $9\sqrt[3]{16} + \dfrac{8}{\sqrt[3]{4}}$

$$9\sqrt[3]{16} + \frac{8}{\sqrt[3]{4}} = 9\sqrt[3]{16} + \frac{8}{\sqrt[3]{4}} \cdot \frac{\sqrt[3]{2}}{\sqrt[3]{2}} = 9\sqrt[3]{8\cdot 2} + \frac{8\sqrt[3]{2}}{\sqrt[3]{8}} = 9\sqrt[3]{8}\sqrt[3]{2} + \frac{8\sqrt[3]{2}}{2}$$

$$= 9\cdot 2\sqrt[3]{2} + \frac{8\sqrt[3]{2}}{2} = 18\sqrt[3]{2} + 4\sqrt[3]{2} = 22\sqrt[3]{2}$$

(*f*) $\dfrac{\sqrt{ab}}{a} - 2\sqrt{\dfrac{b}{a}}$

$$\frac{\sqrt{ab}}{a} - 2\sqrt{\frac{b}{a}} = \frac{\sqrt{ab}}{a} - 2\frac{\sqrt{b}}{\sqrt{a}} = \frac{\sqrt{ab}}{a} - 2\frac{\sqrt{b}}{\sqrt{a}}\cdot\frac{\sqrt{a}}{\sqrt{a}} = \frac{\sqrt{ab}}{a} - \frac{2\sqrt{ab}}{a} = \frac{\sqrt{ab} - 2\sqrt{ab}}{a}$$

$$= \frac{(1-2)\sqrt{ab}}{a} = \frac{-\sqrt{ab}}{a}$$

(*g*) $3y^2\sqrt{48x^7y^6} - 4x\sqrt{108x^5y^{10}}$

$$3y^2\sqrt{48x^7y^6} - 4x\sqrt{108x^5y^{10}} = 3y^2\sqrt{16\cdot 3x^6xy^6} - 4x\sqrt{36\cdot 3x^4xy^{10}}$$

$$= 3y^2\sqrt{16x^6y^6}\sqrt{3x} - 4x\sqrt{36x^4y^{10}}\sqrt{3x}$$

$$= 3y^2 4x^3y^3\sqrt{3x} - 4x6x^2y^5\sqrt{3x}$$

$$= 12x^3y^5\sqrt{3x} - 24x^3y^5\sqrt{3x} = -12x^3y^5\sqrt{3x}$$

Do supplementary problem 5.13.

5.20 Perform the indicated operations and simplify. Assume radicands are positive.

(*a*) $\sqrt{3}(\sqrt{2} - \sqrt{3})$

$$\sqrt{3}(\sqrt{2} - \sqrt{3}) = \sqrt{3}\sqrt{2} - \sqrt{3}\sqrt{3} = \sqrt{6} - \sqrt{9} = \sqrt{6} - 3$$

(*b*) $(3\sqrt{2} - 4\sqrt{3})(5\sqrt{2} + 2\sqrt{3})$

Employ the FOIL method to distribute, then combine like terms.

$$(3\sqrt{2} - 4\sqrt{3})(5\sqrt{2} + 2\sqrt{3}) = 3\sqrt{2}\cdot 5\sqrt{2} + 3\sqrt{2}\cdot 2\sqrt{3} - 4\sqrt{3}\cdot 5\sqrt{2} - 4\sqrt{3}\cdot 2\sqrt{3}$$

$$= 15\cdot 2 + 6\sqrt{6} - 20\sqrt{6} - 8\cdot 3 = 30 - 14\sqrt{6} - 24$$

$$= 6 - 14\sqrt{6}$$

(*c*) $(x - \sqrt{5})(2x + 3\sqrt{5})$

$$(x - \sqrt{5})(2x + 3\sqrt{5}) = 2x^2 + 3x\sqrt{5} - 2x\sqrt{5} - 3\cdot 5 = 2x^2 + x\sqrt{5} - 15$$

(*d*) $(\sqrt{s} - \sqrt{t})^2 + (\sqrt{s+t})^2$

Apply the binomial difference special product form to the first term.

$$(\sqrt{s} - \sqrt{t})^2 + (\sqrt{s+t})^2 = (\sqrt{s})^2 - 2\sqrt{s}\sqrt{t} + (\sqrt{t})^2 + s + t$$

$$= s - 2\sqrt{st} + t + s + t = 2s + 2t - 2\sqrt{st}$$

(*e*) $(\sqrt[3]{a} + \sqrt[3]{b})(\sqrt[3]{a^3} - \sqrt[3]{ab} + \sqrt[3]{b^3}) = P$

Distribute term by term.

$$P = (\sqrt[3]{a} + \sqrt[3]{b})(\sqrt[3]{a^2} - \sqrt[3]{ab} + \sqrt[3]{b^2})$$

$$= \sqrt[3]{a}\sqrt[3]{a^2} - \sqrt[3]{a}\sqrt[3]{ab} + \sqrt[3]{a}\sqrt[3]{b^2} + \sqrt[3]{b}\sqrt[3]{a^2} - \sqrt[3]{b}\sqrt[3]{ab} + \sqrt[3]{b}\sqrt[3]{b^2}$$

$$= \sqrt[3]{a^3} - \sqrt[3]{a^2b} + \sqrt[3]{ab^2} + \sqrt[3]{a^2b} - \sqrt[3]{ab^2} + \sqrt[3]{b^3}$$

$$= \sqrt[3]{a^3} + \sqrt[3]{b^3} = a + b$$

Note that we obtained a special product; it is the sum of two cubes: $P = \left(\sqrt[3]{a}\right)^3 + \left(\sqrt[3]{b}\right)^3 = a + b$.

If the steps in the above problems are reversed, we observe that the results can be written as a product of factors with irrational terms. There are situations which require irrational coefficients in some or all of the terms in factors.

See supplementary problem 5.14 to practice multiplying radical expressions.

5.21 Simplify. Assume radicands are positive.

(*a*) $\dfrac{8-\sqrt{32}}{12}$

Simplify the radical, then reduce.

$$\frac{8-\sqrt{32}}{12} = \frac{8-\sqrt{16\cdot 2}}{12} = \frac{8-4\sqrt{2}}{12} = \frac{4(2-\sqrt{2})}{4\cdot 3} = \frac{2-\sqrt{2}}{3}$$

(*b*) $\dfrac{4}{7+\sqrt{3}}$

Rationalize the denominator and simplify.

$$\frac{4}{7+\sqrt{3}} = \frac{4}{\left(7+\sqrt{3}\right)}\cdot\frac{\left(7-\sqrt{3}\right)}{\left(7-\sqrt{3}\right)} = \frac{4\left(7-\sqrt{3}\right)}{49-3} = \frac{4\left(7-\sqrt{3}\right)}{46} = \frac{2\left(7-\sqrt{3}\right)}{23}$$

(*c*) $\dfrac{\sqrt{2}}{3-\sqrt{x}}$

$$\frac{\sqrt{2}}{3-\sqrt{x}} = \frac{\sqrt{2}}{\left(3-\sqrt{x}\right)}\cdot\frac{\left(3+\sqrt{x}\right)}{\left(3+\sqrt{x}\right)} = \frac{\sqrt{2}\left(3+\sqrt{x}\right)}{9-x} = \frac{3\sqrt{2}+\sqrt{2x}}{9-x}$$

(*d*) $\dfrac{\sqrt{3}}{\sqrt{5}+\sqrt{7}}$

$$\frac{\sqrt{3}}{\sqrt{5}+\sqrt{7}} = \frac{\sqrt{3}}{\left(\sqrt{5}+\sqrt{7}\right)}\cdot\frac{\left(\sqrt{5}-\sqrt{7}\right)}{\left(\sqrt{5}-\sqrt{7}\right)} = \frac{\sqrt{3}\sqrt{5}-\sqrt{3}\sqrt{7}}{5-7} = \frac{\sqrt{15}-\sqrt{21}}{-2} = \frac{\sqrt{21}-\sqrt{15}}{2}$$

(*e*) $\dfrac{l-w}{\sqrt{l}+\sqrt{w}}$

$$\frac{l-w}{\sqrt{l}+\sqrt{w}} = \frac{(l-w)}{\left(\sqrt{l}+\sqrt{w}\right)}\cdot\frac{\left(\sqrt{l}-\sqrt{w}\right)}{\left(\sqrt{l}-\sqrt{w}\right)} = \frac{(l-w)\left(\sqrt{l}-\sqrt{w}\right)}{l-w} = \sqrt{l}-\sqrt{w}$$

(*f*) $\dfrac{20}{\sqrt[3]{5}}$

$$\frac{20}{\sqrt[3]{5}} = \frac{20}{\sqrt[3]{5}}\cdot\frac{\sqrt[3]{5^2}}{\sqrt[3]{5^2}} = \frac{20\sqrt[3]{5^2}}{\sqrt[3]{5^3}} = \frac{20\sqrt[3]{25}}{5} = 4\sqrt[3]{25}$$

(*g*) $\dfrac{2}{\sqrt[3]{3}+\sqrt[3]{5}}$

Refer to solved problem 5.20(*e*) for an indication of the appropriate rationalizing factor. We must multiply by a factor such that the product is the cube of the terms in the denominator. Form 6 of the special products given in an earlier section is the appropriate form.

$$\frac{2}{\sqrt[3]{3}+\sqrt[3]{5}} = \frac{2}{\left(\sqrt[3]{3}+\sqrt[3]{5}\right)}\cdot\frac{\left(\sqrt[3]{3^2}-\sqrt[3]{3}\sqrt[3]{5}+\sqrt[3]{5^2}\right)}{\left(\sqrt[3]{3^2}-\sqrt[3]{3}\sqrt[3]{5}+\sqrt[3]{5^2}\right)} = \frac{2\left(\sqrt[3]{3^2}-\sqrt[3]{3}\sqrt[3]{5}+\sqrt[3]{5^2}\right)}{\sqrt[3]{3^3}+\sqrt[3]{5^3}}$$

$$= \frac{2\left(\sqrt[3]{9}-\sqrt[3]{15}+\sqrt[3]{25}\right)}{3+5} = \frac{2\left(\sqrt[3]{9}-\sqrt[3]{15}+\sqrt[3]{25}\right)}{8} = \frac{\sqrt[3]{9}-\sqrt[3]{15}+\sqrt[3]{25}}{4}$$

See supplementary problem 5.15 for similar problems.

5.22 Express each of the following in complex number form. Express the results appropriately in $a + bi$ or $a + ib$ form.

(*a*) 5

$$5 = 5 + 0i$$

(*b*) -88

$$-88 = -88 + 0i$$

(*c*) $-\sqrt{-6}$

$$-\sqrt{-6} = -\sqrt{-1 \cdot 6} = -\sqrt{-1}\sqrt{6} = -i\sqrt{6} = 0 - i\sqrt{6}$$

(*d*) $3 + \sqrt{-2}$

$$3 + \sqrt{-2} = 3 + \sqrt{-1 \cdot 2} = 3 + \sqrt{-1}\sqrt{2} = 3 + i\sqrt{2}$$

(*e*) $4\sqrt{-12}$

$$4\sqrt{-12} = 4\sqrt{-1 \cdot 4 \cdot 3} = 4\sqrt{-1}\sqrt{4}\sqrt{3} = 4i\,(2)\sqrt{3} = 8i\sqrt{3} = 0 + 8i\sqrt{3}$$

Note the order of the factors in the imaginary part.

5.23 Perform the indicated operations.

(*a*) $(2 + 7i) + (3 + i)$

$$(2 + 7i) + (3 + i) = (2 + 3) + (7 + 1)i = 5 + 8i$$

(*b*) $(-2 + 4i) + (3 - 3i)$

$$(-2 + 4i) + (3 - 3i) = (-2 + 3) + (4 - 3)i = 1 + 1i = 1 + i$$

(*c*) $7 + (4 - 2i)$

$$7 + (4 - 2i) = (7 + 4) + (-2i) = 11 - 2i$$

(*d*) $(2 + 7i) - (3 + i)$

$$(2 + 7i) - (3 + i) = (2 - 3) + (7 - 1)i = -1 + 6i$$

(*e*) $(-2 + 4i) - (3 - 3i)$

$$(-2 + 4i) - (3 - 3i) = (-2 - 3) + [4 - (-3)]\,i = -5 + 7i$$

(*f*) $7 - (4 - 2i)$

$$7 - (4 - 2i) = (7 - 4) + [0 - (-2)]\,i = 3 + 2i$$

The results above illustrate we can treat i as a variable and combine like terms when we are either adding or subtracting complex numbers.

5.24 Perform the indicated operations.

(*a*) $(2 + 5i)(3 + 2i)$

Use the FLOI rule to distribute.

$$(2 + 5i)(3 + 2i) = 2 \cdot 3 + (5i)(2i) + 2\,(2i) + 3\,(5i) = 6 + 10i^2 + 4i + 15i$$
$$= 6 - 10 + (4 + 15)i = -4 + 19i$$

(*b*) $(5 - 3i)(2 + i)$

Use the FLOI rule to distribute.

$$(5 - 3i)(2 + i) = 5 \cdot 2 + (-3i)i + 5i + 2(-3i) = 10 - 3i^2 + 5i - 6i = 10 - (-3) + (5 - 6)i = 13 - i$$

(*c*) $3(4 + 7i)$

Think of i as a variable and distribute in the usual manner.

$$3(4 + 7i) = 3 \cdot 4 + 3(7i) = 12 + 21i$$

(*d*) $-i(2 - 9i)$

Distribute in the usual manner.

$$-i(2 - 9i) = -2i + 9i^2 = -2i + 9(-1) = -9 - 2i$$

(*e*) $\sqrt{-3}(2 - \sqrt{-2})$

Rewrite $\sqrt{-3}$ and $\sqrt{-2}$ first. Then distribute.

$$\sqrt{-3}(2 - \sqrt{-2}) = i\sqrt{3}(2 - i\sqrt{2}) = 2i\sqrt{3} - i^2\sqrt{6} = 2i\sqrt{3} - (-1)\sqrt{6}$$

$$= 2i\sqrt{3} + \sqrt{6} = \sqrt{6} + 2i\sqrt{3}$$

(*f*) $(5 - i)^2$

Apply the special product form for the square of a binomial difference.

$$(5 - i)^2 = 5^2 - 2 \cdot 5i + i^2 = 25 - 10i + (-1) = 24 - 10i$$

(*g*) $(4 + 3i)(4 - 3i)$

Use the appropriate special product form.

$$(4 + 3i)(4 - 3i) = 4^2 - (3i)^2 = 16 - 9i^2 = 16 - 9(-1) = 16 + 9 = 25$$

Part (*g*) illustrates a special situation. Note that the product is a real number. The factors are complex conjugates of each other.

5.25 Express the following in $a + bi$ or $a + ib$ form.

(*a*) $\dfrac{7 - 8i}{3}$

$$\frac{7 - 8i}{3} = \frac{7}{3} - \frac{8}{3}i$$

(*b*) $\dfrac{11 - 20i}{-5}$

$$\frac{11 - 20i}{-5} = \frac{11}{-5} - \frac{20}{-5}i = \frac{-11}{5} - (-4)i = \frac{-11}{5} + 4i$$

(*c*) $\dfrac{5 + \sqrt{-2}}{2i}$

$$\frac{5 + \sqrt{-2}}{2i} = \frac{5 + i\sqrt{2}}{2i} = \frac{5 + i\sqrt{2}}{2i} \cdot \frac{i}{i} = \frac{5i + i^2\sqrt{2}}{2i^2} = \frac{5i - \sqrt{2}}{-2} = \frac{5i}{-2} - \frac{\sqrt{2}}{-2}$$

$$= \frac{-5i}{2} + \frac{\sqrt{2}}{2} = \frac{\sqrt{2}}{2} - \frac{5}{2}i$$

(*d*) $\frac{7+i}{-i}$

$$\frac{7+i}{-i} = \frac{7+i}{-i}\cdot\frac{i}{i} = \frac{7i+i^2}{-i^2} = \frac{7i-1}{-(-1)} = \frac{-1+7i}{1} = -1+7i$$

(*e*) $\frac{4}{3-i}$

$$\frac{4}{3-i} = \frac{4}{3-i}\cdot\frac{3+i}{3+i} = \frac{12+4i}{3^2-i^2} = \frac{12+4i}{9-(-1)} = \frac{12+4i}{10} = \frac{12}{10}+\frac{4}{10}i = \frac{6}{5}+\frac{2}{5}i$$

(*f*) $\frac{5+2i}{4+3i}$

$$\frac{5+2i}{4+3i} = \frac{5+2i}{4+3i}\cdot\frac{4-3i}{4-3i} = \frac{20-6i^2-15i+8i}{4^2-(3i)^2} = \frac{20+6-15i+8i}{16+9} = \frac{26-7i}{25} = \frac{26}{25}-\frac{7}{25}i$$

(*g*) $\frac{6}{\sqrt{-3}}$

$$\frac{6}{\sqrt{-3}} = \frac{6}{i\sqrt{3}} = \frac{6}{i\sqrt{3}}\cdot\frac{i}{i} = \frac{6i}{i^2\sqrt{3}} = \frac{6i}{-\sqrt{3}} = \frac{-6i}{\sqrt{3}}\cdot\frac{\sqrt{3}}{\sqrt{3}} = \frac{-6i\sqrt{3}}{3} = -2i\sqrt{3}$$

(*h*) $\frac{3+\sqrt{-25}}{5-\sqrt{-9}}$

$$\frac{3+\sqrt{-25}}{5-\sqrt{-9}} = \frac{3+i\sqrt{25}}{5-i\sqrt{9}} = \frac{3+5i}{5-3i} = \frac{3+5i}{5-3i}\cdot\frac{5+3i}{5+3i} = \frac{15+15i^2+9i+25i}{5^2-(3i)^2}$$

$$= \frac{15-15+34i}{25+9} = \frac{34i}{34} = i$$

Answers to division problems may be checked by multiplying the quotient by the divisor to compare with the dividend. We illustrate the process by checking the answer to solved problem 5.25(*f*). We determined that $\frac{5+2i}{4+3i} = \frac{26}{25}-\frac{7}{25}i$ above. The check follows.

$$\left(\frac{26}{25}-\frac{7}{25}i\right)(4+3i) = \left(\frac{26}{25}\right)4 + \left(\frac{-7}{25}i\right)(3i) + \left(\frac{26}{25}\right)(3i) + \left(\frac{-7}{25}i\right)4$$

$$= \frac{104}{25} - \frac{21}{25}i^2 + \frac{78}{25}i - \frac{28}{25}i = \frac{104+21}{25} + \frac{78-28}{25}i$$

$$= \frac{125}{25} + \frac{50}{25}i = 5+2i$$

The quotient times the divisor does equal the dividend. The check is complete.

See supplementary problem 5.16 to practice operations on complex numbers.

SUPPLEMENTARY PROBLEMS

5.1 Perform the indicated operations and simplify.

(*a*) 4^{-2} (*b*) $2\cdot 4^{-2}$ (*c*) $\frac{1}{3\cdot 9^{-2}}$ (*d*) $\frac{2^{-1}}{5}$

(*e*) $(-2)^{-3}$ (*f*) $\left(\frac{1}{4\cdot 8^{-2}}\right)^2$ (*g*) $\left(\frac{2}{7}\right)^0 - \frac{2}{5}$ (*h*) $\left(\frac{5^{-3}}{4^{-2}}\right)^{-1}$

5.2 Perform the indicated operations and simplify. Assume that no variable is zero.

(*a*) $10 \cdot 5^{-2}$

(*b*) -4^{-3}

(*c*) $\left(\dfrac{2}{3 \cdot 6^{-2}}\right)^2$

(*d*) $2^{-1} + 3^{-2}$

(*e*) $n^{-2} \cdot n^9 \cdot n^{-3}$

(*f*) $(x^{-3}y)(x^0 y^{-4})$

(*g*) $\left(\dfrac{3}{a^{-3}b^{-2}}\right)^{-2}$

(*h*) $(-2s^{-2}t^2)^0(5s^{-4}t^3)^{-2}$

(*i*) $\dfrac{x^0 x^{-4} x^5}{x^{-2}x^2}$

(*j*) $\left(\dfrac{x^{-2} + y^{-1}}{xy^{-2}}\right)^{-1}$

(*k*) $\left(\dfrac{b^n}{b^{n-1}}\right)^{-2}$

(*l*) $\left(\dfrac{xy^{-4}}{2^{-1}x^{-2}}\right)^3 \left(\dfrac{8x^{-2}y^0}{3^{-1}xy^{-3}}\right)^{-2}$

5.3 Express the following in scientific notation.

(*a*) 93,000,000 (*b*) 100,000,000 (*c*) 0.000034

(*d*) 20.005 (*e*) 0.0007090 (*f*) 0.008

5.4 Express the following in standard (decimal) form.

(*a*) 5.87×10^5 (*b*) 6.23×10^{-4} (*c*) 7×10^7

(*d*) 2.03×10^{-3} (*e*) 4×10^{-6} (*f*) 9.82×10^0

5.5 Convert to the appropriate factored form of each number and perform the indicated operations. Express the result in standard form.

(*a*) $\dfrac{700 \times 810}{9{,}000 \times 2.8}$ (*b*) $\dfrac{0.075 \times 0.068}{0.00034 \times 0.0015}$ (*c*) $\dfrac{560 \times 0.0423}{1410 \times 0.080}$

5.6 Determine the number of significant digits and the accuracy of the following numbers.

(*a*) 1.27×10^3 (*b*) 1.270×10^3 (*c*) 127,000 (*d*) 0.0060

(*e*) 30,200 (*f*) 5.001 (*g*) 4×10^{-3} (*h*) 90.0×10^{-3}

5.7 Change to radical notation.

(*a*) $b^{\frac{1}{2}}$ (*b*) $x^{\frac{2}{5}}$ (*c*) $3y^{\frac{5}{4}}$

(*d*) $(5t)^{\frac{-1}{3}}$ (*e*) $5t^{\frac{-1}{3}}$ (*f*) $(l + w)^{\frac{1}{2}}$

5.8 Perform the indicated operations and simplify. Assume all variables represent positive-real numbers.

(*a*) $c^{\frac{1}{2}}c^{\frac{1}{5}}$

(*b*) $x^{\frac{-2}{5}}x^{\frac{2}{3}}$

(*c*) $\dfrac{4t^{\frac{1}{2}}}{t^{\frac{2}{3}}}$

(*d*) $\left(a^{\frac{1}{3}}b^{\frac{-1}{2}}\right)^6$

(*e*) $\left(\dfrac{y^{\frac{1}{5}}y^{\frac{-1}{3}}}{y}\right)^{-15}$

(*f*) $\left(\dfrac{x^{\frac{1}{6}}y}{x^{-2}y^{\frac{1}{6}}}\right)^3$

(*g*) $b^2\left(b^{\frac{1}{2}} - b^{\frac{1}{4}}\right)$

(*h*) $\left(s^{\frac{1}{2}} - t^{\frac{1}{2}}\right)^2$

(*i*) $\left(x^{\frac{3}{2}} - 2y^{\frac{1}{2}}\right)\left(7x^{\frac{1}{2}} + y^{\frac{3}{2}}\right)$

5.9 Express in exponential form.

(a) $\sqrt[3]{7^4}$ (b) $\sqrt{st}$ (c) $a\sqrt[3]{b^2}$

(d) $\sqrt{x} - \sqrt{y}$ (e) $\sqrt[4]{9}\,\sqrt[5]{xy^3}$ (f) $\dfrac{9}{\sqrt{a+b}}$

5.10 Use the properties of radicals to express the following in simplest radical form. For simplicity, assume all variables represent positive numbers.

(a) $\sqrt{50}$ (b) $\sqrt[3]{250x^5}$ (c) $\sqrt[4]{243s^{10}t^{17}}$

(d) $3\sqrt{2a^3b^5}\,\sqrt{32ab^2}$ (e) $\sqrt{\dfrac{500x^3}{4y^4}}$ (f) $\sqrt{\dfrac{3a^3}{b^4}}\,\sqrt{\dfrac{6a}{b^2}}$

(g) $\sqrt[3]{(x-y)^{10}}$ (h) $\sqrt[3]{s^3 - t^3}$

5.11 Simplify. Assume all radicands are positive numbers.

(a) $\dfrac{7}{\sqrt{6}}$ (b) $\dfrac{10\sqrt{10}}{\sqrt{5}}$ (c) $\dfrac{4}{\sqrt[3]{9}}$ (d) $\dfrac{11}{\sqrt[3]{k^2}}$

(e) $\sqrt{\dfrac{2x}{7y}}$ (f) $\dfrac{3}{\sqrt[4]{3t^3}}$ (g) $\dfrac{1}{3x-4}$ (h) $\dfrac{20}{4-\sqrt{6}}$

(i) $\dfrac{6}{2+\sqrt{w}}$

5.12 Simplify. Assume radicands are positive numbers

(a) $\sqrt[4]{9}$ (b) $\sqrt[6]{x^6y^4}$ (c) $\sqrt[12]{64\,a^3b^6}$

(d) $\sqrt[4]{6}\,\sqrt[3]{6}$ (e) $\dfrac{\sqrt{k}}{\sqrt[4]{k}}$ (f) $\dfrac{2\sqrt{x}\sqrt[4]{x}}{\sqrt[6]{x^2}}$

5.13 Simplify. Assume variables represent positive numbers.

(a) $5\sqrt{7} - 2\sqrt{7}$ (b) $3\sqrt{6} - (9\sqrt{6} - 2\sqrt{6})$

(c) $7 - \sqrt[3]{k^2} + 4\sqrt[3]{k^2} - 3$ (d) $2\sqrt[3]{32} - 7\sqrt[3]{108}$

(e) $\sqrt{\dfrac{1}{7}} + 3\sqrt{7}$ (f) $3\sqrt{\dfrac{2}{x}} - \dfrac{\sqrt{2x}}{5}$

(g) $4xy\sqrt{45x^7y^6} + \sqrt{5x^9y^{8\,8}}$ (h) $4\sqrt[3]{\dfrac{1}{3}} - \dfrac{10}{\sqrt[3]{3}} + 7\sqrt[3]{9}$

5.14 Simplify. Assume radicands are positive.

(a) $\sqrt{5}\,(3 + \sqrt{2})$ (b) $(\sqrt{x} - \sqrt{y})\sqrt{x}$ (c) $(\sqrt{7} + \sqrt{3})(\sqrt{7} - \sqrt{3})$

(d) $(2\sqrt{7} + 3\sqrt{5})(5 - \sqrt{5})$ (e) $(\sqrt{a} + \sqrt{b})^2$ (f) $(\sqrt[3]{s} + t)^3$

5.15 Simplify. Assume radicands are positive.

(a) $\dfrac{\sqrt{27} - 6}{3}$ (b) $\dfrac{5}{\sqrt{3} - 4}$ (c) $\dfrac{\sqrt{6}}{4 + \sqrt{y}}$ (d) $\dfrac{\sqrt{2}}{\sqrt{6} + \sqrt{5}}$

(e) $\dfrac{t-2}{\sqrt{t} + \sqrt{2}}$ (f) $\dfrac{9}{\sqrt[3]{x^2}}$ (g) $\dfrac{1}{\sqrt[4]{x^2}}$ (h) $\dfrac{8}{\sqrt[3]{4} - \sqrt[3]{2}}$

5.16 Perform the indicated operations.

(*a*) $10 + (-3 + 4i)$ (*b*) $(-4 + 6i) + (5 - 2i)$ (*c*) $(-3 - 5i) + 3i$

(*d*) $10 - (-3 + 4i)$ (*e*) $(-4 + 6i) - (5 - 2i)$ (*f*) $(-3 - 5i) - 3i$

(*g*) $(6 - 3i)(-2 + i)$ (*h*) $-5(3 + 2i)$ (*i*) $(2 - 6i)\,i$

(*j*) $(3 + 7i)(3 - 7i)$ (*k*) $\sqrt{-16}\,(3 + \sqrt{-1})$ (*l*) $(6 + 2i)^2$

(*m*) $(4 + \sqrt{-2})(-5 - \sqrt{-6})$ (*n*) $\dfrac{-10 + 16i}{2}$ (*o*) $\dfrac{12 - 28i}{4i}$

(*p*) $\dfrac{6}{2 + i}$ (*q*) $\dfrac{4 - 3i}{2 - 7i}$ (*r*) $\dfrac{9i}{\sqrt{-3}}$

(*s*) $\dfrac{2 - \sqrt{-9}}{1 + \sqrt{-16}}$ (*t*) $\dfrac{4 - 2i}{i(5 + 2i)}$

ANSWERS TO SUPPLEMENTARY PROBLEMS

5.1 (*a*) $\frac{1}{16}$ (*b*) $\frac{1}{8}$ (*c*) 27 (*d*) $\frac{1}{10}$

(*e*) $\frac{-1}{8}$ (*f*) 256 (*g*) $\frac{3}{15}$ (*h*) $\frac{125}{16}$

5.2 (*a*) $\frac{2}{5}$ (*b*) $\frac{-1}{64}$ (*c*) 576 (*d*) $\frac{11}{18}$

(*e*) n^4 (*f*) $\dfrac{1}{x^3y^3}$ (*g*) $\dfrac{1}{9a^6b^4}$ (*h*) $\dfrac{s^8}{25t^6}$

(*i*) x (*j*) $\dfrac{x^3}{y(y + x^2)}$ (*k*) $\dfrac{1}{b^2}$ (*l*) $\dfrac{x^{15}}{72y^{18}}$

5.3 (*a*) 9.3×10^7 (*b*) 1×10^8 (*c*) 3.4×10^{-5}

(*d*) 2.0005×10 (*e*) 7.090×10^{-4} (*f*) 8×10^{-3}

5.4 (*a*) 587,000 (*b*) 0.000623 (*c*) 70,000,000

(*d*) 0.00203 (*e*) 0.000004 (*f*) 9.82

5.5 (*a*) 22.5 (*b*) 10,000 (*c*) 0.21

5.6 (*a*) 3, nearest ten (*b*) 4, nearest unit (*c*) 3, nearest thousand

(*d*) 2, nearest ten-thousandth (*e*) 3, nearest hundred (*f*) 4, nearest thousandth

(*g*) 1, nearest thousandth (*h*) 3, nearest ten-thousandth

5.7 (*a*) $\sqrt{b}$ (*b*) $\left(\sqrt[5]{x}\right)^2$ *or* $\sqrt[5]{x^2}$ (*c*) $3\left(\sqrt[4]{y}\right)^5$ *or* $3\sqrt[4]{y^5}$

(*d*) $\dfrac{1}{\sqrt[3]{5t}}$ (*e*) $\dfrac{5}{\sqrt[3]{t}}$ (*f*) $\sqrt{l + w}$

5.8 (*a*) $c^{\frac{7}{10}}$ (*b*) $x^{\frac{4}{15}}$ (*c*) $\dfrac{4}{t^{\frac{1}{6}}}$

(*d*) $\dfrac{a^2}{b^3}$ (*e*) y^{17} (*f*) $x^{\frac{13}{2}}y^{\frac{5}{2}}$

(*g*) $b^{\frac{5}{2}} - b^{\frac{9}{4}}$ (*h*) $s - 2s^{\frac{1}{2}}t^{\frac{1}{2}} + t$ (*i*) $7x^2 + x^{\frac{3}{2}}y^{\frac{3}{2}} - 14x^{\frac{1}{2}}y^{\frac{1}{2}} - 2y^2$

5.9 (*a*) $7^{\frac{4}{3}}$ (*b*) $(st)^{\frac{1}{2}}$ (*c*) $ab^{\frac{2}{3}}$

(*d*) $x^{\frac{1}{2}} - y^{\frac{1}{2}}$ (*e*) $9^{\frac{1}{4}}(xy^3)^{\frac{1}{5}}$ (*f*) $\dfrac{9}{(a+b)^{\frac{1}{2}}}$ or $9(a+b)^{\frac{-1}{2}}$

5.10 (*a*) $5\sqrt{2}$ (*b*) $5x\sqrt[3]{2x^2}$ (*c*) $3s^2t^4\sqrt[4]{3s^2t}$

(*d*) $24a^2b^3\sqrt{b}$ (*e*) $\dfrac{5x\sqrt{5x}}{y^2}$ (*f*) $\dfrac{3a^2\sqrt{2}}{b^3}$

(*g*) $(x-y)^3\sqrt[3]{x-y}$ (*h*) Cannot be simplified further.

5.11 (*a*) $\dfrac{7\sqrt{6}}{6}$ (*b*) $10\sqrt{2}$ (*c*) $\dfrac{4\sqrt[3]{3}}{3}$

(*d*) $\dfrac{11\sqrt[3]{k}}{k}$ (*e*) $\dfrac{\sqrt{14xy}}{7y}$ (*f*) $\dfrac{\sqrt[4]{27t}}{t}$

(*g*) $\dfrac{\sqrt{3x-4}}{3x-4}$ (*h*) $8+2\sqrt{6}$ (*i*) $\dfrac{6(2-\sqrt{w})}{4-w}$

5.12 (*a*) $\sqrt{3}$ (*b*) $x\sqrt[3]{y^2}$ (*c*) $\sqrt[4]{4ab^2}$

(*d*) $\sqrt[12]{6^7}$ (*e*) $\sqrt[4]{k}$ (*f*) $2\sqrt[12]{x^5}$

5.13 (*a*) $3\sqrt{7}$ (*b*) $-4\sqrt{6}$ (*c*) $4 + 3\sqrt[3]{k^2}$

(*d*) $-17\sqrt[3]{4}$ (*e*) $\dfrac{22\sqrt{7}}{7}$ (*f*) $\dfrac{(15-x)\sqrt{2x}}{5x}$

(*g*) $13x^4y^4\sqrt{5x}$ (*h*) $5\sqrt[3]{9}$

5.14 (*a*) $3\sqrt{5} + \sqrt{10}$ (*b*) $x - \sqrt{xy}$ (*c*) 4

(*d*) $10\sqrt{7} - 2\sqrt{35} + 15\sqrt{5} - 15$ (*e*) $a + 2\sqrt{ab} + b$ (*f*) $s + t$

5.15 (*a*) $\sqrt{3} - 2$ (*b*) $\dfrac{-5(\sqrt{3}+4)}{13}$ (*c*) $\dfrac{4\sqrt{6}-\sqrt{6y}}{16-y}$

(*d*) $2\sqrt{3} - \sqrt{10}$ (*e*) $\sqrt{t} - \sqrt{2}$ (*f*) $\dfrac{9\sqrt[3]{x}}{x}$

(*g*) $\dfrac{\sqrt[4]{x^2}}{x} = \dfrac{\sqrt{x}}{x}$ (*h*) $4(2\sqrt[3]{2} + 2 + \sqrt[3]{4})$

5.16 (*a*) $7 + 4i$ (*b*) $1 + 4i$ (*c*) $-3 - 2i$

(*d*) $13 - 4i$ (*e*) $-9 + 8i$ (*f*) $-3 - 8i$

(*g*) $-9 + 12i$ (*h*) $-15 - 10i$ (*i*) $6 + 2i$

(*j*) 58 (*k*) $-4 + 12i$ (*l*) $32 + 24i$

(*m*) $(-20 + 2\sqrt{3}) - (4\sqrt{6} + 5\sqrt{2})i$ (*n*) $-5 + 8i$ (*o*) $-7 - 3i$

(*p*) $\dfrac{12}{5} - \dfrac{6}{5}i$ (*q*) $\dfrac{29}{53} + \dfrac{22}{53}i$ (*r*) $3\sqrt{3}$

(*s*) $\dfrac{-10}{17} - \dfrac{11}{17}i$ (*t*) $\dfrac{-18}{29} - \dfrac{16}{29}i$

Second-Degree Equations and Inequalities

6.1 Solving by Factoring and Square Root Methods

A *second-degree equation* is a polynomial equation of degree two. Second-degree equations in one variable are commonly referred to as *quadratic equations*. The *standard form* of a quadratic equation is $ax^2 + bx + c = 0$ where $a > 0$. The *solutions, roots,* or *zeros* of a quadratic equation are the values of the variable for which the equation is true. There are normally two solutions or roots of a quadratic equation, although there occasionally is one or no solution to the equation.

The variable terms may vanish as we solve certain equations. The equation $x^2 + 1 = x^2 - 2$ has no solution, since the x^2 terms vanish if we subtract x^2 from both sides and obtain $1 = -2$. This statement is a contradiction. There is no solution to the equation.

If a given equation contains a variable in the denominator of a fraction, we multiply by the LCD to clear all fractions. The resulting equation may have solutions that do not satisfy the original equation. This situation occurs for values of the variable which produce a zero value in the original denominator. Equivalently the LCD may contain a factor whose value is zero for a particular value of the variable. Recall that equivalent equations are obtained only when we multiply both sides of an equation by a nonzero quantity. The apparent solutions to the resulting equation which do not satisfy the original equation are called *extraneous roots*. We are obligated to check for extraneous roots when we multiply by factors which involve a variable. It is part of the solution process.

Factor Method

The reader should review factoring polynomials in Chapter 2 before continuing. We shall employ the zero factor property to solve quadratic equations.

Zero Factor Property: If $a \cdot b = 0$, then $a = 0$ or $b = 0$.

The zero factor property states that if the product of two quantities is zero, then at least one of the quantities is zero.

See solved problem 6.1.

Square Root Method

If $b = 0$ in the quadratic equation $ax^2 + bx + c = 0$, the equation becomes $ax^2 + c = 0$. Furthermore if c is subtracted from both sides and both sides are divided by a, then $x^2 = -c/a$. Replace $-c/a$ by the single letter d. The equation then has the form $x^2 = d$. The values of x for which the equation is true are those numbers we can square to obtain d. The possibilities are $\sqrt{d}$ and $-\sqrt{d}$. A common way to express $\sqrt{d}$ and $-\sqrt{d}$ is $\pm\sqrt{d}$. The results of the above discussion are stated in the theorem which follows.

THEOREM 1. If $x^2 = d$, then $x = \pm\sqrt{d}$. The solution set is $\{\pm\sqrt{d}\}$.

If the first-degree term is missing in a quadratic equation, we may find the roots by applying the above theorem. This technique is referred to as the square root method or simply as the extraction of roots.

The square root method may also be employed to solve equations which involve a perfect square equaling a real number. We illustrate on p. 179.

See solved problem 6.2.

6.2 Completing the Square and the Quadratic Formula

Quadratic equations are best solved by factoring, but not all quadratic expressions factor. Consider $x^2 + x + 1$, for example. Additionally, the square root method only works if the equation can be written in the form $(x + k)^2 = d$.

We now develop a technique that works for all quadratic equations. The process is rather cumbersome, although it is important in a variety of applications you may encounter in subsequent courses. The process is called *completing the square*.

We first observe some useful relationships when a binomial of the form $x + k$ is squared. The result is a perfect square trinomial.

$$(x + k)^2 = x^2 + 2kx + k^2$$

Note the following relationships on the right side of the equation.

1. The coefficient of the squared term, x^2, is one.
2. The coefficient of the linear term, x, is $2k$.
3. The constant term, k^2, is the square of one-half the coefficient of x.

We shall make use of the above relationships to solve quadratic equations by completing the square.

Let us first illustrate how to form perfect square trinomials.

See solved problem 6.3.

Let us now employ what we have learned and the square root method to solve quadratic equations by completing the square.

Completing the Square

Our equation must be in the form $(x + k)^2 = d$ in order for us to solve by the square root method. Hence, our first objective will be to use the above skills to write the equation in the appropriate form. Our next step is to extract roots as was illustrated in Section 6.1. We illustrate on p. 181.

See solved problem 6.4.

The steps involved in solving a quadratic equation by completing the square are summarized below.

1. Isolate the variable terms.
2. Make the coefficient of the squared term one if it is other than one.
3. Determine the square of one-half the coefficient of the linear term and add to both sides.
4. Factor the perfect square trinomial.
5. Extract square roots and simplify if necessary.
6. Isolate the variable.
7. State the solution set. A check is recommended.

Work supplementary problem 6.4 to improve your skills.

Solving quadratic equations by completing the square is rather laborious. We now make use of the technique to develop a more efficient method that reduces some of the effort. The method employs the quadratic formula which we now derive.

The Quadratic Formula

The quadratic formula is developed through solving a quadratic equation by completing the square. We begin with the standard form of the equation $ax^2 + bx + c = 0$.

1. Isolate the variable terms. $ax^2 + bx = -c$

2. Divide both sides by a. $x^2 + \frac{b}{a}x = \frac{-c}{a}$

3. Determine $k^2 = \left[\frac{1}{2}\left(\frac{b}{a}\right)\right]^2 = \frac{b^2}{4a^2}$

   Add to both sides and simplify.
   $$x^2 + \frac{b}{a}x + \frac{b^2}{4a^2} = \frac{-c}{a} + \frac{b^2}{4a^2}$$
   $$= \frac{-c}{a} \cdot \frac{4a}{4a} + \frac{b^2}{4a^2}$$
   $$= \frac{b^2 - 4ac}{4a^2}$$

4. Factor the left side. $\left(x + \frac{b}{2a}\right)^2 = \frac{b^2 - 4ac}{4a^2}$

5. Extract square roots and simplify.
   $$x + \frac{b}{2a} = \pm\sqrt{\frac{b^2 - 4ac}{4a^2}} = \pm\frac{\sqrt{b^2 - 4ac}}{\sqrt{4a^2}}$$
   $$= \pm\frac{\sqrt{b^2 - 4ac}}{2a}$$

6. Isolate the variable. $x = \frac{-b}{2a} \pm \sqrt{\frac{b^2 - 4ac}{2a}} = \frac{-b \pm \sqrt{b^2 - 4ac}}{2a}$

The last equation is called the quadratic formula. We restate the result for reference.

The Quadratic Formula

The solutions to $ax^2 + bx + c = 0, a \neq 0$, are given by

$$x = \frac{-b \pm \sqrt{b^2 - 4ac}}{2a}.$$

The formula expresses the solutions to any quadratic equation which can be written in standard form in terms of the coefficients a, b, and c. We merely substitute the appropriate values into the formula and simplify. The process may be employed to solve *every* quadratic equation that can be written in standard form. The formula is used often. Learn it and remember it!

See solved problem 6.5.

In summary, to solve quadratic equations:

1. Use the factor method if the expression can be factored rather easily.
2. If the first-degree term is missing, employ the square root method.
3. The quadratic formula may be utilized to solve other quadratic equations. It may be employed to solve all quadratic equations, although it is less efficient for the types suggested in steps 1 and 2 above.
4. Completing the square may be used without exception, although it is the most cumbersome and prone to error. Completing the square is an important technique that we need for reasons other than solving equations.

Many applications in the real world involve quadratic equations with coefficients other than integers. The coefficients are often decimal numerals or irrational numbers. We often employ the quadratic formula and a calculator to approximate the solutions in those instances. Work through the problems on p. 183 with your calculator to become familiar with the process.

See solved problem 6.6.

6.3 Equations Involving Radicals

If a given equation contains one or more radical expressions, we shall attempt to eliminate the radicals to solve the equation. The property which facilitates the process is stated below.

Property 1. If $a = b$, then $a^2 = b^2$.

Essentially the property says if two expressions are equal, then their squares are equal. The statement cannot be reversed, however. That is, the converse is not true. We'll make use of the fact that the solutions of the equation $a = b$ are included in the solutions of $a^2 = b^2$. The squared version sometimes has solutions which are not solutions of the original equation. In other words, squaring both sides of an equation sometimes introduces extraneous roots. We are obligated to check for extraneous roots when Property 1 is used; it is part of the solution process. We now illustrate the procedure.

See solved problem 6.7.

Property 1 can be generalized as follows.

Property 2. If $a = b$, then $a^p = b^p$ where p is a rational number.

In other words, rational powers of equal expressions are equal. If the rational exponent is in reduced form and its numerator is even, a check is required as part of the solution process. Extraneous solutions may have been introduced. Property 2 is used to solve equations which contain higher order roots.

See solved problem 6.8.

6.4 Quadratic Form Equations

Some equations can be transformed into a quadratic equation by substitution. We shall let the variable u represent an appropriate expression, then substitute and solve an equation of the form $au^2 + bu + c = 0$. Once values for u are obtained, we will find values for the original variable which satisfy the given equation. Quadratic form equations are encountered in equations in which the variable factor in one term is the square of the variable factor in another term and no other terms contain variable factors. Watch for this type of equation.

See solved problem 6.9.

6.5 Applications

The procedures suggested in Section 1.7 for translating verbal statements into equations should be reviewed at this time. In the current section, the equations obtained will be quadratic. It may be that none, one, or both solutions of the equation are meaningful for the physical situation at hand. The solutions obtained to equations should always be checked in the original statement.

See solved problem 6.10.

6.6 Graphs of Second-Degree Equations

The graph of $y = ax^2 + bx + c, a \neq 0$, is a curve called a *parabola*. Since the curve is not a straight line, we must plot more points in order to establish the pattern than is necessary when graphing first-degree equations. The shape of a parabola will become apparent as we graph the equations on p. 197.

See solved problem 6.11.

We now discuss how we can determine the more significant parts of the graph. The highest or lowest point on the curve can be very important. The highest or lowest point on the curve is called the *maximum* or *minimum point*, respectively. We employ the process of completing the square to find the maximum or minimum point. See the problem on p. 198. Review Section 6.2 before proceeding.

See solved problem 6.12.

Refer to the expression in solved problem 6.12(*a*). Think of the general form $y = ax^2 + bx + c$. In $y = x^2 - 6x + 5, a = 1, b = -6$ and $c = 5$. The *x*-coordinate of the minimum point was $x = 3$. Note that $\frac{-b}{2a} = \frac{-(-6)}{2(1)} = \frac{6}{2} = 3$. The *y*-coordinate of the minimum point was $y = -4$. Observe that $\frac{4ac - b^2}{4a} = \frac{4(1)(5) - (-6)^2}{4(1)} = \frac{20 - 36}{4} = \frac{-16}{4} = -4$. Similar results occur if we apply the same procedures to the expression in solved problem 6.12(*b*). The reader should verify the conclusions. In general,

> The maximum or minimum point on the graph of $y = ax^2 + bx + c, a \neq 0$, occurs at $x = -b/(2a)$. The maximum or minimum point is called the *vertex* of the parabola. Evaluate $y = ax^2 + bx + c$ at $x = -b/(2a)$ to find the *y*-coordinate of the vertex.

A line parallel to the *y*-axis that passes through the vertex divides the parabola into parts that are mirror images of each other. This line is called the *axis of symmetry* of the parabola. The idea of symmetry can be employed when graphing the curve. Recall that the *y*-intercept of a graph is found if we let $x = 0$ and solve for *y*. Similarly if $y = 0$, the value(s) of *x* are the *x*-intercepts. In solved problem 6.12 (*a*), the *y*-intercept is 5 and the *x*-intercepts are 1 and 5. The intercepts of a parabola should usually be found. Sometimes the *x*-intercepts are irrational and therefore cumbersome to determine. We normally omit *x*-intercepts if this is the case.

The ideas discussed above are summarized below. To graph parabolas with the equation $y = ax^2 + bx + c$,

1. If $a > 0$, the parabola opens upward. If $a < 0$, the parabola opens downward.
2. Find the *x*-coordinate of the vertex. It is given by $x = -b/(2a)$.
3. Find the *y*-coordinate of the vertex. Evaluate $y = ax^2 + bx + c$ for $x = -b/(2a)$.
4. Let $x = 0$ and evaluate $y = ax^2 + bx + c$ to determine the *y*-intercept.
5. Let $y = 0$ and solve $y = ax^2 + bx + c$ for *x* to determine the *x*-intercepts, if any. This step is optional if the expression does not factor.
6. Plot the points found and one or two additional points. Use symmetry to complete the graph.

See solved problem 6.13.

If the equation includes a second-degree expression in *y*, that is, $x = ay^2 + by + c$ where $a \neq 0$, the roles of *x* and *y* are interchanged. The curve is a parabola that opens to the right if $a > 0$ and the parabola opens to the left if $a < 0$. The vertex has *y*-coordinate $-b/(2a)$. A line parallel to the *x*-axis that passes through the vertex is the axis of symmetry. The intercepts are found in the usual manner.

The procedure used to graph a parabola of the form $x = ay^2 + by + c$ is analogous to the procedure stated previously. To plot arbitrary points, pick values for *y* and solve for *x*.

See solved problem 6.14.

Graphs on the Calculator

Second-degree equations of the form $y = ax^2 + bx + c, a \neq 0$, can be easily graphed on a graphing calculator. We must first determine an appropriate graphing window to display the graph. We know that the vertex occurs at $x = -b/(2a)$, so set the range of x values so that $-b/(2a)$ is near the middle of the interval. The appropriate range of y values is determined partially by observing whether the parabola opens upward or downward. If the y-coordinate of the vertex is not determined, an acceptable range can be found through trial and error. Enter the expression and graph it after the graphing window has been identified. Graphing windows are not unique. Acceptable windows may vary slightly from those indicated in the problems which follow on p. 202.

See solved problem 6.15.

The zeros or x-intercepts and the vertex are often very useful in applications. The approximate coordinates of these points can be determined using the *trace* function on your graphing calculator. The calculator displays either x-coordinates or y-coordinates (or both) of points on a curve as the cursor is moved along the curve.

The first steps are to graph the expression and then set the calculator to trace the path. The cursor begins at the left of the x-range on many calculators. The zeros occur when $y = 0$, so set the calculator to display the y-coordinates of points and press the direction arrow until y is as close to zero as possible. Move the cursor back and forth through zero until you are satisfied the best value has been obtained. Next display the x-coordinate of the point and record it. The value obtained is an approximate zero of the expression. Display the y-coordinate once more and move the cursor along the parabola until the second zero, if there is one, is obtained.

The approximate coordinates of the vertex can be obtained similarly. Simply display the y-coordinates of points and trace the curve until the maximum or minimum value is obtained. Record the y-value, display the corresponding x-value and record it. These recorded values are the approximate coordinates of the vertex. The results obtained may vary slightly as the graphing window is altered. Remember that results obtained using the above technique are approximate.

The *zoom* feature on a calculator, if it has one, can be used to improve the approximations discussed above. This feature essentially alters the graphing window being utilized. The zoom feature allows us to magnify a part of the graph to obtain more refined estimates of significant coordinates. It should be employed if a high degree of accuracy is required.

Many graphing calculators have built-in algorithms that will find the zeros and maximum and minimum points easily and quickly. Check your calculator manual to find if, and how, your calculator may do this.

See solved problem 6.16.

6.7 Quadratic and Rational Inequalities

Quadratic Inequalities

A quadratic inequality of the form $ax^2 + bx + c < 0$ with $a > 0$ is in *standard form*. (Note that $<$ may be replaced by $>, \leq$, or $\geq$.)

In order to solve the inequality $x^2 - x - 6 < 0$, we must find the values of x for which $x^2 - x - 6$ is negative. We begin by solving the equation $x^2 - x - 6 = 0$. The factor method works well. $x^2 - x - 6 = (x + 2)(x - 3) = 0$ if $x = -2$ or $x = 3$.

The value(s) of the variable for which an expression is either zero or is undefined are called the *critical value(s)* for the expression. Thus, -2 and 3 are the critical values for the expression $x^2 - x - 6$.

The inequality $x^2 - x - 6 < 0$ can be written as $(x + 2)(x - 3) < 0$. The product of the factors is negative when one factor is positive and the other factor is negative. We shall examine the signs of the factors $x + 2$ and $x - 3$ for various arbitrary values in the intervals established by the critical values -2 and 3.

We illustrate the relevant information in Figure 6.1. We call the figure shown in Figure 6.1 a *sign diagram*. The signs of the factors and their product for various values of x in the intervals formed by graphing the critical values on a number line are shown. The value for x (test value) used in a given interval is arbitrary.

(Choose an integer value to test in general.) We also note that a factor has the same sign for all values of the variable in a particular interval. We can determine the solution set by reading the results indicated from the sign diagram.

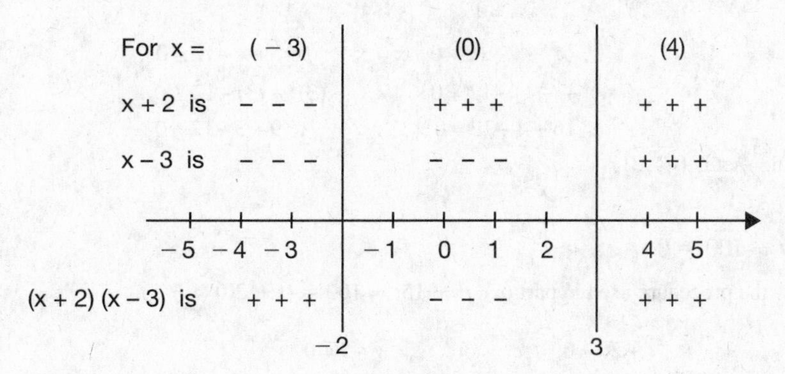

Figure 6.1

In our illustration it is apparent that $x^2 - x - 6 < 0$ for all x in the interval $(-2, 3)$.

We now summarize the procedure employed above.

Procedure for Solving Quadratic Inequalities

1. Write the inequality in standard form.
2. Find the critical values by forming a quadratic equation of the form $ax^2 + bx + c = 0$. Solve by factoring or use the quadratic formula.
3. Construct a sign diagram for the factors by choosing arbitrary test values for the variable in the intervals determined by the critical values. Find the signs of the factors and their product in each interval.
4. Identify and record the solution set for the inequality.

See solved problem 6.17.

Rational Inequalities

We can employ the same method to solve rational inequalities. The inequality must be written in the form $P(x)/Q(x) < 0$ to apply the method. (The $<$ symbol may be replaced by $>, \leq$, or $\geq$.) We illustrate on p. 205.

See solved problem 6.18.

General Inequalities

After the inequality is in standard form and fully factored, the power of the factor which generates a critical value determines the solution behavior for the interval on either side of it. An odd power indicates inclusion of exactly one of the two intervals, while an even power indicates inclusion of both or neither of the intervals.

See solved problem 6.19.

SOLVED PROBLEMS

6.1 Solve the following quadratic equations by the factor method. Check your answers.

(*a*) $x^2 + x - 12 = 0$

Factor the quadratic expression and apply the zero factor property: $x^2 + x - 12 = (x + 4)(x - 3)$. Now solve the resulting linear equations.

$$4 = 0 \quad \text{or} \quad x - 3 = 0$$
$$x = -4 \qquad x = 3$$

Check: $x = -4$;

$$x^2 + x - 12 = 0$$
$$(-4)^2 + (-4) - 12 \overset{?}{=} 0$$
$$16 - 4 - 12 = 0$$

$x = 3$;

$$x^2 + x - 12 = 0$$
$$(3)^2 + (3) - 12 \overset{?}{=} 0$$
$$9 + 3 - 12 = 0$$

The solution set is $\{-4, 3\}$.

(*b*) $t^2 + 15t - 100 = 0$

Follow the procedure used in part (*a*). $t^2 + 15t - 100 = (t + 20)(t - 5)$.

$$t + 20 = 0 \quad \text{or} \quad t - 5 = 0$$
$$t = -20 \qquad t = 5$$

Check:

$t = -20$;

$$t^2 + 15t - 100 = 0$$
$$(-20)^2 + 15(-20) - 100 \overset{?}{=} 0$$
$$400 - 300 - 100 = 0$$

$t = 5$;

$$t^2 + 15t - 100 = 0$$
$$(5)^2 + 15(5) - 100 \overset{?}{=} 0$$
$$25 + 75 - 100 = 0$$

The solution set is $\{-20, 5\}$.

(*c*) $w^2 - 2w + 1 = 0$

$w^2 - 2w + 1 = (w - 1)(w - 1)$

$$w - 1 = 0 \quad \text{or} \quad w - 1 = 0$$
$$w = 1 \qquad w = 1$$

Check: $w = 1$; $w^2 - 2w + 1 = 0$

$$1^2 - 2(1) + 1 \overset{?}{=} 0$$
$$1 - 2 + 1 = 0$$

The equation has only one solution, since the expression has only one factor. The solution set is $\{1\}$. Note the factor $w - 1$ occurs twice; hence, 1 is called a *double root* or a *root of multiplicity two*.

(*d*) $(2r + 7)(3r - 8) = 0$

Apply the zero factor property immediately.

$$2r + 7 = 0 \quad \text{or} \quad 3r - 8 = 0$$
$$2r = -7 \qquad 3r = 8$$
$$r = \frac{-7}{2} \qquad r = \frac{8}{3}$$

Check:

$r = \frac{-7}{2}$;

$$(2r + 7)(3r - 8) = 0$$
$$\left[2\left(\frac{-7}{2}\right) + 7\right]\left[3\left(\frac{-7}{2}\right) - 8\right] \overset{?}{=} 0$$
$$[-7 + 7]\left[\frac{-21}{2} - 8\right] \overset{?}{=} 0$$
$$0\left[\frac{-21}{2} - 8\right] = 0$$

$r = \frac{8}{3}$;

$$(2r + 7)(3r - 8) = 0$$
$$\left[2\left(\frac{8}{3}\right) + 7\right]\left[3\left(\frac{8}{3}\right) - 8\right] \overset{?}{=} 0$$
$$\left[\frac{16}{3} + 7\right][8 - 8] \overset{?}{=} 0$$
$$\left[\frac{16}{3} + 7\right]0 = 0$$

The solution set is $\left\{\frac{-7}{2}, \frac{8}{3}\right\}$.

(*e*) $4y^2 + 2y = 0$

$$4y^2 + 2y = 2y(2y+1)$$

$$2y = 0 \quad \text{or} \quad 2y+1=0$$

$$y = \frac{0}{2} \qquad 2y = -1$$

$$y = 0 \qquad y = \frac{-1}{2}$$

Check:

$y = 0;\quad 4y^2 + 2y = 0$	$y = \frac{-1}{2};\quad 4y^2 + 2y = 0$
$4(0)^2 + 2(0) \stackrel{?}{=} 0$	$4\left(\frac{-1}{2}\right)^2 + 2\left(\frac{-1}{2}\right) \stackrel{?}{=} 0$
$0 + 0 = 0$	$4\left(\frac{1}{4}\right) + 2\left(\frac{-1}{2}\right) \stackrel{?}{=} 0$
	$1 - 1 = 0$

The solution set is $\left\{0, \frac{-1}{2}\right\}$.

(*f*) $2p(p-3) = 5p^2 - 7p$

Distribute and write the equation in standard form first. Then factor.

$$2p^2 - 6p = 5p^2 - 7p$$

$$0 = 3p^2 - p = p(3p-1)$$

$$p = 0 \quad \text{or} \quad 3p - 1 = 0$$

$$3p = 1$$

$$p = \frac{1}{3}$$

Check:

$p = 0;\quad 2p^2 - 6p = 5p^2 - 7p$	$p = \frac{1}{3};\quad 2p^2 - 6p = 5p^2 - 7p$
$2(0)^2 - 6(0) \stackrel{?}{=} 5(0)^2 - 7(0)$	$2\left(\frac{1}{3}\right)^2 - 6\left(\frac{1}{3}\right) \stackrel{?}{=} 5\left(\frac{1}{3}\right)^2 - 7\left(\frac{1}{3}\right)$
$0 - 0 \stackrel{?}{=} 0 - 0$	$2\left(\frac{1}{9}\right) - 2 \stackrel{?}{=} 5\left(\frac{1}{9}\right) - \frac{7}{3}$
$0 = 0$	$\frac{2}{9} - \frac{18}{9} \stackrel{?}{=} \frac{5}{9} - \frac{21}{9}$
	$\frac{-16}{9} = \frac{-16}{9}$

The solution set is $\left\{0, \frac{1}{3}\right\}$.

(*g*) $\frac{2}{x^2} + 1 = \frac{3}{x}$

Multiply both sides by the LCD to clear fractions. Then write the equation in standard form and solve. Check for extraneous solutions.

$$x^2\left(\frac{2}{x^2} + 1\right) = x^2\left(\frac{3}{x}\right)$$

$$2 + x^2 = 3x$$

$$x^2 - 3x + 2 = 0$$

$$(x-1)(x-2) = 0$$

$$x - 1 = 0 \quad \text{or} \quad x - 2 = 0$$

$$x = 1 \qquad x = 2$$

Check :

$x = 1;$

$$\frac{2}{x^2}+1=\frac{3}{x}$$
$$\frac{2}{1^2}+1\stackrel{?}{=}\frac{3}{1}$$
$$2+1\stackrel{?}{=}3$$
$$3=3$$

$x = 2;$

$$\frac{2}{x^2}+1=\frac{3}{x}$$
$$\frac{2}{2^2}+1\stackrel{?}{=}\frac{3}{2}$$
$$\frac{2}{4}+1\stackrel{?}{=}\frac{3}{2}$$
$$\frac{3}{2}=\frac{3}{2}$$

The solution set is $\{1, 2\}$.

(*h*) $\dfrac{5}{t-3}-2=\dfrac{30}{t^2-9}$

Follow the procedure employed in part (*g*): $t^2 - 9 = (t + 3)(t - 3) = \text{LCD}$.

$$\frac{5}{t-3}-2=\frac{30}{t^2-9}$$
$$(t+3)(t-3)\left[\frac{5}{t-3}-2\right]=(t+3)(t-3)\left[\frac{30}{t^2-9}\right]$$
$$5(t+3)-2(t+3)(t-3)=30$$
$$5t+15-2(t^2-9)=30$$
$$5t+15-2t^2+18=30$$
$$0=2t^2-5t-3$$
$$0=(2t+1)(t-3)$$
$$2t+1=0 \quad \text{or} \quad t-3=0$$
$$2t=-1 \qquad t=3$$
$$t=\frac{-1}{2}$$

We must check for extraneous solutions.

Check :

$t = \dfrac{-1}{2};$

$$\frac{5}{t-3}-2=\frac{30}{t^2-9}$$
$$\frac{5}{\frac{-1}{2}-3}-2\stackrel{?}{=}\frac{30}{\left(\frac{-1}{2}\right)^2-9}$$
$$\frac{(5)}{\left(\frac{-1}{2}-3\right)}\cdot\frac{2}{2}-2\stackrel{?}{=}\frac{(30)}{\left(\frac{1}{4}-9\right)}\cdot\frac{4}{4}$$
$$\frac{10}{-1-6}-2\stackrel{?}{=}\frac{120}{1-36}$$
$$\frac{-10}{7}-\frac{14}{7}\stackrel{?}{=}\frac{120}{-35}$$
$$\frac{-24}{7}=\frac{-24}{7}$$

$t = 3;$

$$\frac{5}{t-3}-2=\frac{30}{t^2-9}$$
$$\frac{5}{3-3}-2\stackrel{?}{=}\frac{30}{3^2-9}$$
$$\frac{5}{0}-2\stackrel{?}{=}\frac{30}{0} \text{ is undefined.}$$

$t = 3$ is not a solution. The solution set is $\left\{\dfrac{-1}{2}\right\}$. $t = 3$ is an extraneous root. There is only one solution.

We can check our possible solutions by the calculator using the method we employed for solved problem 4.4(*e*) as shown below for solved problem 6.1 (*h*):

Calculator check for $\frac{5}{t-3} - 2 = \frac{30}{t^2-9}$:

$t = \frac{-1}{2}$; [(-)] [1] [÷] [2] [sto▶] [Alpha] [t]

[5] [÷] [(] [t] [−] [3] [)] [−] [2] [=] ⇒ Result: −3.42857

[30] [÷] [(] [t] [^] [2] [−] [9] [)] [=] ⇒ Result: −3.42857

The results are the same for both sides so $\frac{-1}{2}$ is a true solution. Whereas

$t = 3$; [3] [sto▶] [Alpha] [t]

[5] [÷] [(] [t] [−] [3] [)] [−] [2] [=] ⇒ Result: *undef*

[30] [÷] [(] [t] [^] [2] [−] [9] [)] [=] ⇒ Result: *undef*

Since the result for either side was *undefined* (both sides in this case), 3 is not a solution. It was extraneous. (3 would have also been extraneous if the results for the two sides would have simply been different numbers.)

Practice solving quadratic equations using the factor method by working supplementary problem 6.1.

6.2 Use the square root method to solve the following.

(*a*) $x^2 - 49 = 0$

Solve for x^2 and extract roots.

$$x^2 - 49 = 0$$
$$x^2 = 49$$
$$x = \pm\sqrt{49} = \pm 7$$

The solution set is $\{\pm 7\}$. The results may be checked mentally.

(*b*) $y^2 + 20 = 0$

$$y^2 + 20 = 0$$
$$y^2 = -20$$
$$y = \pm\sqrt{-20} = \pm\sqrt{-1 \cdot 4 \cdot 5} = \pm 2i\sqrt{5}$$

The solution set is $\{\pm 2i\sqrt{5}\}$.

(*c*) $5t^2 + 1 = 37$

$$5t^2 + 1 = 37$$
$$5t^2 = 36$$
$$t^2 = \frac{36}{5}$$
$$t = \pm\sqrt{\frac{36}{5}} \pm \frac{\sqrt{36}}{\sqrt{5}} = \pm\frac{6}{\sqrt{5}} = \pm\frac{6}{\sqrt{5}} \cdot \frac{\sqrt{5}}{\sqrt{5}} = \pm\frac{6\sqrt{5}}{5}$$

The solution set is $\left\{\pm\frac{6\sqrt{5}}{5}\right\}$.

(*d*) $(2w - 3)^2 = 7$

$$(2w-3)^2 = 7$$
$$2w - 3 = \pm\sqrt{7}$$
$$2w = 3 \pm \sqrt{7}$$
$$w = \frac{3 \pm \sqrt{7}}{2}; \text{ that is, } w = \frac{3+\sqrt{7}}{2} \text{ or } w = \frac{3-\sqrt{7}}{2}$$

The solution set is $\left\{\frac{3 \pm \sqrt{7}}{2}\right\}$.

(*e*) $(5x + 1)^2 + 13 = 0$

$$(5x-1)^2+13=0$$
$$(5x+1)^2=-13$$
$$5x+1=\pm\sqrt{-13}=\pm i\sqrt{13}$$
$$5x=-1\pm i\sqrt{13}$$
$$x=\frac{-1\pm i\sqrt{13}}{5}$$

Check: $x=\dfrac{-1+i\sqrt{13}}{5}$; $\quad (5x+1)^2+13=0$

$$\left[5\left(\frac{-1+i\sqrt{13}}{5}\right)+1\right]^2+13\overset{?}{=}0$$
$$[-1+i\sqrt{13}+1]^2+13\overset{?}{=}0$$
$$\left[i\sqrt{13}\right]^2+13\overset{?}{=}0$$
$$i^2(\sqrt{13})^2+13\overset{?}{=}0$$
$$-1\cdot 13+13=0$$

$x=\dfrac{-1-i\sqrt{13}}{5}$; $\quad (5x+1)^2+13=0$

$$\left[5\left(\frac{-1-i\sqrt{13}}{5}\right)+1\right]^2+13\overset{?}{=}0$$
$$\left[-1-i\sqrt{13}+1\right]^2+13\overset{?}{=}0$$
$$\left[-i\sqrt{13}\right]^2+13\overset{?}{=}0$$
$$(-i)^2(\sqrt{13})^2+13\overset{?}{=}0$$
$$-1\cdot 13+13=0$$

The solution set is $\left\{\dfrac{-1\pm i\sqrt{13}}{5}\right\}$.

See supplementary problem 6.2 for similar exercises.

6.3 Determine the number k such that if k^2 is added to the given expression, the result is a perfect square trinomial. Write the result as the square of a binomial.

(*a*) $x^2 + 8x$

The coefficient of x is $8 = 2k$, so $k = \dfrac{8}{2} = 4$. Add $k^2 = 4^2 = 16$.

$$x^2 + 8x + 4^2 = x^2 + 8x + 16 = (x + 4)^2$$

(*b*) $x^2 - 8x$

The coefficient of x is $-8 = 2k$, so $k = \dfrac{-8}{2} = -4$. Add $k^2 = (-4)^2 = 16$.

$$x^2 - 8x + (-4)^2 = x^2 - 8x + 16 = (x - 4)^2$$

(*c*) $t^2 + 20t$

The coefficient of t is $20 = 2k$, so $k = \dfrac{20}{2} = 10$. Add $k^2 = 10^2 = 100$.

$$t^2 + 20t + 10^2 = t^2 + 20t + 100 = (t + 10)^2$$

(*d*) $w^2 - 11w$

The coefficient of w is $-11 = 2k$, so $k = \frac{-11}{2}$. Add $k^2 = \left(\frac{-11}{2}\right)^2 = \frac{121}{4}$

$$w^2 - 11w + \left(\frac{-11}{2}\right)^2 = w^2 - 11w + \frac{121}{4} = \left(w - \frac{11}{2}\right)^2$$

See supplementary problem 6.3.

6.4 Solve by completing the square.

(*a*) $x^2 + 10x - 3 = 0$

Add 3 to both sides. $x^2 + 10x = 3$

Determine $k = \frac{10}{2} = 5$

Square $k = k^2 = 5^2 = 25$

Add $k^2 = 25$ to both sides. $x^2 + 10x + 25 = 3 + 25$

Factor the left side. $(x + 5)^2 = 28$

Take square roots and simplify. $x + 5 = \pm\sqrt{28} = \pm\sqrt{4 \cdot 7} = \pm 2\sqrt{7}$

Solve for x. $x = -5 \pm 2\sqrt{7}$

The solution set is $\{-5 \pm 2\sqrt{7}\}$.

(*b*) $t^2 - 7t + 2 = 0$

Subtract 2 from both sides. $t^2 - 7t = -2$

Determine $k = \frac{-7}{2}$

Square $k = k^2 = \left(\frac{-7}{2}\right)^2 = \frac{49}{4}$

Add $\frac{49}{4}$ to both sides. $t^2 - 7t + \frac{49}{4} = -2 + \frac{49}{4} = \frac{-8 + 49}{4} = \frac{41}{4}$

Factor the left side. $\left(t - \frac{7}{2}\right)^2 = \frac{41}{4}$

Take square roots and simplify. $t - \frac{7}{2} = \pm\sqrt{\frac{41}{4}} = \frac{\sqrt{41}}{\sqrt{4}} = \pm\frac{\sqrt{41}}{2}$

Solve for t. $t = \frac{7}{2} \pm \frac{\sqrt{41}}{2} = \frac{7 \pm \sqrt{41}}{2}$

The solution set is $\left\{\frac{7 \pm \sqrt{41}}{2}\right\}$.

(*c*) $56y^2 - 3y = 4$

The coefficient of y^2 is not one, so begin by dividing both sides by six. Then proceed as in parts (*a*) and (*b*).

$$\frac{6y^2 - 3y}{6} = \frac{4}{6} \text{ or } y^2 - \frac{y}{2} = \frac{2}{3}$$

Determine $k = \frac{1}{2}\left(\frac{-1}{2}\right) = \frac{-1}{4}$

Therefore, $k^2 = \left(\frac{-1}{4}\right)^2 = \frac{1}{16}$

Add $\frac{1}{16}$ to both sides. $\quad y^2 - \frac{y}{2} + \frac{1}{16} = \frac{2}{3} + \frac{1}{16} = \frac{35}{48}$

Factor the left side. $\quad \left(y - \frac{1}{4}\right)^2 = \frac{35}{48}$

Take square roots and simplify. $\quad y - \frac{1}{4} = \pm\sqrt{\frac{35}{48}} = \pm\frac{\sqrt{35}}{\sqrt{48}} = \pm\frac{\sqrt{35}}{\sqrt{16\cdot 3}}$

$$= \pm\frac{\sqrt{35}}{\sqrt{16}\sqrt{3}} = \pm\frac{\sqrt{35}}{4\sqrt{3}}$$

Solve for y. $\quad y = \frac{1}{4} \pm \frac{\sqrt{35}}{4\sqrt{3}} = \frac{1}{4} \pm \frac{\sqrt{35}}{4\sqrt{3}}\cdot\frac{\sqrt{3}}{\sqrt{3}}$

$$= \frac{1}{4} \pm \frac{\sqrt{105}}{12} = \frac{3 \pm \sqrt{105}}{12}$$

The solution set is $\left\{\frac{3 \pm \sqrt{105}}{12}\right\}$.

6.5 Use the quadratic formula to solve the following.

(*a*) $x^2 - 3x + 1 = 0$

The equation is in the appropriate form. Therefore, the coefficients can readily be identified: $a = 1, b = -3$ and $c = 1$. Substitute the values into the formula and simplify.

$$x = \frac{-b \pm \sqrt{b^2 - 4ac}}{2a} = \frac{-(-3) \pm \sqrt{(-3)^2 - 4(1)(1)}}{2(1)} = \frac{3 \pm \sqrt{9-4}}{2} = \frac{3 \pm \sqrt{5}}{2}$$

The solution set is $\left\{\frac{3 \pm \sqrt{5}}{2}\right\}$.

(*b*) $3x^2 + 4x - 5 = 0$

The equation is in the appropriate form. Therefore, the coefficients can readily be identified; $a = 3, b = 4$ and $c = -5$. Substitute the values into the formula and simplify.

$$x = \frac{-b \pm \sqrt{b^2 - 4ac}}{2a} = \frac{-4 \pm \sqrt{4^2 - 4(3)(-5)}}{2(3)} = \frac{-4 \pm \sqrt{16+60}}{6} = \frac{-4 \pm \sqrt{76}}{6}$$

$$= \frac{-4 \pm \sqrt{4\cdot 19}}{6} = \frac{-4 \pm \sqrt{4}\sqrt{19}}{6} = \frac{-4 \pm 2\sqrt{19}}{6} = \frac{2\left(-2 \pm \sqrt{19}\right)}{2\cdot 3} = \frac{-2 \pm \sqrt{19}}{3}$$

The solution set is $\left\{\frac{-2 \pm \sqrt{19}}{3}\right\}$.

(*c*) $3t^2 = t - 2$

We must first write the equation in the standard form $3t^2 - t + 2 = 0$. Now $a = 3, b = -1$ and $c = 2$. Substitute into the formula and simplify.

$$x = \frac{-b \pm \sqrt{b^2 - 4ac}}{2a} = \frac{1 \pm \sqrt{(-1)^2 - 4(3)(2)}}{2(3)} = \frac{1 \pm \sqrt{1-24}}{6} = \frac{1 \pm \sqrt{-23}}{6} = \frac{1 \pm i\sqrt{23}}{6}$$

The solution set is $\left\{\frac{1 \pm i\sqrt{23}}{6}\right\}$.

(*d*) $w^2 + 3 = -8w$

Proceed as in part (*c*). The standard form is $w^2 + 8w + 3 = 0$. $a = 1, b = 8$ and $c = 3$.

$$x = \frac{-b \pm \sqrt{b^2 - 4ac}}{2a} = \frac{-8 \pm \sqrt{8^2 - 4(1)(3)}}{2(1)} = \frac{-8 \pm \sqrt{64 - 12}}{2} = \frac{-8 \pm \sqrt{-52}}{2} = \frac{-8 \pm \sqrt{4 \cdot 13}}{2}$$

$$= \frac{-8 \pm \sqrt{4}\sqrt{13}}{2} = \frac{-8 \pm 2\sqrt{13}}{2} = \frac{2\left(-4 \pm \sqrt{13}\right)}{2} = -4 \pm \sqrt{13}$$

The solution set is $\{-4 \pm \sqrt{13}\}$.

Work supplementary problem 6.5 for practice.

6.6 Use your calculator to solve the following. Verify the intermediate values.

(*a*) $2.36x^2 + 4.02x + 1.18 = 0$

Use the quadratic formula; $a = 2.36$, $b = 4.02$, and $c = 1.18$.

$$x = \frac{-b \pm \sqrt{b^2 - 4ac}}{2a} = \frac{-4.02 \pm \sqrt{4.02^2 - 4(2.36)(1.18)}}{2(2.36)}$$

$$= \frac{-4.02 \pm \sqrt{16.1604 - 11.1392}}{4.7200} = \frac{-4.02 \pm \sqrt{5.0212}}{4.7200} = \frac{-4.02 \pm 2.2408}{4.7200}$$

$$x = \frac{-4.02 + 2.2408}{4.7200} = \frac{-1.7792}{4.7200} \approx -0.3769 \approx -0.38 \quad \text{or}$$

$$x = \frac{-4.02 - 2.2408}{4.7200} = \frac{-6.2608}{4.7200} \approx -1.3264 \approx -1.33$$

The solution set accurate to two decimal places is $\{-0.38, -1.33\}$.

(*b*) $1.18x^2 - 2.35x + 4.03 = 0$

$$a = 1.18, b = -2.35 \text{ and } c = 4.03$$

$$x = \frac{-b \pm \sqrt{b^2 - 4ac}}{2a} = \frac{-(-2.35) \pm \sqrt{(-2.35)^2 - 4(1.18)(4.03)}}{2(1.18)} = \frac{2.35 \pm \sqrt{5.5225 - 19.0216}}{2.3600}$$

$$= \frac{2.35 \pm \sqrt{-13.4991}}{2.3600} = \frac{2.35 \pm i\sqrt{13.4991}}{2.3600} = \frac{2.35 \pm 3.6741i}{2.3600}$$

$$x \approx 0.9958 \pm 1.5568i \approx 1.00 \pm 1.56i$$

The solution set is $\{1.00 \pm 1.56i\}$.

You were asked to verify the intermediate values in the above problems. Refer to your owner's manual if the values obtained by you were different. Normally it is unnecessary to record the intermediate values in the calculations. They were included for your reference.

The given equations contained coefficients accurate to two decimal places. The answers given are approximate because of round-off, and should be expressed accurate to two decimal places also. In order to accomplish two decimal place accuracy, the intermediate values are given to four decimal place accuracy.

Try supplementary problem 6.6 for additional drill.

6.7 Solve the following. Check for extraneous roots.

(*a*) $\sqrt{3x} = 9$

Apply Property 1 and solve the resulting equation.

$$\sqrt{3x} = 9$$

$$\left(\sqrt{3x}\right)^2 = 9^2$$

$$3x = 81$$

$$x = \frac{81}{3} = 27$$

Check:

$$x = 27; \qquad \sqrt{3x} = 9$$
$$\sqrt{3(27)} \stackrel{?}{=} 9$$
$$\sqrt{81} \stackrel{?}{=} 9$$
$$9 = 9$$

The solution set is $\{27\}$.

(*b*) $\sqrt{-5x} = -15$

Apply Property 1 and solve the resulting equation.

$$\left(\sqrt{-5x}\right)^2 = (-15)^2$$
$$-5x = 225$$
$$x = \frac{225}{-5} = -45$$

Check:

$$x = -45; \qquad \sqrt{-5x} = -15$$
$$\sqrt{-5(-45)} \stackrel{?}{=} -15$$
$$\sqrt{225} \stackrel{?}{=} -15$$
$$15 \neq -15$$

The solution set is { } = Ø. There is no solution.

(*c*) $\sqrt{2t+1} = 5$

Apply Property 1 and solve.

$$\sqrt{2t+1} = 5$$
$$\left(\sqrt{2t+1}\right)^2 = 5^2$$
$$2t + 1 = 25$$
$$2t = 24$$
$$t = \frac{24}{2} = 12$$

Check :

$$t = 12; \qquad \sqrt{2t+1} = 5$$
$$\sqrt{2(12)+1} \stackrel{?}{=} 5$$
$$\sqrt{24+1} \stackrel{?}{=} 5$$
$$\sqrt{24} \stackrel{?}{=} 5$$
$$5 = 5$$

The solution set is $\{12\}$.

(*d*) $(x - 10)^{\frac{1}{2}} = 7$

Recall that the one-half power of an expression represents the square root of the expression. Apply Property 1 and solve.

$$(x-10)^{\frac{1}{2}} = 7$$
$$\left[(x-10)^{\frac{1}{2}}\right]^2 = 7^2$$
$$x - 10 = 49$$
$$x = 59$$

Check :

$$x = 59; \quad (x-10)^{\frac{1}{2}} = 7$$
$$(59-10)^{\frac{1}{2}} \stackrel{?}{=} 7$$
$$(49)^{\frac{1}{2}} \stackrel{?}{=} 7$$
$$7 = 7$$

The solution set is $\{59\}$.

(*e*) $4\sqrt{2y-5} - 3 = 2$

Isolate the radical term, then apply Property 1 and solve.

$$4\sqrt{2y-5} - 3 = 2$$
$$4\sqrt{2y-5} = 5$$
$$\left(4\sqrt{2y-5}\right)^2 = 5^2$$
$$16(2y-5) = 25$$
$$32y - 80 = 25$$
$$32y = 105$$
$$y = \frac{105}{32}$$

Check :

$$y = \frac{105}{32}; \quad 4\sqrt{2y-5} - 3 = 2$$
$$4\sqrt{2\left(\frac{105}{32}\right) - 5} - 3 \stackrel{?}{=} 2$$
$$4\sqrt{\frac{105}{16} - 5} - 3 \stackrel{?}{=} 2$$
$$4\sqrt{\frac{105-5(16)}{16}} - 3 \stackrel{?}{=} 2$$
$$4\sqrt{\frac{105-80}{16}} - 3 \stackrel{?}{=} 2$$
$$4\sqrt{\frac{25}{16}} - 3 \stackrel{?}{=} 2$$
$$4\left(\frac{5}{4}\right) - 3 \stackrel{?}{=} 2$$
$$2 = 2$$

The solution set is $\left\{\frac{105}{32}\right\}$.

(*f*) $\sqrt{3k} - \sqrt{4k-5} = 0$

Solution 1: Proceed as before.

$$\sqrt{3k} - \sqrt{4k-5} = 0$$
$$\left(\sqrt{3k} - \sqrt{4k-5}\right)^2 = 0^2$$
$$3k - 2\sqrt{3k}\sqrt{4k-5} + 4k - 5 = 0$$

Radicals remain, what now? Try a different approach.

Solution 2:

$$\sqrt{3k} - \sqrt{4k-5} = 0$$

Add $\sqrt{4k-5}$ to both sides to obtain

$$\sqrt{3k} = \sqrt{4k-5}$$

Now apply Property 1 and solve.

$$\sqrt{3k} = \sqrt{4k-5}$$
$$\left(\sqrt{3k}\right)^2 = \left(\sqrt{4k-5}\right)^2$$
$$3k = 4k - 5$$
$$5 = k$$

Check:

$k = 5;$

$$\sqrt{3k} - \sqrt{4k-5} = 0$$
$$\sqrt{3(5)} - \sqrt{4(5)-5} \stackrel{?}{=} 0$$
$$\sqrt{15} - \sqrt{20-5} \stackrel{?}{=} 0$$
$$\sqrt{15} - \sqrt{15} \stackrel{?}{=} 0$$
$$0 = 0$$

The solution set is $\{5\}$.

(g) $\sqrt{2t+9} - t = 3$

Isolate the radical term, then square and solve.

$$\sqrt{2t+9} - t = 3$$
$$\sqrt{2t+9} = t + 3$$
$$\left(\sqrt{2t+9}\right)^2 = (t+3)^2$$
$$2t + 9 = t^2 + 6t + 9$$
$$0 = t^2 + 4t$$
$$0 = t(t+4)$$

$$t = 0 \quad \text{or} \quad t + 4 = 0$$
$$t = -4$$

Check:

$t = 0;$

$$\sqrt{2t+9} - t = 3$$
$$\sqrt{2(0)+9} - 0 \stackrel{?}{=} 3$$
$$\sqrt{9} - 0 \stackrel{?}{=} 3$$
$$3 - 0 \stackrel{?}{=} 3$$
$$3 = 3$$

$t = -4;$

$$\sqrt{2t+9} - t = 3$$
$$\sqrt{2(-4)+9} - (-4) \stackrel{?}{=} 3$$
$$\sqrt{-8+9} + 4 \stackrel{?}{=} 3$$
$$\sqrt{1} + 4 \stackrel{?}{=} 3$$
$$1 + 4 \stackrel{?}{=} 3$$
$$5 \neq 3$$

$t = -4$ is an extraneous root. The solution set is $\{0\}$.

(*h*) $\sqrt{w+\sqrt{w+2}}=2$

Apply Property 1 twice, isolating the radical each time.

$$\sqrt{w+\sqrt{w+2}}=2$$
$$\left(\sqrt{w+\sqrt{w+2}}\right)^2=2^2$$
$$w+\sqrt{w+2}=4$$
$$\sqrt{w+2}=4-w$$
$$\left(\sqrt{w+2}\right)^2=(4-w)^2$$
$$w+2=16-8w+w^2$$
$$0=14-9w+w^2$$
$$0=(w-7)(w-2)$$

$$w-7=0 \quad \text{or} \quad w-2=0$$
$$w=7 \qquad\qquad w=2$$

Check:

$w=7;$	$\sqrt{w+\sqrt{w+2}}=2$	$w=2;$	$\sqrt{w+\sqrt{w+2}}=2$
	$\sqrt{7+\sqrt{7+2}}\stackrel{?}{=}2$		$\sqrt{2+\sqrt{2+2}}\stackrel{?}{=}2$
	$\sqrt{7+\sqrt{9}}\stackrel{?}{=}2$		$\sqrt{2+\sqrt{4}}\stackrel{?}{=}2$
	$\sqrt{7+3}\stackrel{?}{=}2$		$\sqrt{2+2}\stackrel{?}{=}2$
	$\sqrt{10}\neq 2$		$\sqrt{4}\stackrel{?}{=}2$
			$2=2$

$w=7$ is an extraneous root. The solution set is {2}.

(*i*) $\sqrt{x+3}=3+\sqrt{2-x}$

Apply Property 1 twice.

$$\sqrt{x+3}=3+\sqrt{2-x}$$
$$\left(\sqrt{x+3}\right)^2=\left(3+\sqrt{2-x}\right)^2$$
$$x+3=9+6\sqrt{2-x}+2-x=11-x+6\sqrt{2-x}$$
$$2x-8=6\sqrt{2-x}$$
$$(2x-8)^2=\left(6\sqrt{2-x}\right)^2$$
$$4x^2-32x+64=72-36x$$
$$4x^2+4x-8=0$$
$$4(x+2)(x-1)=0$$

$$x+2=0 \quad \text{or} \quad x-1=0$$
$$x=-2 \qquad\qquad x=1$$

Check:

$x=-2;$	$\sqrt{x+3}=3+\sqrt{2-x}$	$x=1;$	$\sqrt{x+3}=3+\sqrt{2-x}$
	$\sqrt{(-2)+3}\stackrel{?}{=}3+\sqrt{2-(-2)}$		$\sqrt{(1)+3}\stackrel{?}{=}3+\sqrt{2-(1)}$
	$1\neq 5$		$2\neq 4$

$x=-2$ and $x=1$ are extraneous roots. The solution set is $\{\}=\varnothing$. There is no solution.

Reflect on the procedure used in the above problems. Property 1 was applied in each case after a radical term was isolated. The result was then an equation we could solve. The squaring step sometimes introduces extraneous solutions. We must check all potential solutions in the original equation prior to specifying the solution set.

Improve your skills by working supplementary problem 6.7.

6.8 Solve. Remember to check for extraneous roots.

(*a*) $\sqrt[3]{t} = 4$

Apply Property 2. Cubing undoes cube root.

$$\sqrt[3]{t} = 4$$
$$\left(\sqrt[3]{t}\right)^3 = 4^3$$
$$t = 64$$

Check:

$$t = 64; \quad \sqrt[3]{t} = 4$$
$$\sqrt[3]{64} \stackrel{?}{=} 4$$
$$4 = 4$$

The solution set is $\{64\}$.

(*b*) $\sqrt[3]{4 - x} = -2$

Proceed as in part (*a*).

$$\sqrt[3]{4-x} = -2$$
$$\left(\sqrt[3]{4-x}\right)^3 = (-2)^3$$
$$4 - x = -8$$
$$12 = x$$

Check:

$$x = 12; \quad \sqrt[3]{4-x} = -2$$
$$\sqrt[3]{4-12} \stackrel{?}{=} \sqrt[3]{-8}$$
$$\sqrt[3]{-8} \stackrel{?}{=} -2$$
$$-2 = -2$$

The solution set is $\{12\}$.

(*c*) $(y + 5)^{\frac{1}{4}} = -1$

The one-fourth power represents the fourth root of the expression. Apply Property 2.

$$(y+5)^{\frac{1}{4}} = -1$$
$$\left[(y+5)^{\frac{1}{4}}\right]^4 = (-1)^4$$
$$y + 5 = 1$$
$$y = -4$$

Check:

$$y = -4; \quad (y+5)^{\frac{1}{4}} = -1$$
$$(-4+5)^{\frac{1}{4}} \stackrel{?}{=} -1$$
$$(1)^{\frac{1}{4}} \stackrel{?}{=} -1$$
$$1 \neq -1$$

$y = -4$ is an extraneous solution. There is no solution. The solution set is $\{\ \} = \emptyset$.

(*d*) $\sqrt[4]{w-3}+7=10$

Isolate the radical term first.

$$\sqrt[4]{w-3}+7=10$$
$$\sqrt[4]{w-3}=3$$
$$\left(\sqrt[4]{w-3}\right)^4=3^4$$
$$w-3=81$$
$$w=84$$

Check:

$$w=84;\quad \sqrt[4]{w-3}+7=10$$
$$\sqrt[4]{84-3}+7\overset{?}{=}10$$
$$\sqrt[4]{81}+7\overset{?}{=}10$$
$$3+7\overset{?}{=}10$$
$$10=10$$

The solution set is $\{84\}$.

(*e*) $t^{\frac{-1}{3}}=5$

The variable t will be isolated if we raise both sides to the -3 power.

$$t^{\frac{-1}{3}}=5$$
$$\left(t^{\frac{-1}{3}}\right)^{-3}=5^{-3}$$
$$t=\frac{1}{5^3}=\frac{1}{125}$$

Check:

$$t=\frac{1}{125};\quad t^{\frac{-1}{3}}=5$$
$$\left(\frac{1}{125}\right)^{\frac{-1}{3}}\overset{?}{=}5$$
$$\left[(125)^{-1}\right]^{\frac{-1}{3}}\overset{?}{=}5$$
$$125^{\frac{1}{3}}\overset{?}{=}5$$
$$5=5$$

The solution set is $\left\{\frac{1}{125}\right\}$.

(*f*) $p^{\frac{3}{2}}+20=12$

$$p^{\frac{3}{2}}+20=12$$
$$p^{\frac{3}{2}}=-8$$
$$\left(p^{\frac{3}{2}}\right)^{\frac{2}{3}}=(-8)^{\frac{2}{3}}$$
$$p=\left(\sqrt[3]{-8}\right)^2=(-2)^2=4$$

Check:

$$p = 4; \qquad p^{\frac{3}{2}} + 20 = 12$$
$$(4)^{\frac{3}{2}} + 20 \stackrel{?}{=} 12$$
$$\left(\sqrt{4}\right)^3 + 20 \stackrel{?}{=} 12$$
$$2^3 + 20 \stackrel{?}{=} 12$$
$$8 + 20 \stackrel{?}{=} 12$$
$$28 \neq 12$$

Isolate $p^{\frac{3}{2}}$ and then raise each side to the $\frac{2}{3}$ power.

$p = 4$ is an extraneous root. There is no solution. The solution set is $\{\ \} = \emptyset$.

(*g*) $\sqrt[7]{z+4} + 5 = 6$

Isolate the radical term first.

$$\sqrt[7]{z+4} + 5 = 6$$
$$\sqrt[7]{z+4} = 1$$
$$\left(\sqrt[7]{z+4}\right)^7 = 1^7$$
$$z + 4 = 1$$
$$z = -3$$

Check:

$$z = -3; \qquad \sqrt[7]{z+4} + 5 = 6$$
$$\sqrt[7]{-3+4} + 5 \stackrel{?}{=} 6$$
$$\sqrt[7]{1} + 5 \stackrel{?}{=} 6$$
$$1 + 5 \stackrel{?}{=} 6$$
$$6 = 6$$

The solution set is $\{-3\}$.

See supplementary problem 6.8 for similar problems.

6.9 Solve. Check for extraneous roots when necessary.

(*a*) $y^4 - 9y^2 + 20 = 0$

Since y^4 is the square of y^2, let $u = y^2$ so that $u^2 = y^4$. Now substitute to obtain

$$u^2 - 9u + 20 = 0$$
$$(u - 5)(u - 4) = 0$$
$$u = 5 \quad \text{or} \quad u = 4$$

Since $u = y^2$,

$$y^2 = 5 \quad \text{or} \quad y^2 = 4$$
$$y = \pm\sqrt{5} \qquad\qquad y = \pm 2$$

The solution set is $\{\pm\sqrt{5}, \pm 2\}$.

(*b*) $t^6 + 15 = 8t^3$

$t^6 = (t^3)^2$ so let $u = t^3$, so that $u^2 = t^6$. Now substitute and rearrange.

$$u^2 - 8u + 15 = 0$$
$$(u - 3)(u - 5) = 0$$
$$u = 3 \quad \text{or} \quad u = 5$$

Since $u = t^3$,

$$t^3 = 3 \quad \text{or} \quad t^3 = 5$$
$$t = \sqrt[3]{3} \qquad\qquad t = \sqrt[3]{5}$$

The solution set is $\{\sqrt[3]{3}, \sqrt[3]{5}\}$.

(*c*) $x + 13x^{\frac{1}{2}} + 36 = 0$

$x = \left(x^{\frac{1}{2}}\right)^2$ so if $u = x^{\frac{1}{2}}$, then $u^2 = x$.

$$u^2 + 13u + 36 = 0$$
$$(u+9)(u+4) = 0$$

$$u = -9 \quad \text{or} \quad u = -4$$

Since $u = x^{\frac{1}{2}}$,

$$x^{\frac{1}{2}} = -9 \quad \text{or} \quad x^{\frac{1}{2}} = -4$$
$$x = 81 \qquad\qquad x = 16$$

A check is mandatory this time.

Check:

$x = 81$;	$x + 13x^{\frac{1}{2}} + 36 = 0$	$x = 16$;	$x + 13x^{\frac{1}{2}} + 36 = 0$
	$81 + 13(18)^{\frac{1}{2}} + 36 \stackrel{?}{=} 0$		$16 + 13(16)^{\frac{1}{2}} + 36 \stackrel{?}{=} 0$
	$81 + 13(9) + 36 \stackrel{?}{=} 0$		$16 + 13(4) + 36 \stackrel{?}{=} 0$
	$234 \neq 0$		$104 \neq 0$

Both potential solutions are extraneous. There is no solution. The solution set is { } =Ø.

(*d*) $2w^4 + 5w^2 - 12 = 0$

Let $u = w^2$, then $u^2 = w^4$.

$$2u^2 + 5u - 12 = 0$$
$$(2u-3)(u+4) = 0$$

$$u = \frac{3}{2} \quad \text{or} \quad u = -4$$

Since $u = w^2$,

$$w^2 = \frac{3}{2} \quad \text{or} \quad w^2 = -4$$

$$w = \pm\sqrt{\frac{3}{2}} = \pm\frac{\sqrt{3}}{\sqrt{2}} \cdot \frac{\sqrt{2}}{\sqrt{2}} = \pm\frac{\sqrt{6}}{2} \qquad w = \pm\sqrt{-4} = \pm i\sqrt{4} = \pm 2i$$

The solution set is $\left\{\pm\frac{\sqrt{6}}{2}, \pm 2i\right\}$.

(*e*) $s^{\frac{2}{3}} - 2s^{\frac{1}{3}} - 8 = 0$

Let $u = s^{\frac{1}{3}}$, then $u^2 = s^{\frac{2}{3}}$.

$$u^2 - 2u - 8 = 0$$
$$(u-4)(u+2) = 0$$

$$u = 4 \quad \text{or} \quad u = -2$$

Since $u = s^{\frac{1}{3}}$,

$$s^{\frac{1}{3}} = 4 \quad \text{or} \quad s^{\frac{1}{3}} = -2$$
$$s = 4^3 = 64 \qquad s = (-2)^3 = -8$$

A check is required.

$$s = 64; \quad s^{\frac{2}{3}} - 2s^{\frac{1}{3}} - 8 = 0 \qquad\qquad s = -8; \quad s^{\frac{2}{3}} - 2s^{\frac{1}{3}} - 8 = 0$$

$$64^{\frac{2}{3}} - 2(64)^{\frac{1}{3}} - 8 \stackrel{?}{=} 0 \qquad\qquad (-8)^{\frac{2}{3}} - 2(-8)^{\frac{1}{3}} - 8 \stackrel{?}{=} 0$$

$$\left(64^{\frac{1}{3}}\right)^2 - 2(64)^{\frac{1}{3}} - 8 \stackrel{?}{=} 0 \qquad\qquad \left[(-8)^{\frac{1}{3}}\right]^2 - 2(-8)^{\frac{1}{3}} - 8 \stackrel{?}{=} 0$$

$$4^2 - 2(4) - 8 \stackrel{?}{=} 0 \qquad\qquad (-2)^2 - 2(-2) - 8 \stackrel{?}{=} 0$$

$$16 - 8 - 8 \stackrel{?}{=} 0 \qquad\qquad 4 + 4 - 8 \stackrel{?}{=} 0$$

$$0 = 0 \qquad\qquad 0 = 0$$

The solution set is $\{-8, 64\}$.

(*f*) $x^{-2} - x^{-1} - 20 = 0$

Let $u = x^{-1}$, then $u^2 = x^{-2}$.

$$u^2 - u - 20 = 0$$

$$(u + 4)(u - 5) = 0$$

$$u = -4 \quad \text{or} \quad u = 5$$

Since $u = x^{-1}$,

$$x^{-1} = -4 \quad \text{or} \quad x^{-1} = 5$$

$$x = \frac{-1}{4} \qquad\qquad x = \frac{1}{5}$$

The solution set is $\left\{\frac{-1}{4}, \frac{1}{5}\right\}$.

(*g*) $2t^{\frac{2}{5}} + 7t^{\frac{1}{5}} = -3$

Let $u = t^{\frac{1}{5}}$, then $u^2 = t^{\frac{2}{5}}$.

$$2u^2 + 7u = -3$$

$$2u^2 + 7u + 3 = 0$$

$$(2u + 1)(u + 3) = 0$$

$$u = \frac{-1}{2} \quad \text{or} \quad u = -3$$

Since $u = t^{\frac{1}{5}}$,

$$t^{\frac{1}{5}} = \frac{-1}{2} \quad \text{or} \quad t^{\frac{1}{5}} = -3$$

$$t = \left(\frac{-1}{2}\right)^5 = \frac{-1}{32} \qquad\qquad t = (-3)^5 = -243$$

A check is necessary.

$$t = \frac{-1}{32}; \quad 2t^{\frac{2}{5}} + 7t^{\frac{1}{5}} = -3 \qquad\qquad t = -243; \quad 2t^{\frac{2}{5}} + 7t^{\frac{1}{5}} = -3$$

$$2\left(\frac{-1}{32}\right)^{\frac{2}{5}} + 7\left(\frac{-1}{32}\right)^{\frac{1}{5}} \stackrel{?}{=} -3 \qquad\qquad 2(-243)^{\frac{2}{5}} + 7(-243)^{\frac{1}{5}} \stackrel{?}{=} -3$$

$$2\left[\left(\frac{-1}{32}\right)^{\frac{1}{5}}\right]^2 + 7\left(\frac{-1}{32}\right)^{\frac{1}{5}} \stackrel{?}{=} -3 \qquad\qquad 2\left[(-243)^{\frac{1}{5}}\right]^2 + 7(-243)^{\frac{1}{5}} \stackrel{?}{=} -3$$

$$2\left[\frac{-1}{2}\right]^2 + 7\left(\frac{-1}{2}\right) \stackrel{?}{=} -3 \qquad\qquad 2(-3)^2 + 7(-3) \stackrel{?}{=} -3$$

$$2\left(\frac{1}{4}\right) - \frac{7}{2} \stackrel{?}{=} -3 \qquad\qquad 2(9) - 21 \stackrel{?}{=} -3$$

$$\frac{1}{2} - \frac{7}{2} \stackrel{?}{=} -3 \qquad\qquad 18 - 21 \stackrel{?}{=} -3$$

$$-3 = -3 \qquad\qquad -3 = -3$$

The solution set is $\left\{\frac{-1}{32}, -243\right\}$.

$(h)\quad \sqrt[3]{3r-2}+5=\dfrac{6}{\sqrt[3]{3r-2}}$

First clear the fraction, then apply the above process.

$$\sqrt[3]{3r-2}+5=\frac{6}{\sqrt[3]{3r-2}}$$

$$\sqrt[3]{3r-2}\left(\sqrt[3]{3r-2}+5\right)=\sqrt[3]{3r-2}\left(\frac{6}{\sqrt[3]{3r-2}}\right)$$

$$\left(\sqrt[3]{3r-2}\right)^2+5\sqrt[3]{3r-2}=6$$

Let $u=\sqrt[3]{3r-2}$, then $u^2=\left(\sqrt[3]{3r-2}\right)^2$

$$\left(\sqrt[3]{3r-2}\right)^2+5\sqrt[3]{3r-2}=6$$

$$u^2+5u=6$$

$$u^2+5u-6=0$$

$$(u+6)(u-1)=0$$

$$u=-6 \qquad \text{or} \qquad u=1$$

Since $u=\sqrt[3]{3r-2}$,

$$\sqrt[3]{3r-2}=-6 \qquad \text{or} \qquad \sqrt[3]{3r-2}=1$$

$$\left(\sqrt[3]{3r-2}\right)^3=(-6)^3 \qquad\qquad \left(\sqrt[3]{3r-2}\right)^3=1^3$$

$$3r-2=-216 \qquad\qquad 3r-2=1$$

$$r=\frac{-216+2}{3}=\frac{-214}{3} \qquad\qquad r=\frac{1+2}{3}=1$$

Check:

$r=\dfrac{-214}{3}$

$$\sqrt[3]{3r-2}+5=\frac{6}{\sqrt[3]{3r-2}}$$

$$\sqrt[3]{3\left(\frac{-214}{3}\right)-2}+5\stackrel{?}{=}\frac{6}{\sqrt[3]{3\left(\frac{-214}{3}\right)-2}}$$

$$-6+5\stackrel{?}{=}\frac{6}{-6}$$

$$-1=-1$$

$r=1;$

$$\sqrt[3]{3r-2}+5=\frac{6}{\sqrt[3]{3r-2}}$$

$$\sqrt[3]{3(1)-2}+5\stackrel{?}{=}\frac{6}{\sqrt[3]{3(1)-2}}$$

$$1+5\stackrel{?}{=}\frac{6}{1}$$

$$6=6$$

The solution set is $\left\{\dfrac{-214}{3},1\right\}$.

Practice solving quadratic form equations by working supplementary problem 6.9.

6.10 Solve. State the answer in complete sentence form. A calculator is helpful in many instances.

(*a*) The distance d in feet that an object falls in t seconds is approximated by the formula $d=16t^2$. How many seconds elapse if an object falls 2,700 feet?

$d=16t^2$ so solve $2{,}700=16t^2$ for t.

$$2{,}700=16t^2$$

$$\frac{2{,}700}{16}=t^2$$

$$t=\pm\sqrt{\frac{2{,}700}{16}}=\pm\sqrt{168.75}\approx\pm 13$$

Disregard the negative solution, as it is meaningless.

Check:

$$t = 13; \qquad d = 16t^2$$

$$2{,}700 \stackrel{?}{=} 16(13)^2 = 2{,}704$$

$$2{,}700 \approx 2{,}704$$

Approximately 13 seconds elapse as an object falls 2,700 feet.

(*b*) The area of a circle is given by the formula $A = \pi r^2$. Find the radius r of a circular region with area 1,385 square feet.

$A = \pi r^2$ so solve $1{,}385 = \pi r^2$ for r. Divide by π, then extract square roots.

$$1{,}385 = \pi r^2$$

$$r^2 = \frac{1{,}385}{\pi}$$

$$r = \pm\sqrt{\frac{1{,}385}{\pi}} \approx \pm\sqrt{440.8591} \approx \pm 21$$

The negative solution is meaningless.

Check:

$$r = 21; \qquad A = \pi r^2$$

$$= \pi(21)^2 \approx 1{,}385.44 \approx 1{,}385$$

The radius of the circular region is approximately 21 feet.

(*c*) The volume V of a cylindrical container is determined by $V = \pi r^2 h$, where r is the radius and h is the height of the cylinder. Five cubic feet of planting mix are used to fill a cylindrical planter 18 inches in height. What is the radius of the container?

Isolate r^2 and extract square roots. The height must be expressed in feet since the volume is given in cubic feet. Eighteen inches is $\frac{18}{12} = 1.5$ feet. $V = \pi r^2 h$ or $5 = \pi r^2(1.5)$ so

$$r^2 = \frac{5}{\pi(1.5)} \approx 1.0610$$

$$r \approx \pm\sqrt{1.0610} \approx \pm 1.03$$

The negative solution is not applicable.

Check:

$$r = 1.03; \qquad V = \pi r^2 h$$

$$= \pi(1.03)^2(1.5) \approx 4.999 \approx 5$$

The container has a radius of approximately 1.03 feet.

(*d*) The Pythagorean Theorem states that the sum of the squares of the lengths of the legs, a and b, in a right triangle is equal to the square of the length of the hypotenuse c. That is, $a^2 + b^2 = c^2$. Find the length of a leg of a right triangle with hypotenuse $c = 26$ cm and leg $a = 24$ cm in length.

Suppose b is the length in cm of the unknown leg. Then

$$24^2 + b^2 = 26^2$$

$$b^2 = 26^2 - 24^2 = 676 - 576 = 100$$

$$b = \pm\sqrt{100} = \pm 10$$

Reject the negative root since lengths are positive.

Check:

$$b = 10; \qquad a^2 + b^2 = c^2$$

$$24^2 + 10^2 = 576 + 100 = 676 = 26^2$$

The unknown leg is 10 cm in length.

(*e*) The converse of the Pythagorean Theorem can be employed to construct a right angle. The converse states that if $a^2 + b^2 = c^2$, where a, b, and c are the lengths of the sides, then the triangle is a right triangle. A right angle is formed by the two shorter sides of the triangle. Find the appropriate length of the longest side of a triangle whose shorter sides are 9 and 12 feet in length in order for the shorter sides to form a right angle.

We must determine the length c when $a = 9$ and $b = 12$. We know $a^2 + b^2 = c^2$, so substitute and solve for c.

$$c^2 = 9^2 + 12^2 = 81 + 144 = 225$$

$$c = \pm\sqrt{225} = \pm 15$$

Disregard the negative root.

Check:

$$c = 15; \qquad a^2 + b^2 = c^2$$

$$9^2 + 12^2 \stackrel{?}{=} 15^2$$

$$81 + 144 \stackrel{?}{=} 225$$

$$225 = 225$$

The shorter sides form a right angle if the longest side is 15 feet in length.

(*f*) The height h in feet of a baseball above the ground t seconds after it is thrown upward is given by $h = -16t^2 + 80t + 5$. How long after it is thrown does the ball strike the ground?

The ball strikes the ground when the height $h = 0$. Use the quadratic formula to solve $0 = -16t^2 + 80t + 5$ or $16t^2 - 80t - 5 = 0$ for t.

$$t = \frac{-b \pm \sqrt{b^2 - 4ac}}{2a}$$

$$= \frac{-(-80) \pm \sqrt{(-80)^2 - 4(16)(-5)}}{2(16)} = \frac{80 \pm \sqrt{6400 + 320}}{32}$$

$$= \frac{80 \pm \sqrt{6720}}{32} \approx \frac{80 \pm 81.98}{32}$$

$$t \approx \frac{80 + 81.98}{32} = \frac{161.98}{32} \approx 5.06$$

The other root is meaningless since it is negative.

Check:

$$t = 5.06; \qquad 0 = -16t^2 + 80t + 5$$

$$0 = -16(5.06)^2 + 80(5.06) + 5 = 0.1424 \approx 0$$

The expression differs from 0 slightly because of round-off. The ball strikes the ground approximately 5 seconds after it is thrown upward.

(*g*) Find the dimensions of a rectangle whose area is 1,568 m^2 if the length is twice its width. The area of a rectangle is $A = lw$.

Let w be the width of the rectangle in meters. The length is then $2w$ meters. Therefore, solve

$$1{,}568 = (2w)w = 2w^2$$

$$784 = w^2$$

$$w = \pm\sqrt{784} = \pm 28$$

Disregard the negative solution. If $w = 28$, $l = 2w = 2(28) = 56$.

Check:

$$w = 28; \qquad A = lw$$

$$l = 56; \qquad = 56(28) = 1{,}568$$

The dimensions are $l = 56$ m and $w = 28$ m.

(*h*) An airplane requires 4 hrs, 45 min to complete a round trip of 1,100 miles each way. The plane had an 80 mph tailwind going and a 70 mph headwind on the return trip. Find the speed of the plane in still air.

Let s be the speed in mph of the plane in still air. The rate the plane travels with an 80 mph tailwind is then $s + 80$ mph. The rate the plane travels with a 70 mph headwind is $s - 70$ mph. This is a distance, rate, time problem. The quantities are related by the formula $d = rt$. We need the equivalent form $t = d/r$. We summarize the above information in the table below.

	d in miles	r in mph	t in hr
Going	1,100	$s+80$	$\frac{1{,}100}{s+80}$
Returning	1,100	$s-70$	$\frac{1{,}100}{s-70}$

The equation which relates the known and unknown quantities can be formed by relating the times involved.

Total time = time going + time returning

The times must be expressed in the same units in the equation. We choose hours since we desire the speed of the plane in miles per hour.

$4 \text{ hr } 45 \text{ min} = \left(4 + \frac{45}{60}\right) \text{hr} = 4\frac{3}{4} \text{ hr} = 4.75 \text{ hr}$. Write the equation and solve for s.

$$
\begin{aligned}
4.75 &= \frac{1{,}100}{s+80} + \frac{1{,}100}{s-70} \\
4.75(s+80)(s-70) &= 1{,}100(s-70) + 1{,}100(s+80) \\
4.75(s^2+10s-5{,}600) &= 1{,}100s - 77{,}000 + 1{,}100s + 88{,}000 \\
4.75s^2 + 47.5s - 26{,}600 &= 2{,}200s + 11{,}000 \\
4.75s^2 - 2{,}152.5s - 37{,}600 &= 0
\end{aligned}
$$

Use the quadratic formula.

$$
\begin{aligned}
s &= \frac{-b \pm \sqrt{b^2-4ac}}{2a} \\
&= \frac{-(-2{,}152.5) \pm \sqrt{(-2{,}152.5)^2 - 4(4.75)(-37{,}600)}}{2(4.75)} \\
&= \frac{2{,}152.5 \pm \sqrt{5{,}347{,}656.25}}{9.5} = \frac{2{,}152.5 \pm 2{,}312.5}{9.5} \\
s &= \frac{2{,}152.5 + 2{,}312.5}{9.5} = \frac{4{,}465}{9.5} = 470 \text{ mph}
\end{aligned}
$$

s is negative if we evaluate $(2{,}152.5 - 2{,}312.5)/9.5$. Speed is not negative, so disregard the negative root.

Check:

$$
\begin{aligned}
s = 470; \quad 4.75 &= \frac{1{,}100}{s+80} + \frac{1{,}100}{s-70} \\
&= \frac{1{,}100}{470+80} + \frac{1{,}100}{470-70} = \frac{1{,}100}{550} + \frac{1{,}100}{400} \\
&= 2 + 2.75 = 4.75
\end{aligned}
$$

The speed of the plane in still air is 470 mph.

Refer to supplementary problem 6.10 for similar applications.

6.11 Graph the following.

(*a*) $y = x^2$

We begin by choosing arbitrary values for x and determine the corresponding y values. Make a table of values. Plot the points and connect them with a smooth curve.

x	y
0	0
1	1
2	4
3	9
−1	1
−2	4
−3	9

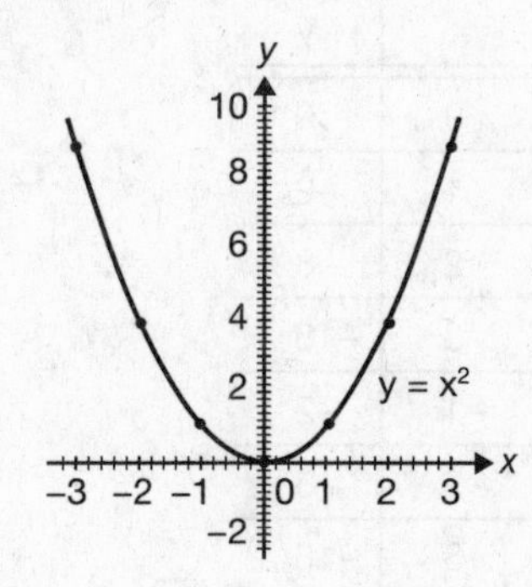

Figure 6.2

(*b*) $y = x^2 + 2$

Proceed as in part (*a*).

x	y
0	2
1	3
2	6
3	11
−1	3
−2	6
−3	11

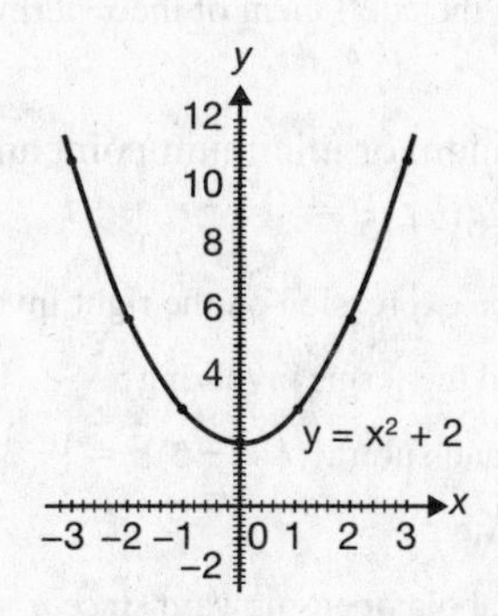

Figure 6.3

(*c*) $y = -x^2 + 3 = 3 - x^2$

Proceed as in part (*a*).

x	y
0	3
1	2
2	−1
3	−6
−1	2
−2	−1
−3	−6

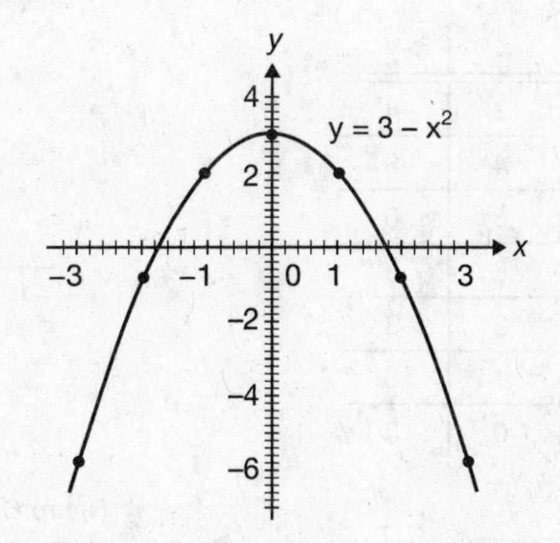

Figure 6.4

(*d*) $y = -2x^2$

Proceed as in part (*a*).

x	y
0	0
1	-2
2	-8
3	−18
-1	-2
-2	-8
-3	−18

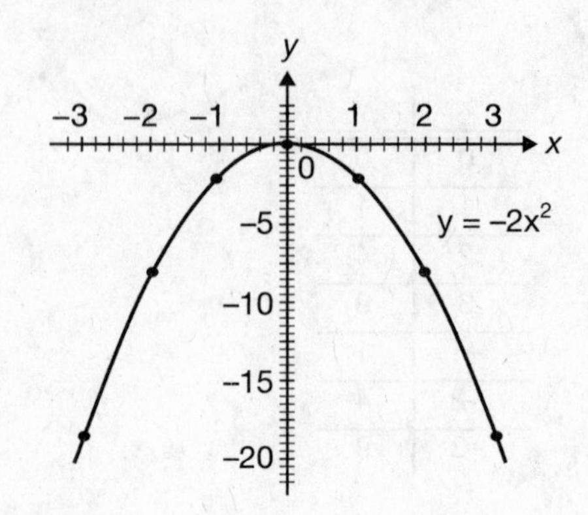

Figure 6.5

Observe that the parabola opens upward when a, the coefficient of x^2, is positive. The parabola opens downward when a is negative. We can therefore predict the direction the parabola opens merely by observing the sign of the coefficient of the x^2 term.

6.12 Find the maximum or minimum point and graph the expression.

(*a*) $y = x^2 - 6x + 5$

Make the expression on the right involve the square of a binomial.

Group the terms involving x. $\quad y = (x^2 - 6x) + 5$

Add and subtract $[\frac{1}{2}(-6)]^2 = [-3]^2 = 9$. $\quad y = (x^2 - 6x + 9) + 5 - 9$

Rewrite. $\quad y = (x - 3)^2 - 4$

The parabola opens upward since $a = 1 > 0$. The smallest value the right side can have occurs when $x = 3$. The value of x which makes the expression in parentheses zero is the x-coordinate of the minimum point. Choose x's equidistant from 3 and evaluate to find additional points on the curve. Plot the points and draw a smooth curve through them.

x	y
3	-4
4	-3
5	0
6	5
2	-3
1	0
0	5

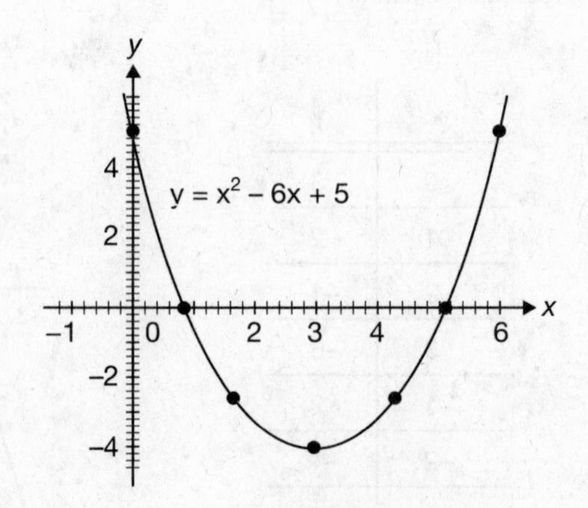

Figure 6.6

(*b*) $y = -x^2 - 4x + 1$

Group the terms involving x.	$y = (-x^2 - 4x) + 1$
Factor out -1 from the group.	$y = -(x^2 + 4x) + 1$
Add and subtract $[\frac{1}{2}(4)]^2 = 4$.	$y = -(x^2 + 4x + 4) + 1 + 4$
(Observe signs carefully.)	
Rewrite.	$y = -(x + 2)^2 + 5$

The parabola opens downward since $a = -1 < 0$. The largest size the right side can have occurs when $x = -2$. The value of x which makes the expression in parentheses zero is the x-coordinate of the maximum point. Choose x's equidistant from -2 and evaluate to find additional points on the parabola. Plot the points and draw a smooth curve through them.

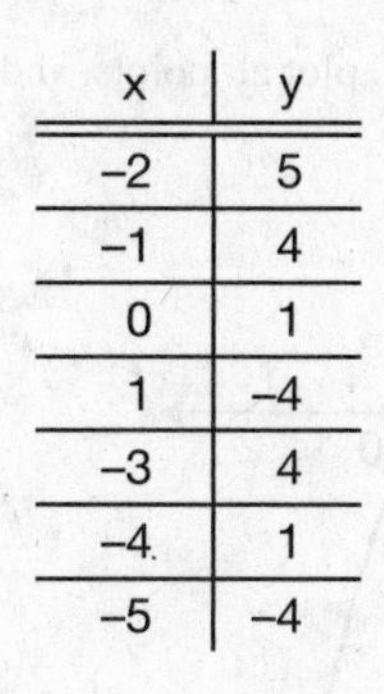

x	y
–2	5
–1	4
0	1
1	–4
–3	4
–4	1
–5	–4

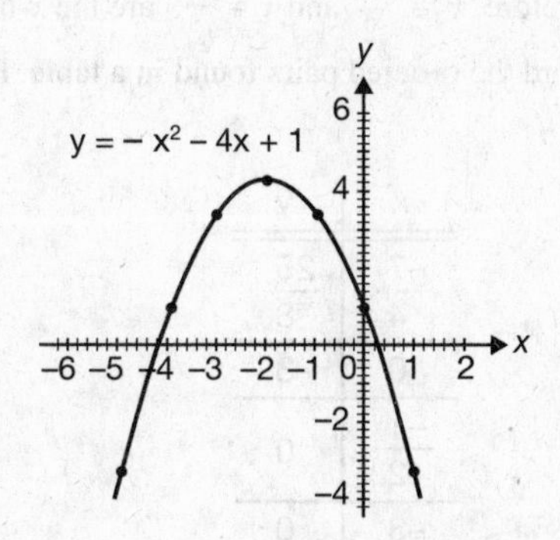

Figure 6.7

6.13 Graph the following. Utilize the procedure stated above.

(*a*) $y = x^2 - 4x$

1. The parabola opens upward since $a = 1 > 0$.
2. The vertex occurs at $x = \frac{-b}{2a} = \frac{-(-4)}{2(1)} = \frac{4}{2} = 2$.
3. The y-coordinate of the vertex is $y = 2^2 - 4(2) = 4 - 8 = -4$.
4. If $x = 0, y = 0^2 - 4(0) = 0 - 0 = 0$. This is the y-intercept.
5. If $y = 0, 0 = x^2 - 4x$ or $x^2 - 4x = 0$. Hence $x(x - 4) = 0$.

   Therefore, $x = 0$ and $x = 4$ are the x-intercepts.
6. We record the ordered pairs found in a table and find a couple of additional points to graph. Plot the points found and draw a smooth curve through them to complete the graph.

x	y
2	–4
0	0
4	0
–1	5
5	5

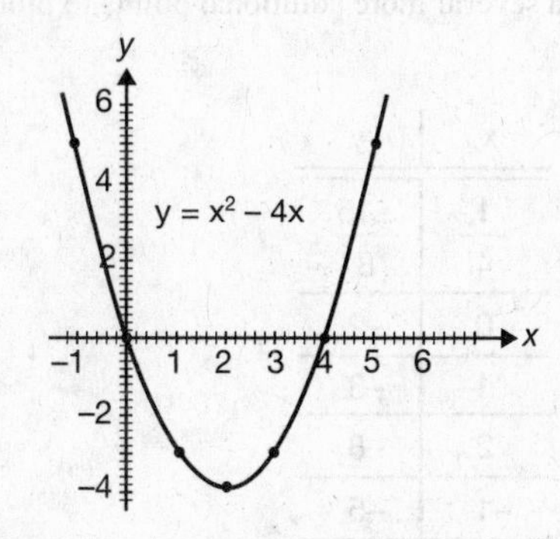

Figure 6.8

(*b*) $y = -2x^2 - 7x - 3$

1. The parabola opens downward since $a = -2 < 0$.
2. The vertex occurs at $x = \frac{-b}{2a} = \frac{-(-7)}{2(-2)} = \frac{7}{-4} = \frac{-7}{4}$.
3. If $x = \frac{-7}{4}, y = -2\left(\frac{-7}{4}\right)^2 - 7\left(\frac{-7}{4}\right) - 3 = -2\left(\frac{49}{16}\right) + \frac{49}{4} - 3$

   $= \frac{-49}{8} + \frac{98}{8} - \frac{24}{8} = \frac{49}{8} - \frac{24}{8} = \frac{25}{8}$.
4. If $x = 0, y = -2(0)^2 - 7(0) - 3 = -3$. This is the y-intercept.
5. If $y = 0, 0 = -2x^2 - 7x - 3$ or $2x^2 + 7x + 3 = 0$ so $(2x + 1)(x + 3) = 0$.

   Therefore, $x = \frac{-1}{2}$ and $x = -3$ are the x-intercepts.
6. Record the ordered pairs found in a table. Find additional points, plot all points, and draw the curve.

x	y
$\frac{-7}{4}$	$\frac{25}{8}$
0	−3
$\frac{-1}{2}$	0
−3	0
1	−12
−4	−7

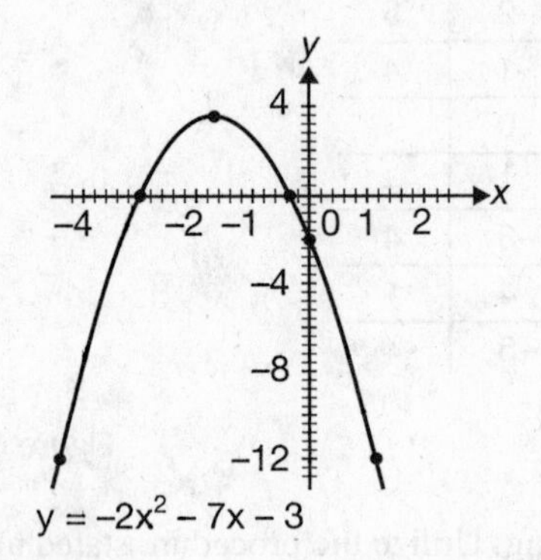

Figure 6.9

(*c*) $y = -2x^2 + x - 2$

1. The parabola opens downward since $a = -2 < 0$.
2. The vertex is at $x = \frac{-b}{2a} = \frac{-1}{2(-2)} = \frac{-1}{-4} = \frac{1}{4}$.
3. If $x = \frac{1}{4}, y = -2\left(\frac{1}{4}\right)^2 + \frac{1}{4} - 2 = -2\left(\frac{1}{16}\right) + \frac{1}{4} - 2 = \frac{-15}{8}$.
4. If $x = 0, y = 2$. This is the y-intercept.
5. There are no x-intercepts since the maximum point has y-coordinate $\frac{-15}{8}$.
6. Find several more additional points to plot and graph the curve.

x	y
$\frac{1}{4}$	$\frac{-15}{8}$
0	−2
1	−3
2	−8
−1	−5
−2	−12

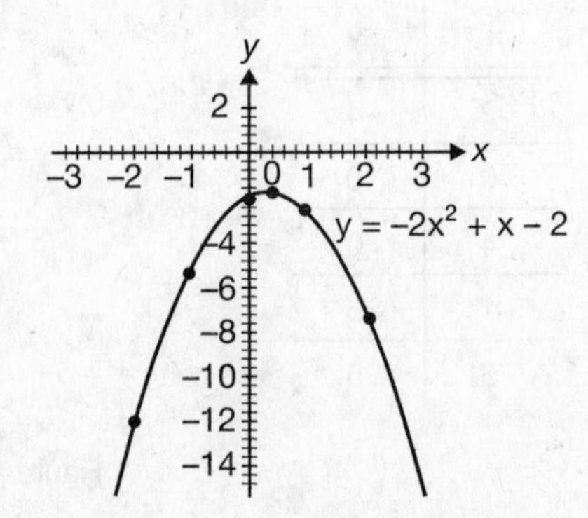

Figure 6.10

6.14 Graph the following.

(*a*) $x = y^2 - 3$

1. The parabola opens to the right since $a = 1 > 0$.
2. The vertex occurs at $y = \dfrac{-b}{2a} = \dfrac{-0}{2(1)} = 0$.
3. The x-coordinate of the vertex is $x = 0^2 - 3 = -3$.
4. If $x = 0, 0 = y^2 - 3$ so $y = \pm\sqrt{3}$ are the y-intercepts.
5. The x-intercept is at the vertex.
6. Record the values in a table and plot the points. Find about two more points and draw a smooth curve through them.

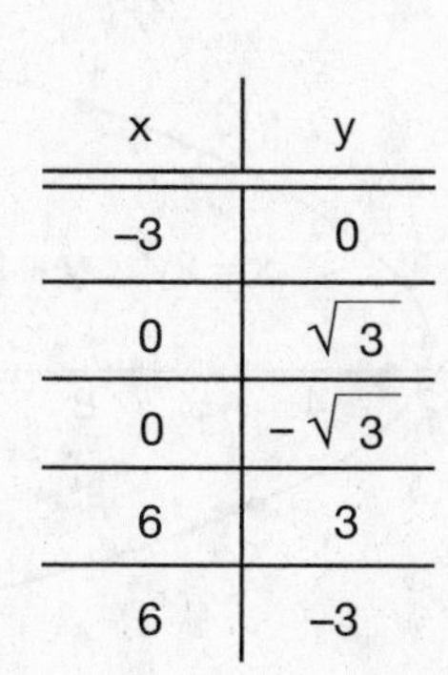

x	y
−3	0
0	$\sqrt{3}$
0	$-\sqrt{3}$
6	3
6	−3

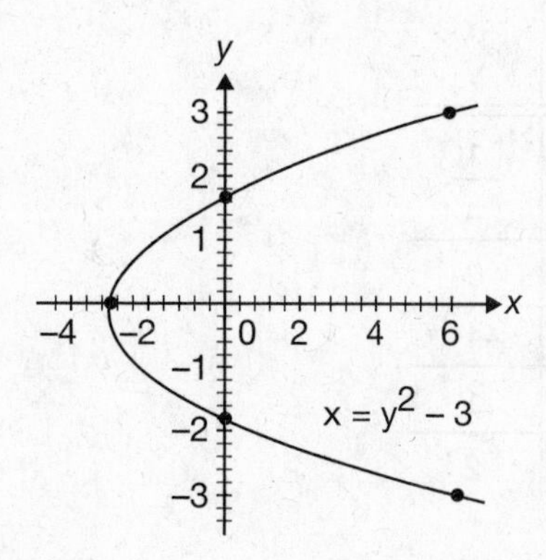

Figure 6.11

(*b*) $x = -y^2 + 5y + 6$

1. The parabola opens to the left since $a = -1 < 0$.
2. The vertex occurs at $y = -b/(2a) = -5/(2(-1)) = -5/(-2) = 5/2$.
3. The x-coordinate of the vertex is $x = -\left(\dfrac{5}{2}\right)^2 + 5\left(\dfrac{5}{2}\right) + 6 = \dfrac{-25}{4} + \dfrac{25}{2} + 6$ $= \dfrac{-25}{4} + \dfrac{50}{4} + \dfrac{24}{4} = \dfrac{49}{4}$.
4. If $x = 0, 0 = -y^2 + 5y + 6 = -(y^2 - 5y - 6) = -(y + 1)(y - 6)$ and $y = -1$ or $y = 6$. These are the y-intercepts.
5. If $y = 0$, then $x = -0^2 + 5(0) + 6 = 6$ is the x-intercept.
6. Make a table of the ordered pairs found above. Pick arbitrary values for y to find additional ordered pairs on the curve. Draw the curve.

x	y
$\dfrac{49}{4}$	$\dfrac{5}{2}$
0	−1
0	6
6	0
−8	7
−8	−2

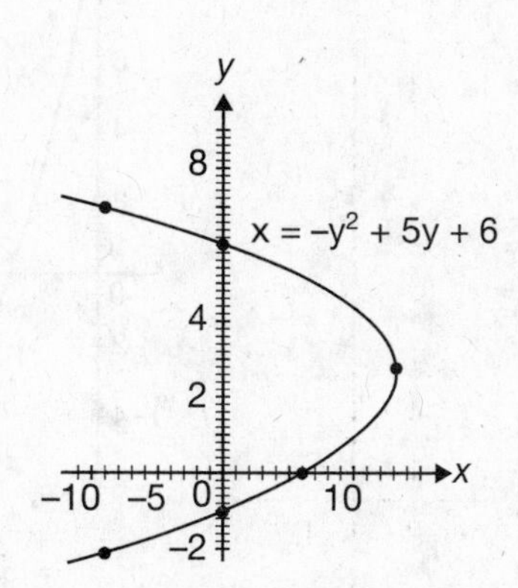

Figure 6.12

(*c*) $x = 2y^2 - y + 2$

1. The parabola opens to the right since $a = 2 > 0$.
2. The vertex occurs at $y = -b/(2a) = -(-1)/(2(2)) = 1/4$.
3. If $y = \frac{1}{4}$, $x = 2\left(\frac{1}{4}\right)^2 - \frac{1}{4} + 2 = \frac{2}{16} - \frac{1}{4} + 2 = \frac{1}{8} - \frac{2}{8} + \frac{16}{8} = \frac{15}{8}$.
4. There are no *y*-intercepts since the vertex lies to the right of the *y*-axis.
5. If $y = 0$, then $x = 2(0)^2 - 0 + 2 = 2$ is the *x*-intercept.
6. Record the known ordered pairs in a table. Find additional ordered pairs to plot and draw the graph.

x	y
$\frac{15}{8}$	$\frac{1}{4}$
2	0
5	−1
3	1
8	2

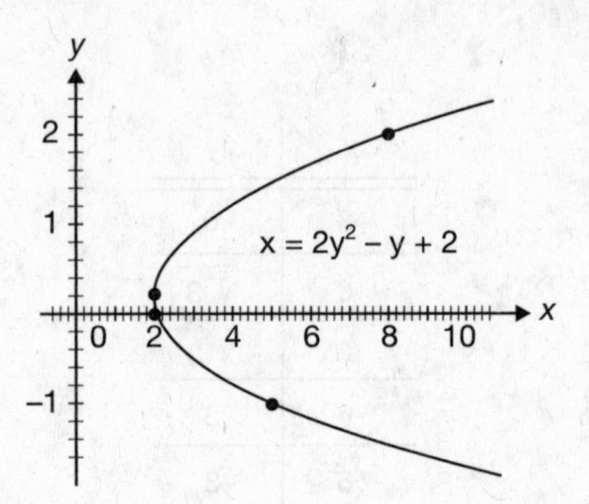

Figure 6.13

Work supplementary problem 6.11 to practice graphing parabolas.

6.15 Graph the following on a graphing calculator.

(*a*) $y = x^2 - 8x + 11$

The parabola opens upward since $a = 1 > 0$. The vertex occurs at $x = -b/(2a) = -(-8)/(2(1)) = 8/2 = 4$. Try an *x*-range of $[-1, 9]$. If $x = 4$, $y = -5$ and the *y*-intercept is 11 so set the *y*-range to $[-6, 12]$. Now enter the equation and graph it.

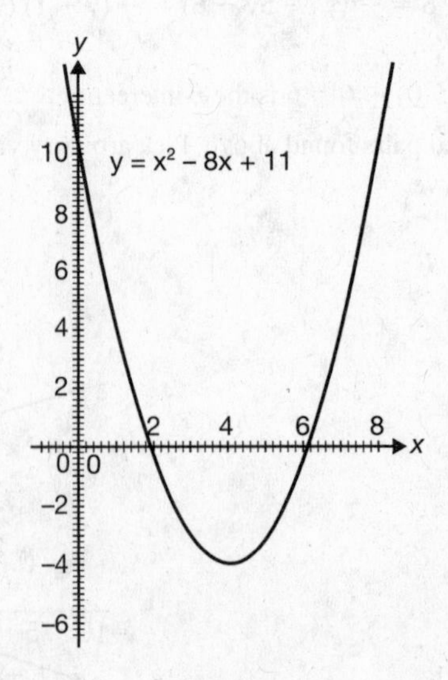

Figure 6.14

(*b*) $y = -x^2 - 6x - 5$

The parabola opens downward since $a = -1 < 0$. The vertex occurs at $x = -b/(2a) = -(-6)/(2(-1)) = 6/(-2) = -3$. Set the x-range to $[-8, 2]$. If $x = -3$, $y = 4$ and the y-intercept is -5 so set the y-range to $[-10, 5]$. Enter the expression and graph it.

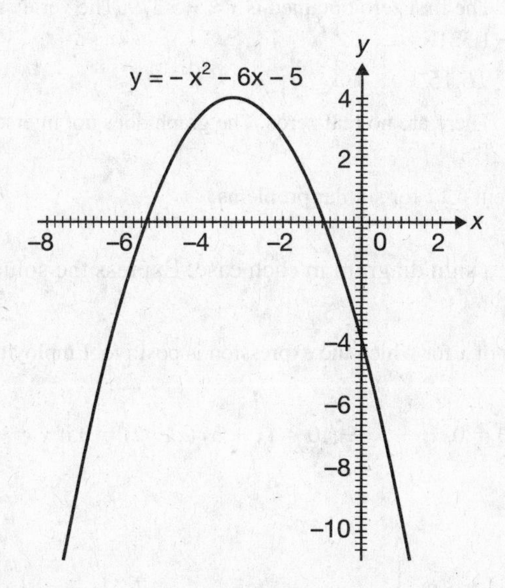

Figure 6.15

(*c*) $y = -0.76\,(x - 4.38)^2 + 7.89$

The parabola opens downward since $a = -0.76 < 0$. The vertex occurs at $x = 4.38$ since the maximum point occurs at the value of x which makes the expression in parentheses zero. The y-value is 7.89 when $x = 4.38$. Set the x-range to $[-2, 10]$ and the y-range to $[-10, 10]$. Graph the expression.

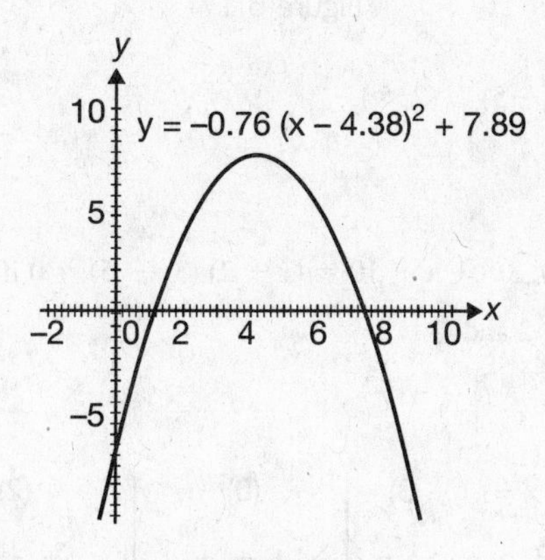

Figure 6.16

6.16 Graph the following on a graphing calculator. Employ the trace feature to approximate the zeros (roots) and vertex of each. Express answers to three decimal place accuracy.

(*a*) $y = x^2 - 4.8x + 2.06$

Graph the expression. Try to fill the graphing window with the part of the curve that includes the zeros and the vertex. The zoom feature can be utilized to magnify the relevant portion of the curve. Display the y-coordinate and instruct the calculator to trace the curve until $y \approx 0$ to determine the first zero. Display the x-coordinate and record it. We obtain $x \approx 0.468$ for the first zero. Display the y-coordinate and continue the trace to obtain the vertex. Record the y-coordinate, then display the x-coordinate and record it. The vertex is approximately $(2.426, -3.699)$. Display the y-coordinate and continue the trace to obtain the other zero. Display the x-coordinate and record it. It is $x \approx 4.298$.

(*b*) $y = -1.2x^2 - 7.92x - 8.368$

Proceed as in part (*a*). The first zero obtained is $x \approx -5.279$. The vertex is approximately $(-3.300, 4.700)$. The second zero is $x \approx -1.321$.

(*c*) $y = 0.77x^2 - 6.70x + 16.15$

Proceed as in part (*a*). There are no real zeros. The graph does not intersect the x-axis. The vertex is approximately $(4.351, 1.575)$.

Refer to supplementary problem 6.12 for similar problems.

6.17 Solve the following. Make a sign diagram in each case. Express the solution set in interval notation.

(*a*) $x^2 + 3x - 10 > 0$

We must find the values of x for which the expression is positive. Employ the procedure stated on p. 175.

1. $x^2 + 3x - 10 > 0$
2. Solve $x^2 + 3x - 10 = 0$. $x^2 + 3x - 10 = (x + 5)(x - 2) = 0$ if $x = -5$ or $x = 2$. The critical values are -5 and 2.
3.

For x = (− 6)
(0)
(3)
x + 5 is− − −
+ + +
+ + +
x − 2 is − − −
− − −
+ + +
− 8
− 6
− 4
− 2
0
4
6
(x + 5) (x − 2) is + + +
− − −
+ + +
− 5
2

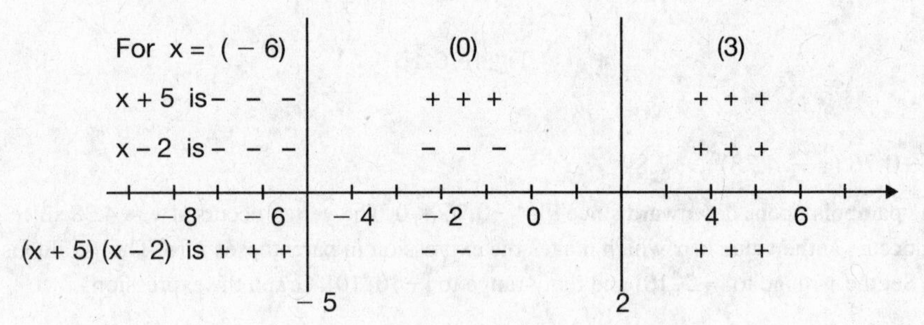

Figure 6.17

4. The solution set is $(-\infty, -5) \cup (2, \infty)$.

(*b*) $3x^2 + x \leq 10$

1. $3x^2 + x - 10 \leq 0$
2. Solve $3x^2 + x - 10 = 0$. $3x^2 + x - 10 = (x + 2)(3x - 5) = 0$ if $x = -2$ or $x = \frac{5}{3}$. The critical values are -2 and $\frac{5}{3}$.
3.

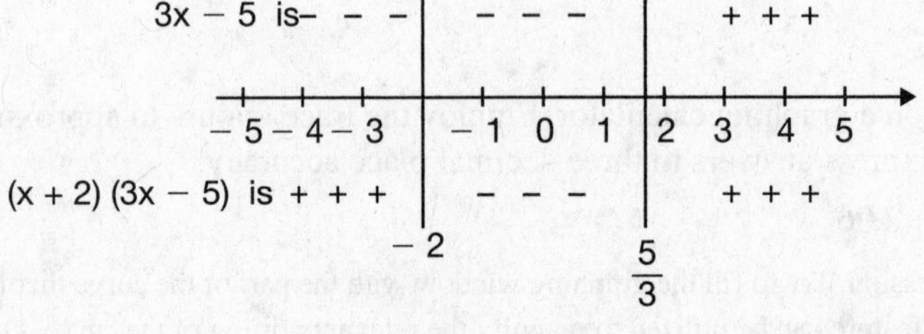

For x = (− 3)
(0)
(2)
x + 2 is − − −
+ + +
+ + +
3x − 5 is− − −
− − −
+ + +
− 5 − 4 − 3
− 1 0 1 2 3 4 5
(x + 2) (3x − 5) is + + +
− − −
+ + +
− 2
5/3

Figure 6.18

4. The endpoints of the interval in the solution set are included since the critical values satisfy the equality relation. The solution set is $\left[-2, \frac{5}{3}\right]$.

(*c*) $2x^2 + x \geq 3$

1. $2x^2 + x - 3 \geq 0$
2. Solve $2x^2 + x - 3 = 0$. $2x^2 + x - 3 = (2x + 3)(x - 1) = 0$ if $x = \frac{-3}{2}$ or $x = 1$.
   The critical values are $\frac{-3}{2}$ and 1.
3.

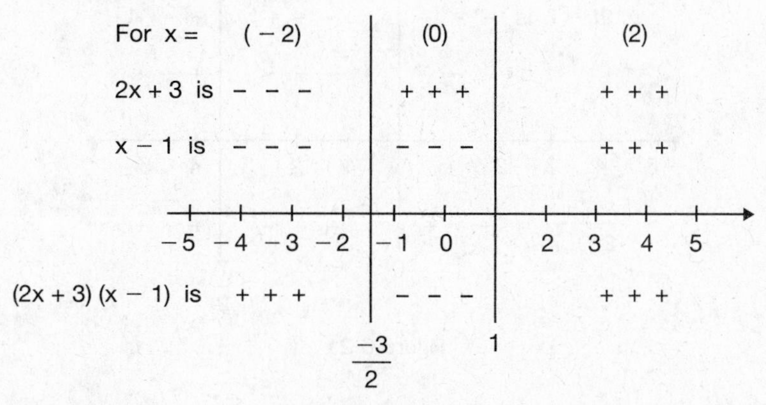

Figure 6.19

4. The endpoints of the intervals in the solution set are included since the critical values satisfy the equality relation. The solution set is $(-\infty, -3/2] \cup [1, \infty)$.

6.18 Solve the following. Make a sign diagram in each case. Express the solution set in interval notation.

(*a*) $\frac{x+3}{x-5} < 0$

1. $\frac{x+3}{x-5} < 0$
2. The critical values are the values for which the expression is zero or undefined. The expression is zero if the numerator is zero or if $x = -3$. The expression is undefined if the denominator is zero or if $x = 5$. The critical values are -3 and 5.
3.

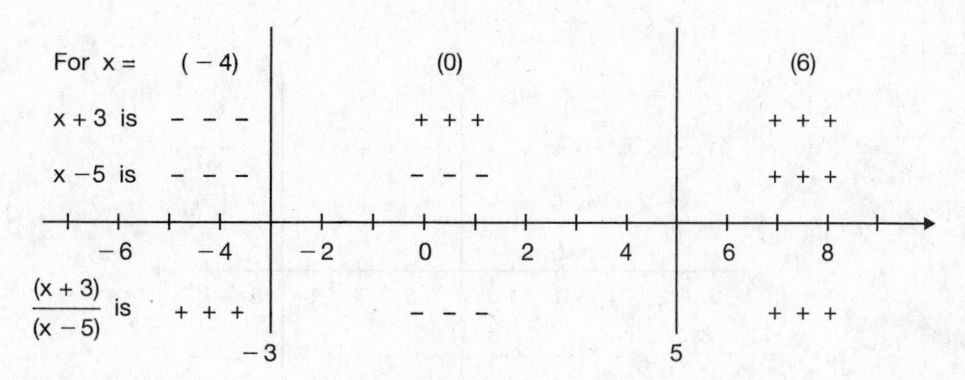

Figure 6.20

4. The solution set is $(-3, 5)$.

(*b*) $\dfrac{t}{2t-7} \geq 0$

1. $\dfrac{t}{2t-7} \geq 0$
2. The numerator is zero if $t = 0$. The denominator is zero if $t = \frac{7}{2}$. The critical values are 0 and $\frac{7}{2}$.
3.

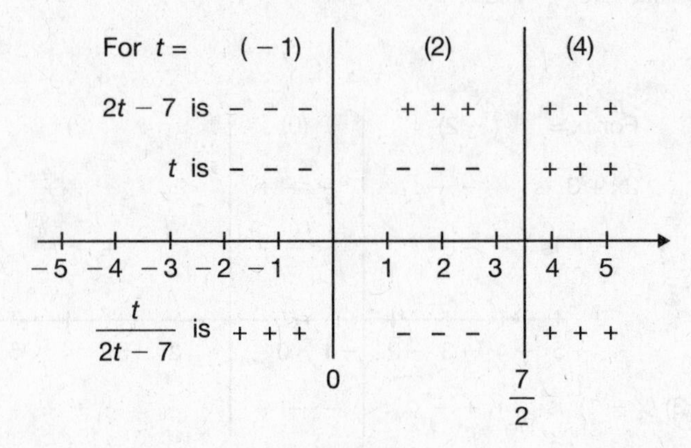

Figure 6.21

4. The solution set is $(-\infty, 0] \cup \left(\frac{7}{2}, \infty\right)$. Note that $t = \frac{7}{2}$ is excluded since the expression is undefined at $t = \frac{7}{2}$.

(*c*) $\dfrac{s+5}{s-2} \leq 3$

1. We must first rewrite the expression with one side of the inequality equal to zero.

$$\frac{s+5}{s-2} - 3 \leq 0$$

$$\frac{s+5}{s-2} - \frac{3(s-2)}{s-2} = \frac{s+5-3(s-2)}{s-2} = \frac{11-2s}{s-2} \text{ so solve } \frac{11-2s}{s-2} \leq 0.$$

2. The numerator is zero if $11 - 2s = 0$ or if $s = \frac{11}{2}$. The denominator is zero if $s - 2 = 0$ or if $s = 2$. The critical values are $\frac{11}{2}$ and 2.
3.

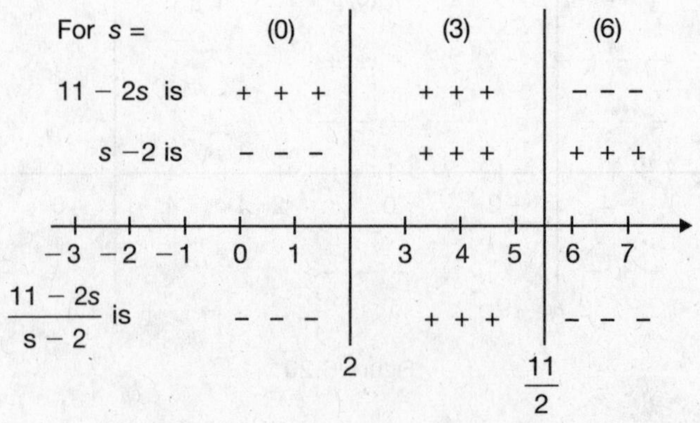

For s = (0) (3) (6)

11 − 2s is + + + + + + − − −

s −2 is − − − + + + + + +

− 3 − 2 − 1 0 1 3 4 5 6 7

$\dfrac{11-2s}{s-2}$ is − − − + + + − − −

2 $\frac{11}{2}$

Figure 6.22

4. The solution set is $(-\infty, 2) \cup \left[\frac{11}{2}, \infty\right)$. It can be verified that values of s in the indicated intervals satisfy $\frac{s+5}{s-2} \le 3$.

See supplementary problem 6.13 to practice solving quadratic and rational inequalities.

6.19 $(x-3)^3 (x+2)^5 (x-1)^4 > 0$

The critical values are 3, −2, and 1. They break the number line into four intervals: $(-\infty, -2)$, $(-2, 1)$, $(1, 3)$, and $(3, \infty)$. We must determine the solution behavior of each interval. We use the test value of zero to start the process:

$x = 0$ gives $(-)^3 (+)^5 (-)^4 = (-)$ which is not greater than zero as the inequality requires.

Therefore, the interval containing zero, $(-2, 1)$, is excluded from the solution. See Figure 6.23 below.

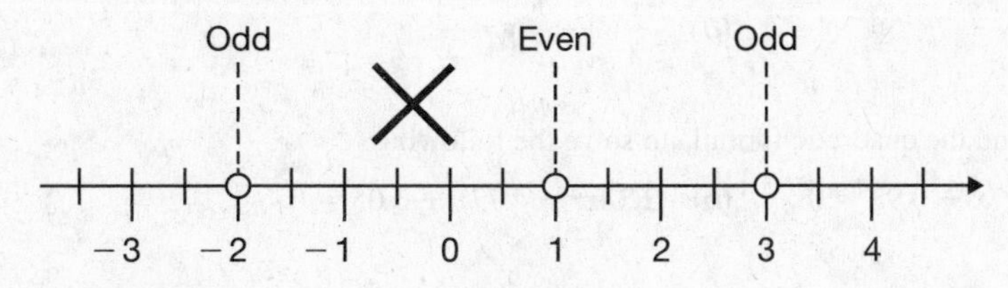

Figure 6.23

$(-2, 1)$ excluded requires	$(1, 3)$ to be excluded	(since 1 came from an even power)
$(1, 3)$ excluded requires	$(3, \infty)$ to be included	(since 3 came from an odd power)
$(-2, 1)$ excluded requires	$(-\infty, -2)$ to be included	(since −2 came from an odd power)

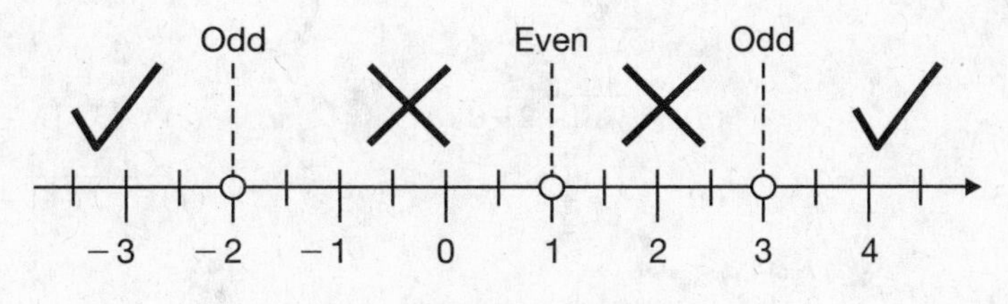

Figure 6.24

The solution set is $(-\infty, -2) \cup (3, \infty)$.

SUPPLEMENTARY PROBLEMS

6.1 Use the factor method to solve the following. Check for extraneous roots where applicable.

(*a*) $y^2 - 8y + 15 = 0$ (*b*) $t^2 - 3t = 10$ (*c*) $2s^2 - 7s + 5 = 0$

(*d*) $2p^2 - 5p = p$ (*e*) $(x + 10)(2x - 7) = 0$ (*f*) $\frac{7}{10w} = \frac{1}{10} + \frac{1}{w^2}$

(*g*) $4 + \frac{3r}{r-2} = \frac{r^2+2}{r-2}$

6.2 Use the square root method to solve the following.

(*a*) $t^2 = 10$ (*b*) $3y^2 + 1 = 28$ (*c*) $2x^2 + 26 = 0$

(*d*) $(2s + 1)^2 = 7$ (*e*) $(3w - 4)^2 + 8 = 0$

6.3 Determine the term that would complete the square for each of the following.

(*a*) $x^2 + 18x$ (*b*) $y^2 - 9y$

6.4 Solve by completing the square.

(*a*) $x^2 - 4x - 11 = 0$ (*b*) $y^2 + 6y = -2$

(*c*) $3t^2 - 6t - 2 = 0$ (*d*) $4p^2 + 7p + 2 = 3$

6.5 Use the quadratic formula to solve the following.

(*a*) $x^2 + 5x - 10 = 0$ (*b*) $t^2 - 17 = 2t$

(*c*) $2y^2 + 2 = 2y + 7$ (*d*) $\frac{1}{2}w^2 + \frac{1}{2}w = \frac{-3}{4}$

6.6 Use a calculator and the quadratic formula to solve the following.

(*a*) $5.21x^2 + 1.17x - 0.98 = 0$ (*b*) $2.38x^2 - 4.77x + 3.05 = 0$

6.7 Solve the following. Remember to check for extraneous solutions.

(*a*) $\sqrt{x-3} = 8$ (*b*) $\sqrt{5y} - 4 = 1$ (*c*) $9\sqrt{4t+3} = 36$

(*d*) $(z+5)^{\frac{1}{2}} = 3$ (*e*) $\sqrt{6w-3} - \sqrt{4w+1} = 0$ (*f*) $s + 5 = 3\sqrt{s+5}$

(*g*) $\sqrt{2x+1} + 7 = x$ (*h*) $\sqrt{2 - \sqrt{x+7}} = 1$

6.8 Solve.

(*a*) $x^{\frac{1}{3}} = -3$ (*b*) $\sqrt[4]{y+1} + 2 = 0$ (*c*) $-3 = (w-2)^{\frac{1}{4}} - 8$

(*d*) $\sqrt[3]{6-t} + 5 = 3$ (*e*) $x^{\frac{-1}{4}} = \frac{1}{2}$ (*f*) $\sqrt[3]{2y+1} + 5 = 8$

6.9 Solve.

(*a*) $y^6 - 8y^3 + 12 = 0$ (*b*) $x^4 - 7x^2 - 18 = 0$

(*c*) $t + 12t^{\frac{1}{2}} = -27$ (*d*) $12w^4 + 5w^2 - 2 = 0$

(*e*) $l^{\frac{2}{3}} - 2l^{\frac{1}{3}} = 15$ (*f*) $2x^{-2} - 3x^{-1} - 20 = 0$

(*g*) $3t^{\frac{4}{5}} = 10t^{\frac{2}{5}} + 25$ (*h*) $8y^{\frac{1}{2}} - 5y^{\frac{1}{4}} = 3$

6.10 Solve. State the answer in complete sentence form. Employ a calculator when necessary.

(*a*) The height in feet of a ball thrown vertically upward is given by $h = 48t - 16t^2$, where t is the time in seconds after the throw. How long will it take the ball to reach a height of 32 ft?

(*b*) Find the radius of a circular region with area 6,940 ft^2.

(*c*) Find the radius of a cylindrical container 8 meters high that contains 1,608 m^3 of material.

(*d*) A diagonal brace is necessary on a gate that is 3 feet by 5 feet. How long must the brace be if it is attached at the corners of the gate?

(*e*) A square with a diagonal of length 17 feet is to be constructed. How long must the sides of the square be?

(*f*) The cost C in dollars of producing n gadgets is given by $C = 1{,}200 + 12n - 6n^2$. How many gadgets can be produced at a cost of \$192?

(*g*) Find the dimensions of a rectangle with area 4,500 cm^2 if the width is 80% of the length.

(*h*) A jet requires 7 hr 10 min to complete a round trip of 1,785 miles each way. The plane had a 70 mph headwind going and a 60 mph tailwind on the return trip. Find how fast the jet travels in still air.

6.11 Graph the following.

(*a*) $y = x^2 - 4$ (*b*) $y = 2 - \frac{1}{2}x^2$

(*c*) $y = x^2 - 6x + 7$ (*d*) $y = -x^2 - 4x + 5$

(*e*) $x = y^2 + 1$ (*f*) $x = y^2 - y - 12$

(*g*) $x = -y^2 - 4y - 5$ (*h*) $y = \frac{1}{2}x^2 - \frac{1}{4}x - \frac{3}{2}$

6.12 Graph the following on a graphing calculator. Employ the trace feature to approximate the zeros (roots) and vertex of each. Express answers to three decimal place accuracy.

(*a*) $y = x^2 - 3.6x - 1.1$ (*b*) $y = -0.8x^2 - 6.6x - 10.7$

(*c*) $y = -1.1x^2 - 5.5x - 7.9$

6.13 Solve the following. (Use a sign diagram to help in identifying the solution in each case.) Express the solution set in interval notation.

(*a*) $x^2 - 3x - 4 < 0$ (*b*) $2x^2 - x - 15 \geq 0$ (*c*) $4x^2 + 4x \leq 35$

(*d*) $\frac{t-2}{t+4} < 0$ (*e*) $\frac{3s+6}{2s} \geq 0$ (*f*) $\frac{w-4}{2w+5} \leq -1$

ANSWERS TO SUPPLEMENTARY PROBLEMS

6.1 (*a*) $\{5, 3\}$ (*b*) $\{5, -2\}$ (*c*) $\left\{\frac{5}{2}, 1\right\}$

(*d*) $\{0, 3\}$ (*e*) $\left\{-10, \frac{7}{2}\right\}$ (*f*) $\{2, 5\}$

(*g*) $\{5\}$

6.2 (*a*) $\{\pm\sqrt{10}\}$ (*b*) $\{\pm 3\}$ (*c*) $\{\pm i\sqrt{13}\}$

(*d*) $\left\{\frac{-1 \pm \sqrt{7}}{2}\right\}$ (*e*) $\left\{\frac{4 \pm 2i\sqrt{2}}{3}\right\}$

6.3 (*a*) 81 (*b*) $\frac{81}{4}$

6.4 (*a*) $\{2 \pm \sqrt{15}\}$ (*b*) $\{-3 \pm \sqrt{7}\}$ (*c*) $\left\{\frac{3 \pm \sqrt{15}}{3}\right\}$ (*d*) $\left\{\frac{-7 \pm \sqrt{65}}{8}\right\}$

6.5 (*a*) $\left\{\frac{-5 \pm \sqrt{65}}{2}\right\}$ (*b*) $\{1 \pm 3\sqrt{2}\}$ (*c*) $\left\{\frac{1 \pm \sqrt{11}}{2}\right\}$ (*d*) $\left\{\frac{-1 \pm i\sqrt{5}}{2}\right\}$

6.6 (*a*) $\{-0.56, 0.34\}$ (*b*) $\{1.00 \pm 0.53i\}$

6.7 (*a*) $\{67\}$ (*b*) $\{5\}$ (*c*) $\left\{\frac{13}{4}\right\}$

(*d*) $\{4\}$ (*e*) $\{2\}$ (*f*) $\{-5, 4\}$

(*g*) $\{12\}$; $x = 4$ is an extraneous root. (*h*) $\{-6\}$

6.8 (*a*) $\{-27\}$ (*b*) $\{\} = \varnothing$ (*c*) $\{627\}$

(*d*) $\{14\}$ (*e*) $\{16\}$ (*f*) $\{13\}$

6.9 (*a*) $\{\sqrt[3]{2}, \sqrt[3]{6}\}$ (*b*) $\{\pm 3, \pm i\sqrt{2}\}$ (*c*) $\{\} = \varnothing$

(*d*) $\left\{\pm\frac{1}{2}, \pm\frac{i\sqrt{6}}{3}\right\}$ (*e*) $\{-27, 125\}$ (*f*) $\left\{-\frac{2}{5}, \frac{1}{4}\right\}$

(*g*) $\left\{\pm 25\sqrt{5}, \pm\frac{25i\sqrt{15}}{27}\right\}$ (*h*) $\left\{\frac{81}{4{,}096}, 1\right\}$

6.10 (*a*) The ball reaches a height of 32 ft 1 sec after it is thrown on the way up and 2 sec after it is thrown on the way down.

(*b*) The radius of the circular region is 47 ft.

(*c*) The radius of the container is approximately 8 m.

(*d*) The brace must be approximately 5 ft 10 in. long.

(*e*) The sides of the square must be 12 ft long.

(*f*) Fourteen gadgets can be produced at a cost of $192.

(*g*) The dimensions of the rectangle are 75 cm long and 60 cm wide.

(*h*) The jet travels approximately 511 mph in still air.

6.11 (*a*) See Figure 6.25. (*b*) See Figure 6.26.

(*c*) See Figure 6.27. (*d*) See Figure 6.28.

(*e*) See Figure 6.29. (*f*) See Figure 6.30.

(*g*) See Figure 6.31. (*h*) See Figure 6.32.

(*a*)

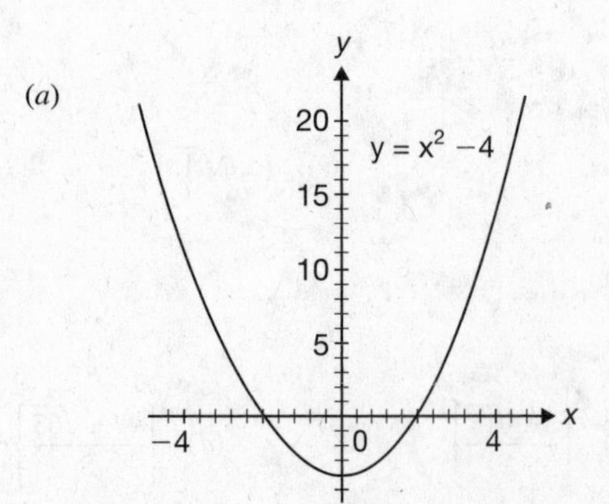

Figure 6.25

(*b*)

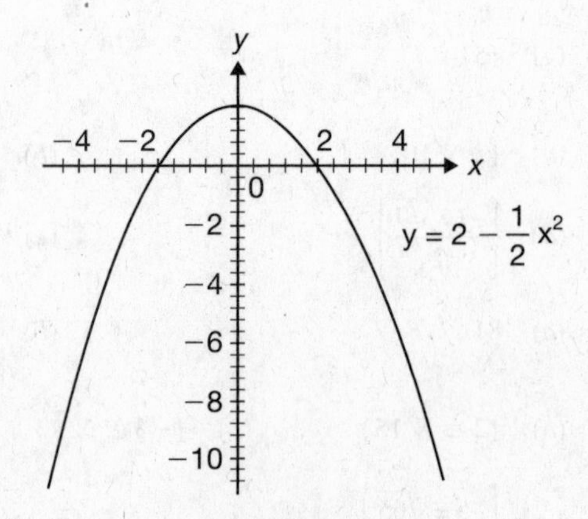

Figure 6.26

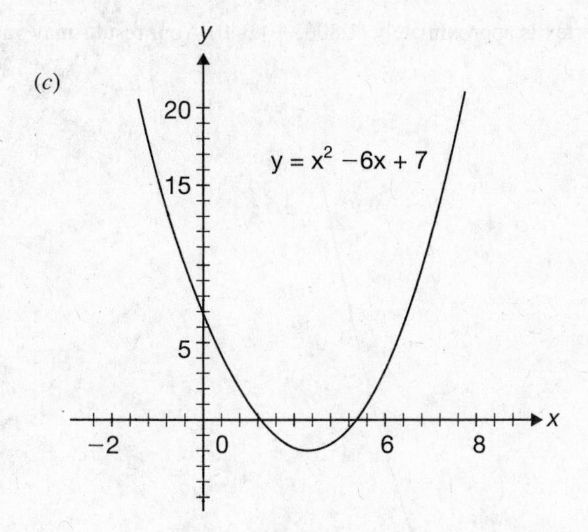

Figure 6.27

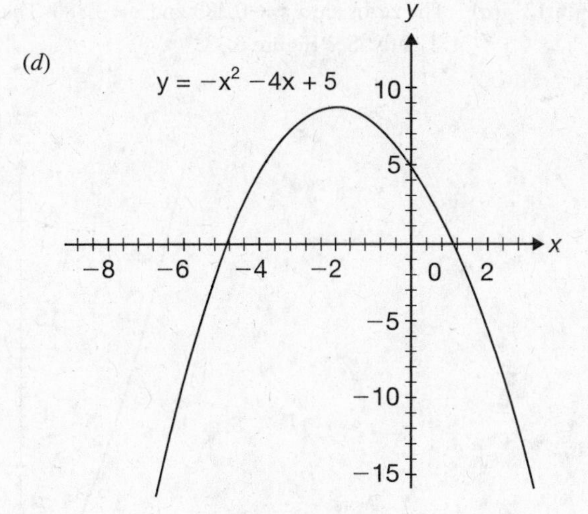

Figure 6.28

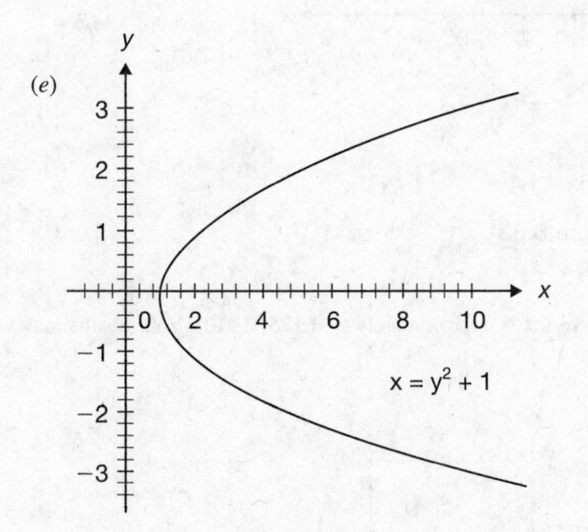

Figure 6.29

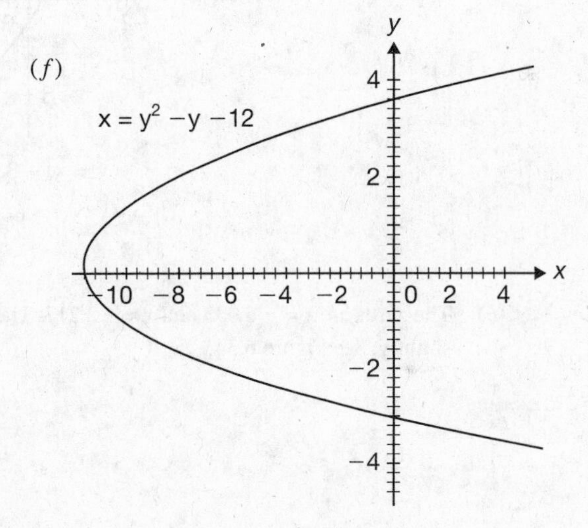

Figure 6.30

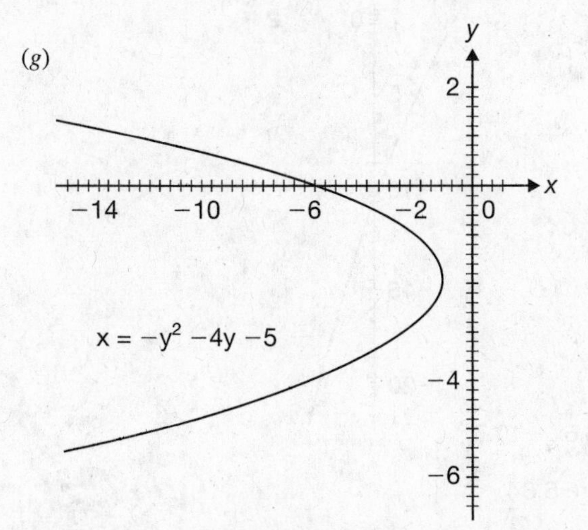

Figure 6.31

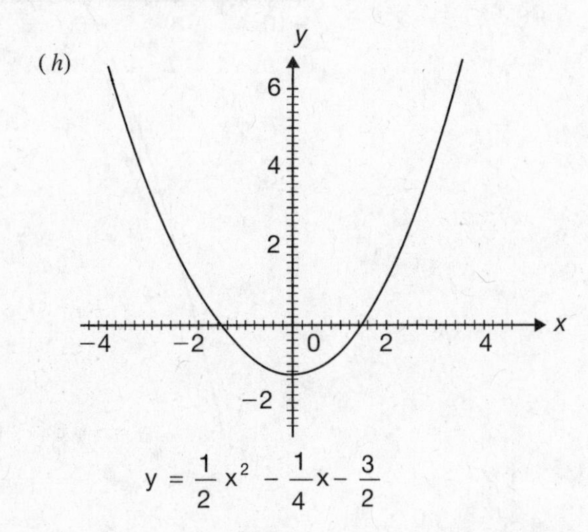

Figure 6.32

6.12 (*a*) The zeros are $x \approx -0.283$ and $x \approx 3.883$. The vertex is approximately $(1.800, -4.340)$. Your results may vary slightly. See Figure 6.33.

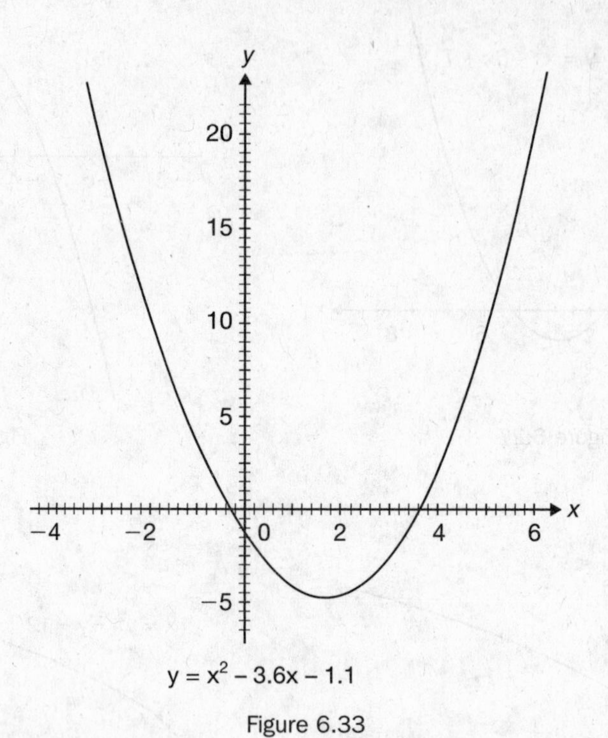

Figure 6.33

(*b*) The zeros are $x \approx -6.033$ and $x \approx -2.217$. The vertex is approximately $(-4.125, 2.912)$. Your results may vary slightly. See Figure 6.34.

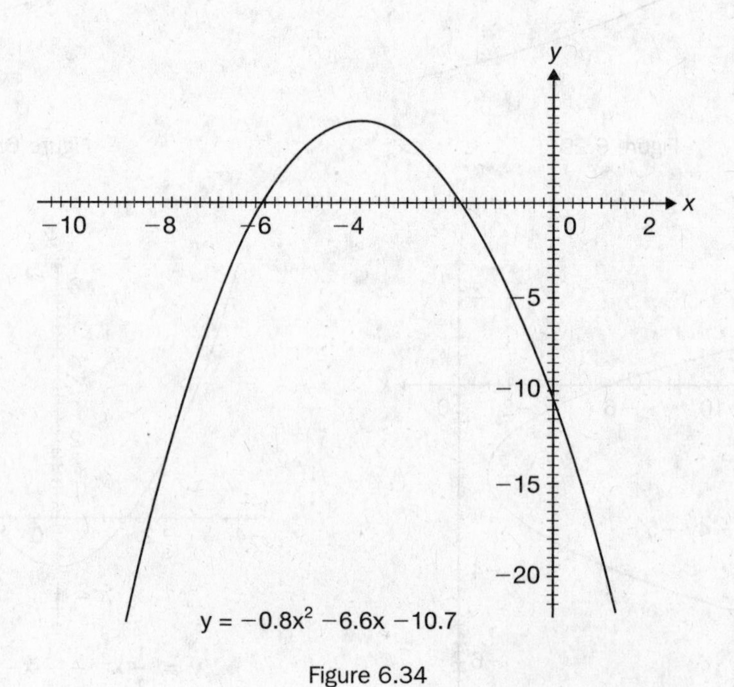

Figure 6.34

(*c*) There are no real zeros. The vertex is approximately (−2.500, −1.025). Your results may vary slightly. See Figure 6.35.

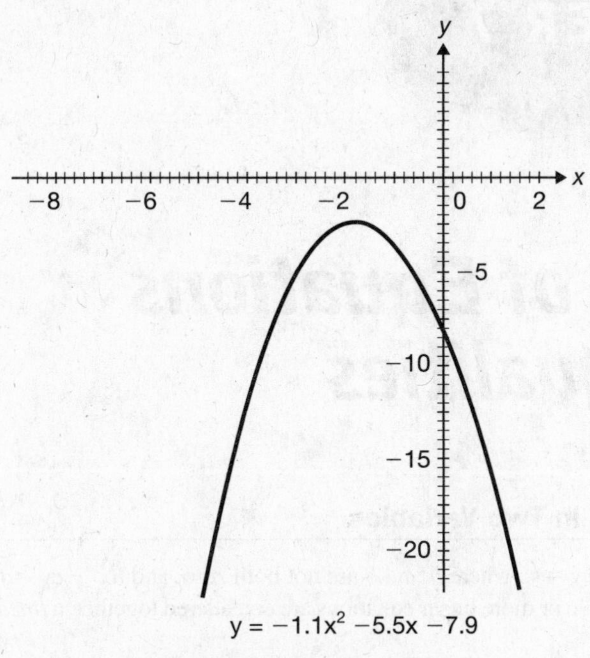

Figure 6.35

6.13 (*a*) $(-1,4)$ (*b*) $\left(-\infty,\frac{-5}{2}\right]\cup[3,\infty)$ (*c*) $\left[\frac{-7}{2},\frac{5}{2}\right]$

(*d*) $(-4,2)$ (*e*) $(-\infty,-2]\cup[0,\infty)$ (*f*) $\left(\frac{-5}{2},\frac{-1}{3}\right]$

Systems of Equations and Inequalities

7.1 Linear Systems in Two Variables

Equations of the form $ax + by = c$, where a and b are not both zero, and $dx + ey = f$, where d and e are not both zero, are linear equations. If two or more linear equations are considered together, a *linear system* is formed. We write

$$\begin{cases} ax + by = c \\ dx + ey = f \end{cases}$$

to represent the system.

Sometimes one of the equations may have the form $y = ax + b$. In that case the system may have the form

$$\begin{cases} y = ax + b \\ cx + dy = f \end{cases}.$$

The brace tells us to consider the equations together. The system is a 2×2 (read 2 by 2) system, since there are two equations and two variables.

Solving a system of equations consists of finding all the ordered pairs, if any, which satisfy each of the equations in the system.

How many ordered pairs can satisfy a system of two equations and two unknowns? The answer is clear if we look at graphs which represent the several possibilities.

Case 1:

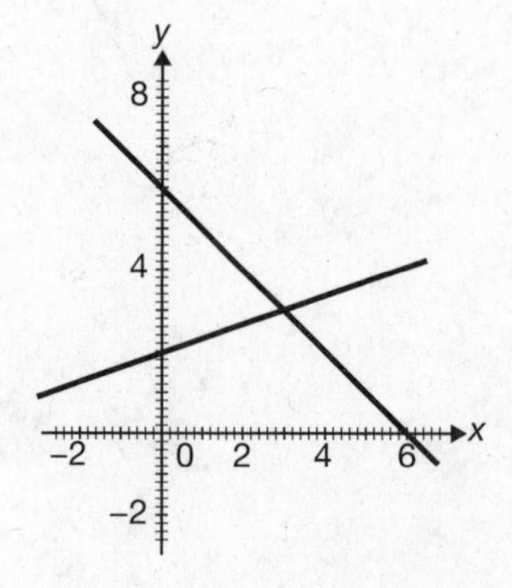

Figure 7.1

The lines intersect in exactly one point. The ordered pair that specifies the coordinates of the point of intersection of the lines is the solution to the system. The solution set contains that ordered pair as its only element. This system is said to be *consistent* and *independent*.

Case 2:

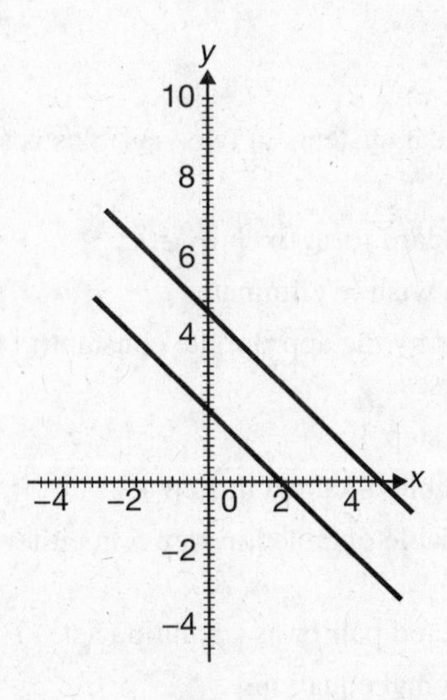

Figure 7.2

The lines are parallel; they do not intersect. There is no solution to the system. The solution set is the empty set Ø. This system is said to be *inconsistent*.

Case 3:

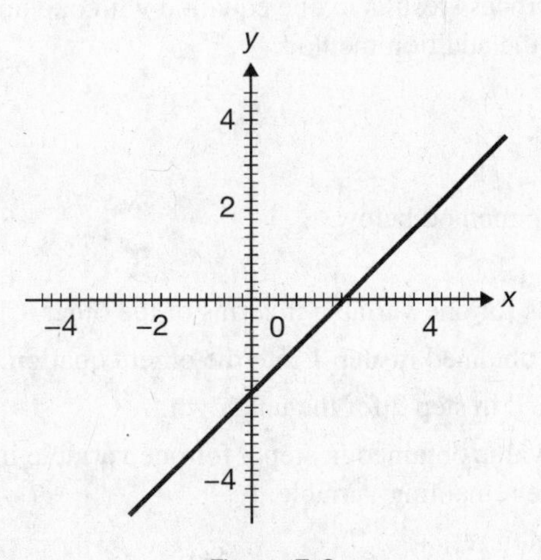

Figure 7.3

The equations represent the same line. All the points on one line are also on the other. There are infinitely many solutions to the system. The solution set contains infinitely many ordered pairs. This system is said to be *dependent*.

We now discuss algebraic methods that allow us to determine the solution(s), if any, to a linear system. The methods employed essentially eliminate unknowns until one equation in one unknown is obtained that we can solve. The result is then used to find the remaining unknown.

The Addition Method

The addition method involves the manipulation of the equations to obtain one equation in one unknown when equations are added. We ordinarily multiply one or both equations by suitable factors to accomplish the task. The appropriate factors are those which result in additive inverse coefficients of one variable in both equations. The process is illustrated in solved problem 7.1.

See solved problem 7.1.

The addition method of solving linear systems in two variables is summarized below.

1. Write the equations in the standard form $ax + by = c$.
2. Determine which variable you wish to eliminate.
3. Multiply one or both equations by the appropriate constant(s) to obtain coefficients of the chosen variable that are additive inverses.
4. Add the equations obtained in step 3.
5. Solve the equation in one variable obtained in step 4.
6. Substitute the value of the variable obtained in step 5 in either of the original equations and solve for the other variable.
7. Express the solution as an ordered pair or as a solution set.
8. Check your solution in the original equations.

Refer to supplementary problem 7.1 for similar exercises.

The Substitution Method

The substitution method entails solving one equation for one variable in terms of the other. We can avoid fractions if we solve for a variable that has a coefficient of $+1$ or -1. We then substitute the expression obtained into the remaining equation. The process results in one equation with one unknown that we can solve. We then proceed in a manner analogous to the addition method.

See solved problem 7.2.

We summarize the substitution method below.

1. Solve one of the equations for one variable in terms of the other. Clear fractions first if necessary.
2. Substitute the expression obtained in step 1 into the other equation.
3. Solve the equation obtained in step 2 for the unknown.
4. Substitute the numerical value obtained in step 3 for one variable into either equation in two variables and solve for the remaining variable.
5. State the solution or solution set.
6. Check the solution in each of the original equations.

Although we can use either of the methods discussed to solve linear systems of two equations in two unknowns, circumstances dictate which method may be the most efficient. As a general rule, if no variable has a coefficient of $+1$ or -1 in either equation, the addition method is preferred. Both techniques have advantages and disadvantages. Experience will facilitate your decision in this matter.

See supplementary problem 7.2 for additional drill.

Applications

We can apply the techniques of solving systems of equations to verbal problems. This capability provides us with more flexibility in how we set up the problem than was the case when only one variable was available. We illustrate.

See solved problem 7.3.

7.2 Linear Systems in Three Variables

A solution of an equation in three variables, such as $x - 2y + 3z = -4$, is an ordered triple of real numbers (x, y, z). Values for each of the three variables must be substituted into the equation before we can decide whether the result is a true statement. The values of the components in the ordered triple are listed in alphabetical order. Thus, $(3, 2, -1)$ and $(-4, 0, 0)$ are solutions of the equation, while $(1,0,1)$ is not. There are infinitely many ordered triples in this solution set.

The solution set of a system of three linear equations in three unknowns, such as

$$\begin{cases} x - 2y + 3z = -4 \\ 2x - y - z = 5 \\ 3x + 2y + z = 12 \end{cases}$$

is the intersection of the solution sets of the individual equations in the system.

The system is a 3×3 (read 3 by 3) system since it consists of three equations in three unknowns. The graph of a first-degree equation in two variables is a straight line in a two-dimensional coordinate system. The graph of a first-degree equation in three variables is a plane in a three-dimensional coordinate system. Each equation in a 3×3 system therefore represents a plane in three-dimensional space. Three planes may intersect at a unique point, may intersect at infinitely many points, or may not intersect at any points common to all three planes. There are therefore three possibilities for the solution set of a 3×3 system;

1. There is a unique solution of the system. It is consistent.
2. There are infinitely many solutions of the system. It is dependent.
3. There is no solution of the system. It is inconsistent.

We employ methods similar to those used to solve systems of linear equations in two variables to solve a 3×3 system. We eliminate variables from pairs of equations until an equation in one unknown is obtained. We then employ "back substitution" to find remaining unknowns. We illustrate by solving the system given above.

$$\begin{cases} x - 2y + 3z = -4 & (1) \\ 2x - y - z = 5 & (2) \\ 3x + 2y + z = 12 & (3) \end{cases}$$

Look at the system and choose a variable to eliminate first. Sometimes one variable can be eliminated more easily than the others. We shall eliminate y first using equations (1) and (3).

$$\begin{array}{ll} x - 2y + 3z = -4 & \\ 3x + 2y + z = 12 & \text{Add} \\ \hline 4x \qquad + 4z = 8 & (4) \end{array}$$

Now choose two different equations and eliminate y again. We shall use (2) and (3).

$$\begin{array}{lrl} \text{2 times (2)} \rightarrow & 4x - 2y - 2z = 10 & \\ & 3x + 2y + z = 12 & \text{Add} \\ \hline & 7x - z = 22 & (5) \end{array}$$

We now form a 2×2 system that consists of equations (4) and (5). Use the methods of the previous section to solve the system.

$$\begin{cases} 4x + 4z = 8 & (4) \\ 7x - z = 22 & (5) \end{cases}$$

$$\begin{array}{lrl} \frac{1}{4} \text{ times (4)} \rightarrow & x + z = 2 & \\ & 7x - z = 22 & \text{Add} \\ \hline & 8x + 0 = 24 \Rightarrow x = 3 & \end{array}$$

"Back substitute" by substituting $x = 3$ into (4) or (5) to find $z = -1$. Now substitute $x = 3$ and $z = -1$ into one of the original equations to find y. We choose equation (2) and find $y = 2$.

The solution is $x = 3, y = 2$, and $z = -1$ or $(3, 2, -1)$. The solution set is $\{(3, 2, -1)\}$. Check: Verify that $(3, 2, -1)$ satisfies each of the original equations.

The procedure used is summarized below. To solve a linear system in three variables:

1. Eliminate any variable from any pair of equations. Try to choose the most convenient variable to eliminate if there is one.
2. Use appropriate steps to eliminate the same variable as in step 1 from a different pair of the original equations.
3. Solve the resulting system of two equations in two variables.
4. Substitute the values obtained in step 3 into one of the original equations. Solve for the remaining variable.
5. Check the solution in each of the original equations.

If the resulting equation vanishes or yields a contradiction at any step in the process, the system contains dependent equations or else two or three inconsistent equations. The system then has infinitely many solutions or no solution, respectively.

See solved problem 7.4.

7.3 Determinants and Cramer's Rule

Determinants

The process of solving systems of linear equations is rather cumbersome. Fortunately the procedure is repetitive in nature. We shall introduce a technique that facilitates the process through the use of determinants. A *determinant* is a number that is associated with a square array of numbers.

Definition 1. A 2×2 *determinant*, designated by $\begin{vmatrix} a & b \\ c & d \end{vmatrix}$, has the value $ad - bc$. It is called a *second-order determinant*.

The expression used to obtain the value of the determinant seems arbitrary. There are logical reasons for using the stated expression which we choose to omit at this point.

See solved problem 7.5.

A third-order determinant has three rows and three columns. Its value is defined as follows.

Definition 2. A 3×3 determinant, designated by $\begin{vmatrix} a_1 & b_1 & c_1 \\ a_2 & b_2 & c_2 \\ a_3 & b_3 & c_3 \end{vmatrix}$ has value $a_1b_2c_3 + b_1c_2a_3 + c_1a_2b_3 - a_3b_2c_1 - b_3c_2a_1 - c_3a_2b_1$.

Because this definition is very cumbersome and difficult to remember, we evaluate third-order determinants by a method called *expansion by minors*. The *minor of an element* (number or letter) of a 3×3 determinant is the 2×2 determinant that remains after you delete the row and column which contain the element. Therefore, to evaluate a 3×3 determinant by expansion by minors of elements in the first column, apply the following relationship.

$$\begin{vmatrix} a_1 & b_1 & c_1 \\ a_2 & b_2 & c_2 \\ a_3 & b_3 & c_3 \end{vmatrix} = a_1\begin{vmatrix} b_2 & c_2 \\ b_3 & c_3 \end{vmatrix} - a_2\begin{vmatrix} b_1 & c_1 \\ b_3 & c_3 \end{vmatrix} + a_3\begin{vmatrix} b_1 & c_1 \\ b_2 & c_2 \end{vmatrix}.$$

Evaluate the 2×2 determinants as illustrated previously. Notice that the signs of the coefficients of the terms alternate. The result is unchanged if we expand the determinant by minors of elements in any column or row. The appropriate signs of the terms in the expansion are displayed by position in the array below.

$$\begin{matrix} + & - & + \\ - & + & - \\ + & - & + \end{matrix}$$

See solved problems 7.6–7.8.

Cramer's Rule

Cramer's rule may be employed to solve systems of linear equations. His rule expresses the solution for each variable as the quotient of two determinants. This allows us to readily employ computers to solve systems of linear equations. We first address systems of linear equations in two unknowns.

Cramer's Rule for 2 × 2 Systems

The solution to $\begin{cases} a_1x + b_1y = k_1 \\ a_2x + b_2y = k_2 \end{cases}$ is given by $x = D_x/D$ and $y = D_y/D$ where

$$D = \begin{vmatrix} a_1 & b_1 \\ a_2 & b_2 \end{vmatrix} \text{ and } D \neq 0,\ D_x = \begin{vmatrix} k_1 & b_1 \\ k_2 & b_2 \end{vmatrix} \text{ and } D_y = \begin{vmatrix} a_1 & k_1 \\ a_2 & k_2 \end{vmatrix}.$$

In practice, find D first since there is no unique solution if $D = 0$.

See solved problem 7.9.

Cramer's rule can be extended to 3×3 systems of linear equations.

Cramer's Rule for 3 × 3 Systems

The solution to $\begin{cases} a_1x + b_1y + c_1z = k_1 \\ a_2x + b_2y + c_2z = k_2 \\ a_3x + b_3y + c_3z = k_3 \end{cases}$ is given by $x = D_x/D$, $y = D_y/D$ and $z = D_z/D$,

$$D = \begin{vmatrix} a_1 & b_1 & c_1 \\ a_2 & b_2 & c_2 \\ a_3 & b_3 & c_3 \end{vmatrix} \text{ and } D \neq 0, D_x = \begin{vmatrix} k_1 & b_1 & c_1 \\ k_2 & b_2 & c_2 \\ k_3 & b_3 & c_3 \end{vmatrix},$$

$$D_y = \begin{vmatrix} a_1 & k_1 & c_1 \\ a_2 & k_2 & c_2 \\ a_3 & k_3 & c_3 \end{vmatrix} \text{ and } D_z = \begin{vmatrix} a_1 & b_1 & k_1 \\ a_2 & b_2 & k_2 \\ a_3 & b_3 & k_3 \end{vmatrix}.$$

See solved problem 7.10.

If $D = 0$, there is no unique solution to the system. The system is either inconsistent or dependent. If at least one, but not all, of D_x, D_y, or D_z as well as D is zero, the system is inconsistent. If D, D_x, D_y, and D_z are all zero, the system is dependent.

Refer to supplementary problem 7.6 to practice applying Cramer's rule.

7.4 Matrix Methods

We previously employed the addition (elimination) method to solve systems of linear equations. In that process, various operations were performed to alter the coefficients of variables in equations. Since our attention was primarily on the coefficients of the variables, we can streamline the process by writing these coefficients only in an orderly array. The array we shall employ is called a matrix. A *matrix* is a rectangular array of elements (entries). The *elements* are usually numbers or letters. The elements or entries are displayed in rows and columns within brackets or parentheses. The following illustrate the symbolism normally used.

$$\begin{bmatrix} 2 & 1 & 0 \\ -3 & 5 & -2 \\ 1 & 8 & 9 \end{bmatrix}, \quad \begin{bmatrix} 1 & -2 & -4 \\ 3 & 5 & -6 \end{bmatrix} \quad \text{and} \quad \begin{bmatrix} 4 \\ -7 \end{bmatrix}$$

The *size*, *dimension*, or *order* of a matrix is specified by stating the number of rows followed by the number of columns. The size, dimension, or order of the above matrices is 3×3 (read 3 by 3), 2×3 (read 2 by 3), and 2×1 (read 2 by 1), respectively. A *square matrix* has the same number of rows and columns.

We now illustrate how matrices are used to solve a system of linear equations. Consider the system

$$\begin{cases} 3s + \ \ t = 7 \\ 2s - 5t = -1 \end{cases}.$$

The matrix $\begin{bmatrix} 3 & 1 \\ 2 & -5 \end{bmatrix}$ is called the *coefficient matrix* of the system. It simply consists of the coefficients of the variables in the equations. The matrix $\begin{bmatrix} 7 \\ -1 \end{bmatrix}$ is called the *constant matrix* of the system. It is composed of the constants in the right members of the equations. The matrix $\begin{bmatrix} 3 & 1 & 7 \\ 2 & -5 & -1 \end{bmatrix}$ or $\left[\begin{array}{cc|c} 3 & 1 & 7 \\ 2 & -5 & -1 \end{array}\right]$ is called the *augmented matrix* of the system. It consists of the coefficient matrix of the system with the constant matrix of the system annexed on the right.

See solved problem 7.11.

Recall that equivalent equations have the same solution. *Equivalent systems of equations* likewise have the same solution(s). The associated augmented matrices of equivalent systems are *row-equivalent matrices*. Row-equivalent matrices are obtained by the following operations.

Elementary Row Operations

1. Interchange any two rows.
2. Multiply each element of any row by a nonzero constant.
3. Add a multiple of one row to another row.

We now illustrate the operations stated above and introduce notation that will help us represent those operations.

See solved problem 7.12.

The next concept needed is that of *row echelon form*. The following matrices are in row echelon form.

$$\begin{bmatrix} 1 & 2 & -1 \\ 0 & 1 & -7 \end{bmatrix}, \begin{bmatrix} 1 & 2 & -3 & 5 \\ 0 & 1 & 4 & -2 \\ 0 & 0 & 1 & 3 \end{bmatrix}, \begin{bmatrix} 1 & 2 & 3 & 2 & 5 \\ 0 & 0 & 1 & 5 & -6 \\ 0 & 0 & 0 & 1 & 2 \end{bmatrix} \text{ and } \begin{bmatrix} 1 & -1 & 0 & 4 \\ 0 & 1 & 1 & 3 \\ 0 & 0 & 0 & 0 \end{bmatrix}.$$

A matrix having the following characteristics is in row echelon form.

Row Echelon Form

1. All rows that contain only zeros are positioned at the bottom of the matrix.
2. If a row is not all zeros, the first nonzero element is a 1.
3. Each leading 1 is at least one column to the right of the leading 1 of the preceding row.

Normally row echelon form is accomplished by first obtaining a 1 in row 1, column 1. Use elementary row operations to next transform the remaining entries in column 1 to zeros. Then obtain a 1 in row 2, column 2 and zeros in the rows below row 2 in column 2. Obtain a 1 in row 3, column 3 and zeros below row 3 in column 3. Continue until row echelon form is obtained. Roughly speaking the lower left portion of a matrix in row echelon form is all zeros, while entries on the main diagonal, which begins at row 1, column 1, and ends at row n, column n, consist of ones.

We now illustrate the techniques employed to transform a matrix to row echelon form.

See solved problem 7.13.

We now employ the concepts given above to solve systems of linear equations using matrices. The process is summarized below. The method is called the *Gaussian elimination method*.

Method for Solving Linear Systems Using Matrices

1. Form the augmented matrix of the system.
2. Use elementary row operations to transform the augmented matrix to row echelon form.
3. Write the system of equations that corresponds to the echelon form matrix.
4. Use back substitution to solve the system.
5. Check the solution in the original equations.

See solved problem 7.14.

7.5 Nonlinear Systems

A nonlinear equation has degree two or more. The graph is not a straight line hence the term nonlinear. A *nonlinear system* has at least one nonlinear equation. Our concentration here is on solving nonlinear systems. The techniques employed eliminate variables, one by one, until an equation in one variable that we can solve remains. There are two methods used primarily: the substitution method and the addition method.

Substitution Method

The substitution method works very well when one of the equations in the system is linear. We simply solve the linear equation explicitly for one variable and substitute into the other equation. The resulting equation must be an equation we can solve.

See solved problem 7.15.

Addition Method

If both equations in a system are second-degree in both variables, the addition (elimination) method often works well.

See solved problem 7.16.

Graphical Method

A graphing calculator may be employed to solve nonlinear systems also. We graph the equations of the system and use the trace feature to approximate the points of intersection of the graphs, if any. The values thus obtained are the approximate solutions to the system.

See solved problem 7.17.

7.6 Systems of Inequalities

Systems of inequalities can be solved graphically. The solution set of a system of inequalities is simply the intersection of the solution sets of the individual inequalities. We first graph each inequality separately. Finally, the graphs are combined to display the solution set of the system.

See solved problem 7.18.

SOLVED PROBLEMS

7.1 Use the addition method to solve the following systems.

(*a*) $\begin{cases} x+y=-3 \\ x-y=11 \end{cases}$

The coefficients of *y* in the equations are additive inverses. Therefore *y* will be eliminated if the equations are added.

$$\begin{array}{rl} x+y=-3 & \\ x-y=11 & \text{Add} \\ \hline 2x+0=8 \Rightarrow x=4 & \end{array}$$

Now substitute 4 for *x* in either of the original equations and solve for *y*. We choose $x + y = -3$, since *y* can be readily isolated in the equation.

$$x + y = -3$$
$$4 + y = -3$$
$$y = -7$$

Check: We substitute $x = 4$ and $y = -7$ into both equations.

$$x + y = -3 \qquad\qquad x - y = 11$$
$$4 + (-7) \stackrel{?}{=} -3 \qquad\qquad 4 - (-7) \stackrel{?}{=} 11$$
$$-3 = -3 \qquad\qquad 11 = 11$$

The solution to the system is $(4, -7)$. The solution set is $\{(4, -7)\}$. The equations are consistent and independent.

(*b*) $\begin{cases} 3s + t = 7 \\ 2s - 5t = -1 \end{cases}$

If the first equation is multiplied by 5, the coefficients of t are additive inverses. The variable t is eliminated when the equations are added.

$$\begin{cases} 3s + t = 7 \\ 2s - 5t = -1 \end{cases} \quad \text{becomes} \quad \begin{cases} 15s + 5t = 35 \\ 2s - 5t = -1 \end{cases}$$

$$\begin{array}{r} 15s + 5t = 35 \\ 2s - 5t = -1 \\ \hline 17s + 0 = 34 \end{array} \quad \text{Add} \quad \Rightarrow s = 2$$

Now substitute 2 for s in either of the original equations and solve for t. We choose $3s + t = 7$, since t can be easily isolated.

$$3s + t = 7$$
$$3(2) + t = 7$$
$$6 + t = 7 \Rightarrow t = 1$$

Check: We substitute $s = 2$ and $t = 1$ into both equations.

$$\begin{aligned} 3s + t &= 7 \\ 3(2) + 1 &\stackrel{?}{=} 7 \\ 6 + 1 &\stackrel{?}{=} 7 \\ 7 &= 7 \end{aligned} \qquad \begin{aligned} 2s - 5t &= -1 \\ 2(2) - 5(1) &\stackrel{?}{=} -1 \\ 4 - 5 &\stackrel{?}{=} -1 \\ -1 &= -1 \end{aligned}$$

The solution to the system is (2, 1). The solution set is {(2, 1)}. The equations are consistent and independent.

(*c*) $\begin{cases} 2x - y = 4 \\ y = 6 \end{cases}$

The second equation states the value of y directly. We need only to substitute $y = 6$ into the first equation and solve for x.

$$2x - y = 4$$
$$2x - 6 = 4$$
$$2x = 10 \Rightarrow x = 5$$

Check: Substitute into both equations.

$$\begin{aligned} 2x - y &= 4 \\ 2(5) - 6 &\stackrel{?}{=} 4 \\ 10 - 6 &\stackrel{?}{=} 4 \\ 4 &= 4 \end{aligned} \qquad \begin{aligned} y &= 6 \\ 6 &= 6 \end{aligned}$$

The solution is (5, 6). The solution set is {(5, 6)}. The equations are consistent and independent.

(*d*) $\begin{cases} 3v - 5w = -1 \\ v + 2w = 7 \end{cases}$

Multiply the second equation by (-3) and add.

$$\begin{cases} 3v - 5w = -1 \\ v + 2w = 7 \end{cases} \quad \text{becomes} \quad \begin{cases} 3v - 5w = -1 \\ -3v - 6w = -21 \end{cases}$$

$$\begin{array}{r} 3v - 5w = -1 \\ -3v - 6w = -21 \\ \hline 0 - 11w = -22 \end{array} \quad \text{Add} \quad \Rightarrow w = 2$$

Now substitute $w = 2$ into either of the original equations and solve for v. We choose $v + 2w = 7$.

$$v + 2w = 7$$
$$v + 2(2) = 7$$
$$v + 4 = 7$$
$$v = 3$$

Check: Substitute into both equations.

$$\begin{aligned} 3v - 5w &= -1 \\ 3(3) - 5(2) &\stackrel{?}{=} -1 \\ 9 - 10 &\stackrel{?}{=} -1 \\ -1 &= -1 \end{aligned} \qquad \begin{aligned} v + 2w &= 7 \\ 3 + 2(2) &\stackrel{?}{=} 7 \\ 3 + 4 &\stackrel{?}{=} 7 \\ 7 &= 7 \end{aligned}$$

The solution is $(3, 2)$. The solution set is $\{(3, 2)\}$. The equations are consistent and independent.

(*e*) $\begin{cases} 2t - 3u = 13 \\ 5t + 2u = 4 \end{cases}$

We choose to eliminate u since the coefficients are opposite in sign. We therefore need not multiply by negative factors. Multiply the first equation by 2 and the second equation by 3 to obtain additive inverse coefficients of u in the equations.

$$\begin{cases} 2t - 3u = 13 \\ 5t + 2u = 4 \end{cases} \quad \text{becomes} \quad \begin{cases} 4t - 6u = 26 \\ 15t + 6u = 12 \end{cases}$$

$$\begin{array}{r} 4t - 6u = 26 \\ 15t + 6u = 12 \\ \hline 19t + 0 = 38 \end{array} \quad \text{Add} \quad \Rightarrow t = 2$$

Now substitute $t = 2$ into either of the original equations and solve for u. We choose $5t + 2u = 4$ since the coefficient of u is positive.

$$5t + 2u = 4$$
$$5(2) + 2u = 4$$
$$10 + 2u = 4$$
$$2u = -6 \Rightarrow u = -3$$

Check: Substitute into both equations.

$$2t - 3u = 13 \qquad 5t + 2u = 4$$

$$2(2) - 3(-3) \stackrel{?}{=} 13 \qquad 5(2) + 2(-3) \stackrel{?}{=} 4$$

$$4 + 9 \stackrel{?}{=} 13 \qquad 10 - 6 \stackrel{?}{=} 4$$

$$13 = 13 \qquad 4 = 4$$

The solution is $(2, -3)$. The solution set is $\{(2, -3)\}$. The equations are consistent and independent.

$$(f) \quad \begin{cases} \frac{1}{2}x - \frac{2}{3}y = \frac{1}{3} \\ \frac{-1}{4}x + \ \ y = \frac{-1}{8} \end{cases}$$

Our approach will be more apparent if we first clear fractions. Multiply the first equation by 6 and the second by 8.

$$\begin{cases} \frac{1}{2}x - \frac{2}{3}y = \frac{1}{3} \\ \frac{-1}{4}x + \ \ y = \frac{-1}{8} \end{cases} \quad \text{becomes} \quad \begin{cases} 3x - 4y = 2 \\ -2x + 8y = -1 \end{cases}$$

The coefficients of y will be additive inverses if we now multiply $3x - 4y = 2$ by 2.

$$\begin{cases} 3x - 4y = 2 \\ -2x + 8y = -1 \end{cases} \quad \text{becomes} \quad \begin{cases} 6x - 8y = 4 \\ -2x + 8y = -1 \end{cases}$$

$$\begin{array}{r} 6x - 8y = 4 \\ -2x + 8y = -1 \\ \hline 4x + \ 0 = 3 \end{array} \quad \text{Add} \quad \Rightarrow x = \frac{3}{4}$$

Now substitute $x = \frac{3}{4}$ into $\frac{-1}{4}x + y = \frac{-1}{8}$ and solve for y.

$$\frac{-1}{4}\left(\frac{3}{4}\right) + y = \frac{-1}{8}$$

$$\frac{-3}{16} + y = \frac{-1}{8}$$

$$y = \frac{-1}{8} + \frac{3}{16} = \frac{-2+3}{16} = \frac{1}{16}$$

Check: Substitute into the original equations.

$$\frac{1}{2}x - \frac{2}{3}y = \frac{1}{3} \qquad \frac{-1}{4}x + y = \frac{-1}{8}$$

$$\frac{1}{2}\left(\frac{3}{4}\right) - \frac{2}{3}\left(\frac{1}{16}\right) \stackrel{?}{=} \frac{1}{3} \qquad \frac{-1}{4}\left(\frac{3}{4}\right) + \frac{1}{16} \stackrel{?}{=} \frac{-1}{8}$$

$$\frac{3}{8} - \frac{1}{24} \stackrel{?}{=} \frac{1}{3} \qquad \frac{-3}{16} + \frac{1}{16} \stackrel{?}{=} \frac{-1}{8}$$

$$\frac{9-1}{24} \stackrel{?}{=} \frac{1}{3} \qquad \frac{-2}{16} \stackrel{?}{=} \frac{-1}{8}$$

$$\frac{8}{24} \stackrel{?}{=} \frac{1}{3} \qquad \frac{-1}{8} = \frac{-1}{8}$$

$$\frac{1}{3} = \frac{1}{3}$$

The solution is $\left(\frac{3}{4}, \frac{1}{16}\right)$. The solution set is $\left\{\left(\frac{3}{4}, \frac{1}{16}\right)\right\}$. The system is consistent and independent.

(g) $\begin{cases} 2x - 5y = 3 \\ 6x - 15y = 11 \end{cases}$

Multiply the first equation by (−3).

$$\begin{cases} 2x - 5y = 3 \\ 6x - 15y = 11 \end{cases} \quad \text{becomes} \quad \begin{cases} -6x + 15y = -9 \\ 6x - 15y = 11 \end{cases}$$

$$\begin{array}{r} -6x + 15y = -9 \\ 6x - 15y = 11 \\ \hline 0 + 0 = 2 \end{array} \quad \text{Add; is false.}$$

There is no solution to the system. The solution set is { } = Ø. The lines are parallel. The system is inconsistent.

(h) $\begin{cases} 2x - 5y = 3 \\ 6x - 15y = 9 \end{cases}$

Multiply the first equation by (−3).

$$\begin{cases} 2x - 5y = 3 \\ 6x - 15y = 9 \end{cases} \quad \text{becomes} \quad \begin{cases} -6x + 15y = -9 \\ 6x - 15y = 9 \end{cases}$$

$$\begin{array}{r} -6x + 15y = -9 \\ 6x - 15y = 9 \\ \hline 0 + 0 = 0 \end{array} \quad \text{Add; is always true.}$$

All of the ordered pairs which satisfy the first equation also satisfy the second. The equations represent the same line. The solution set consists of the infinitely many ordered pairs which satisfy either equation. The equations are dependent.

7.2 Use the substitution method to solve the following.

(a) $\begin{cases} x - 3y = -7 \\ 4x + 3y = 2 \end{cases}$

Solve $x - 3y = -7$ for x in terms of y: $x - 3y = -7 \Rightarrow x = 3y - 7$

Now substitute $3y - 7$ for x into $4x + 3y = 2$ and solve for y.

$$4(3y - 7) + 3y = 2$$
$$12y - 28 + 3y = 2$$
$$15y = 30 \Rightarrow y = 2$$

Now substitute 2 for y into $x = 3y - 7$ and solve for x: $x = 3(2) - 7 = 6 - 7 = -1$

Check:

$$\begin{array}{rl} x - 3y = -7 & \qquad 4x + 3y = 2 \\ -1 - 3(2) \stackrel{?}{=} -7 & \qquad 4(-1) + 3(2) \stackrel{?}{=} 2 \\ -1 - 6 \stackrel{?}{=} -7 & \qquad -4 + 6 \stackrel{?}{=} 2 \\ -7 = -7 & \qquad 2 = 2 \end{array}$$

The solution is $(-1, 2)$. The solution set is $\{(-1, 2)\}$. The system is consistent and independent.

(*b*) $\begin{cases} 3s - 5t = -1 \\ s + 2t = 7 \end{cases}$

Solve $s + 2t = 7$ for s in terms of t: $s + 2t = 7 \Rightarrow s = 7 - 2t$

Now substitute $7 - 2t$ for s into $3s - 5t = -1$ and solve for t.

$$3(7 - 2t) - 5t = -1$$
$$21 - 6t - 5t = -1$$
$$21 - 11t = -1$$
$$-11t = -22 \Rightarrow t = 2$$

Now substitute 2 for t into $s = 7 - 2t$ and solve for s: $s = 7 - 2(2) = 7 - 4 = 3$

Check:

$3s - 5t = -1$	$s + 2t = 7$
$3(3) - 5(2) \stackrel{?}{=} -1$	$3 + 2(2) \stackrel{?}{=} 7$
$9 - 10 \stackrel{?}{=} -1$	$3 + 4 \stackrel{?}{=} 7$
$-1 = -1$	$7 = 7$

The solution is $(3, 2)$. The solution set is $\{(3, 2)\}$. The system is consistent and independent.

(*c*) $\begin{cases} \frac{t}{2} - \frac{w}{4} = \frac{-7}{4} \\ 3t - 2w = -6 \end{cases}$

Clear fractions first to obtain $\begin{cases} 2t - w = -7 \\ 3t - 2w = -6 \end{cases}$. Now solve $2t - w = -7$ for w in terms of t: $2t - w = -7$ $\Rightarrow 2t + 7 = w$. Now substitute $2t + 7$ for w into $3t - 2w = -6$ and solve for t.

$$3t - 2(2t + 7) = -6$$
$$3t - 4t - 14 = -6$$
$$-t = 8 \text{ or } t = -8$$

Next substitute -8 for t into $w = 2t + 7$ and solve for w: $w = 2(-8) + 7 = -16 + 7 = -9$.

Check:

$\frac{t}{2} - \frac{w}{4} = \frac{-7}{4}$	$3t - 2w = -6$
$\frac{-8}{2} - \frac{-9}{4} \stackrel{?}{=} \frac{-7}{4}$	$3(-8) - 2(-9) \stackrel{?}{=} -6$
$\frac{-16}{4} + \frac{9}{4} \stackrel{?}{=} \frac{-7}{4}$	$-24 + 18 \stackrel{?}{=} -6$
$\frac{-7}{4} = \frac{-7}{4}$	$-6 = -6$

The solution is $(-8, -9)$. The solution set is $\{(-8, -9)\}$. The system is consistent and independent.

(*d*) $\begin{cases} 0.1x - 1.2y = 0.3 \\ 0.4x - 4.8y = 1.6 \end{cases}$

Clear fractions by multiplying each equation by 10.

$$\begin{cases} 0.1x - 1.2y = 0.3 \\ 0.4x - 4.8y = 1.6 \end{cases} \quad \text{becomes} \quad \begin{cases} x - 12y = 3 \\ 4x - 48y = 16 \end{cases}$$

Solve $x - 12y = 3$ for x in terms of y: $x - 12y = 3 \Rightarrow x = 12y + 3$

Next substitute $12y + 3$ for x into $4x - 48y = 16$ and solve for y.

$$4(12y + 3) - 48y = 16$$
$$48y + 12 - 48y = 16$$
$$12 = 16 \text{ is false.}$$

If no errors were made, we conclude there is no solution to the system. The solution set is $\{\ \} = \emptyset$. The system is inconsistent. The lines are parallel.

7.3 Use a system of equations to solve the following.

(*a*) An auto parts store ordered a combined total of 40 cases of oil filters and air cleaners that cost a total of $2807.60. Each case of oil filters cost $59.99 while each case of air cleaners cost $83.99. How many cases of oil filters and how many cases of air cleaners were ordered?

We must define variables, write appropriate equations, and form the system to be solved. Use the first letter of a key word to define variables to facilitate keeping the meaning of the variables in mind. Let f be the number of cases of oil filters ordered and c be the number of cases of air cleaners ordered. Since there are two unknowns, we must find two distinct equations that relate them. We know that a combined total of 40 cases were ordered. Therefore $f + c = 40$ is one equation. Now write a cost equation from the information given. It is $59.99f + 83.99c = 2807.60$. Form the system and solve it.

$$\begin{cases} f + c = 40 \\ 59.99f + 83.99c = 2807.60 \end{cases}$$

We solve $f + c = 40$ for f in terms of c: $f = 40 - c$, and substitute into the remaining equation. The details are omitted. We determine $c = 17$ and $f = 23$. Hence 23 cases of oil filters and 17 cases of air cleaners were ordered by the auto parts store.

(*b*) The Cash's have money invested in Best Fund and Secure Fund. The Best Fund account earns 4.2% interest while the Secure Fund account earns 4.0% interest. Their combined investment of $61,500 had total earnings of $2534 last year. How much did the Cash's have invested in each fund last year?

Define variables, form the system, and solve it. Let B represent the amount invested in Best Fund and S represent the amount invested in Secure Fund. The system is

$$\begin{cases} B + S = 61500 \\ 0.042B + 0.0405 = 2534 \end{cases}$$

Use the substitution method to find $B = 37{,}000$ and $S = 24{,}500$. The Cash's had $37,000 invested in Best Fund and $24,500 invested in Secure Fund last year.

(*c*) Mel has two part-time jobs. One week he earned $179.50 by working 10 hours on lawns and 7 hours making deliveries. The previous week he earned $176.00 by working 9 hours on lawn jobs and 8 hours making deliveries. How much does he earn per hour on each job?

Let l be the hourly rate for lawn jobs and let d be the hourly rate making deliveries. The system is

$$\begin{cases} 10l + 7d = 179.50 \\ 9l + 8d = 176.00 \end{cases}$$

Employ the addition method to obtain $l = 12$ and $d = 8.5$. Mel earns $12.00 per hour on lawn jobs and $8.50 per hour making deliveries.

See supplementary problem 7.3 for similar problems.

7.4 Solve the following systems.

$$(a)\quad \begin{cases} 2x + y - z = -1 & (1) \\ x - 3y + z = 7 & (2) \\ 3x + 2y + 2z = 16 & (3) \end{cases}$$

We choose to eliminate z using (1) and (2).

$$\begin{array}{rl} 2x + y - z = -1 & \\ x - 3y + z = 7 & \text{Add} \\ \hline 3x - 2y \quad = 6 & (4) \end{array}$$

Next eliminate z using (1) and (3).

$$\begin{array}{rrl} \text{2 times (1)} \rightarrow & 4x + 2y - 2z = -2 & \\ & 3x + 2y + 2z = 16 & \text{Add} \\ \hline & 7x + 4y \quad = 14 & (5) \end{array}$$

Form and solve the 2 × 2 system using (4) and (5).

$$\begin{array}{rrl} & \begin{cases} 3x - 2y = 6 \\ 7x + 4y = 14 \end{cases} & \begin{array}{l} (4) \\ (5) \end{array} \\ \text{2 times (4)} \rightarrow & 6x - 4y = 12 & \\ & 7x + 4y = 14 & \text{Add} \\ \hline & 13x \quad = 26 \Rightarrow x = 2 & \end{array}$$

Substitute $x = 2$ into (4) or (5) and find $y = 0$. Then substitute $x = 2$ and $y = 0$ into (1) or (2) to find $z = 5$. The solution is $x = 2, y = 0$, and $z = 5$ or $(2, 0, 5)$. The solution set is $\{(2, 0, 5)\}$.

Check: Verify that (2,0,5) satisfies each of the original equations.

$$(b)\quad \begin{cases} -x + 3y + z = -1 & (1) \\ x + 2y - 3z = -13 & (2) \\ 3x - y + 5z = 3 & (3) \end{cases}$$

We use (1) and (2) to eliminate x.

$$\begin{array}{rl} -x + 3y + z = -1 & \\ x + 2y - 3z = -13 & \text{Add} \\ \hline 5y - 2z = -14 & (4) \end{array}$$

Next eliminate x using (1) and (3).

$$\begin{array}{rrl} \text{3 times (1)} \rightarrow & -3x + 9y + 3z = -3 & \\ & 3x - y + 5z = 3 & \text{Add} \\ \hline & 8y + 8z = 0 & (5) \end{array}$$

Form and solve the 2 × 2 system using (4) and (5).

$$\begin{cases} 5y - 2z = -14 & (4) \\ 8y + 8z = 0 & (5) \end{cases}$$

$$\begin{array}{rrl} & 5y - 2z = -14 & \\ \frac{1}{4} \text{ times (5)} \rightarrow & 2y + 2z = 0 & \text{Add} \\ \hline & 7y \quad = -14 \Rightarrow y = -2 & \end{array}$$

Substitute $y = -2$ into (4) or (5) and find $z = 2$. Next substitute $y = -2$ and $z = 2$ into (1) or (2) to find $x = -3$. The solution is $x = -3, y = -2$, and $z = 2$ or $(-3, -2, 2)$. The solution set is $\{(-3, -2, 2)\}$.

Check: Verify that $(-3, -2, 2)$ satisfies each of the original equations.

$$(c) \quad \begin{cases} 2s + 4t + 8u = 12 & (1) \\ \frac{s}{2} + t + 2u = 3 & (2) \\ -s - 2t - 4u = -6 & (3) \end{cases}$$

We use (1) and (3) to eliminate s.

$$\begin{array}{rrl} & 2s + 4t + 8u = 12 & \\ 2 \text{ times (3)} \rightarrow & -2s - 4t - 8u = -12 & \text{Add} \\ \hline & 0 = 0 & \text{is always true} \end{array}$$

Equations (1) and (3) are dependent. The equations are equivalent and represent the same plane. Now eliminate s using (2) and (3).

$$\begin{array}{rrl} 2 \text{ times (2)} \rightarrow & s + 2t + 4u = 6 & \\ & -s - 2t - 4u = -6 & \text{Add} \\ \hline & 0 = 0 & \text{is always true} \end{array}$$

Equations (2) and (3) are dependent also. All three equations are equivalent and represent the same plane. There are infinitely many solutions. The solution set is the set of all ordered triples of real numbers which satisfy any one of the equations. Therefore the solution set is $\{(s, t, u) | 2s + 4t + 8u = 12 \text{ where } s, t, u \in R\}$

$$(d) \quad \begin{cases} 2x - \ y + 3z = -4 & (1) \\ x + 2y - 5z = -6 & (2) \\ 4x - 2y + 6z = -8 & (3) \end{cases}$$

We shall use (1) and (2) to eliminate y.

$$\begin{array}{rrll} 2 \text{ times (1)} \rightarrow & 4x - 2y + 6z = -8 & & \\ & x + 2y - 5z = -6 & \text{Add} & \\ \hline & 5x \quad + z = -14 & & (4) \end{array}$$

Next use (1) and (3) to again eliminate y.

$$\begin{array}{rrl} -2 \text{ times (1)} \rightarrow & -4x + 2y - 6z = 8 & \\ & 4x - 2y + 6z = -8 & \text{Add} \\ \hline & 0 = 0 & \text{is always true} \end{array}$$

Equations (1) and (3) are dependent. They represent the same plane. The three equations in the system actually represent two unique planes. Two planes that are not parallel intersect in a line. Therefore the solution set consists of infinitely many points. The technique employed in this situation is to express the infinitely many ordered triples in terms of one of the variables. Equation (4) can readily be solved for z in terms of x, so we shall express the ordered triples in the solution set in terms of arbitrary x-coordinates.

$$5x + z = -14$$

$$z = -5x - 14 \qquad (5)$$

Now use back substitution in (1) to obtain an equation involving x and y and solve for y in terms of x.

$$2x - y + 3z = -4 \text{ so}$$

$$2x - y + 3(-5x - 14) = -4$$

$$2x - y - 15x - 42 = -4$$

$$-13x - y - 38 = 0$$

$$y = -13x - 38 \qquad (6)$$

Equations (5) and (6) can be used to express the solution set. Suppose x is any real number n. Use (6) to obtain $y = -13n - 38$ and (5) to obtain $z = -5n - 14$. The solution set consists of all ordered triples of the form $(n, -13n - 38, -5n - 14)$. Hence the solution is $\{(n, -13n - 38, -5n - 14), n \in R\}$. Some specific ordered triples in the solution set can be obtained by choosing particular n's. For example:

If $n = 0$, we obtain $(0, -38, -14)$.

If $n = -1$, we obtain $(-1, -25, -9)$.

If $n = 5$, we obtain $(5, -103, -39)$.

Check: These particular points satisfy all of the original equations. We cannot do an all-inclusive check point by point since there are infinitely many ordered triples in the solution set. It is possible to perform a general check algebraically, however, by using $x = n$, $y = -13n$, and $z = -5n - 14$. You should verify that each of the original equations is satisfied by those substitutions (be sure to employ parentheses around each substituted expression).

(*e*) $$\begin{cases} 3a - 2b + c = 4 & (1) \\ 2a + b - c = 4 & (2) \\ -6a + 4b - 2c = 1 & (3) \end{cases}$$

We use (1) and (2) to eliminate c

$$\begin{array}{rl} 3a - 2b + c = 4 & \\ 2a + b - c = 4 & \text{Add} \\ \hline 5a - b \quad = 8 & \qquad (4) \end{array}$$

Next use (1) and (3) to eliminate c

$$\begin{array}{rrl} \text{2 times (1)} \rightarrow & 6a - 4b + 2c = 8 & \\ & -6a + 4b - 2c = 1 & \text{Add} \\ \hline & 0 = 9 & \text{is a contradiction.} \end{array}$$

Equations (1) and (3) are inconsistent. The planes are parallel and do not intersect. There is no solution to the system. The solution set is $\{ \} = \emptyset$.

(*f*)
$$\begin{cases} \dfrac{x}{2} + 5y + \dfrac{z}{4} = \dfrac{-9}{4} & (1) \\ 4x \qquad\qquad = z & (2) \\ \qquad 10y + z = -4 & (3) \end{cases}$$

We begin by clearing fractions and rewriting (2)

$$\begin{array}{rl} \text{4 times (1)} \rightarrow & \begin{cases} 2x + 20y + z = -9 & (4) \\ 4x \qquad - z = 0 & (5) \\ \qquad 10y + z = -4 & (6) \end{cases} \\ \text{Subtract } z \rightarrow & \end{array}$$

Since y does not appear in (5), let's eliminate y from (6) also.

$$\begin{array}{rrl} & 2x + 20y + z = -9 & \\ -2 \text{ times (6)} \rightarrow & -20y - 2z = 8 & \text{Add} \\ \hline & 2x \qquad - z = -1 & (7) \end{array}$$

Form and solve the 2×2 system using (5) and (7).

$$\begin{cases} 4x - z = 0 & (5) \\ 2x - z = -1 & (7) \end{cases}$$

$$\begin{array}{rrl} & 4x - z = 0 & \\ -1 \text{ times (7)} \rightarrow & -2x + z = 1 & \text{Add} \\ \hline & 2x \quad = 1 \Rightarrow x = \frac{1}{2} & \end{array}$$

Substitute $x = \frac{1}{2}$ into (2) to find $z = 2$. Next substitute $z = 2$ into (3) and find $y = \frac{-6}{10} = \frac{-3}{5}$. The solution is $x = \frac{1}{2}, y = \frac{-3}{5}, z = 2$ or $\left(\frac{1}{2}, \frac{-3}{5}, 2\right)$. The solution set is $\left\{\left(\frac{1}{2}, \frac{-3}{5}, 2\right)\right\}$.

Check: Verify that $\left(\frac{1}{2}, \frac{-3}{5}, 2\right)$ satisfies each of the original equations.

See supplementary problem 7.4 for similar problems.

7.5 Evaluate the following.

(*a*) $\begin{vmatrix} 2 & 3 \\ 1 & 5 \end{vmatrix}$

$$\begin{vmatrix} 2 & 3 \\ 1 & 5 \end{vmatrix} = 2(5) - 3(1) = 10 - 3 = 7$$

(*b*) $\begin{vmatrix} -3 & -2 \\ 4 & 5 \end{vmatrix}$

$$\begin{vmatrix} -3 & -2 \\ 4 & 5 \end{vmatrix} = -3(5) - (-2)(4) = -15 - (-8) = -15 + 8 = -7$$

(*c*) $\begin{vmatrix} 5 & 6 \\ 0 & -2 \end{vmatrix}$

$$\begin{vmatrix} 5 & 6 \\ 0 & -2 \end{vmatrix} = 5(-2) - 6(0) = -10 - 0 = -10$$

(*d*) $\begin{vmatrix} 4 & -6 \\ 2 & -3 \end{vmatrix}$

$$\begin{vmatrix} 4 & -6 \\ 2 & -3 \end{vmatrix} = 4(-3) - (-6)(2) = -12 - (-12) = -12 + 12 = 0$$

(*e*) $\begin{vmatrix} 2 & 3 \\ x & y \end{vmatrix}$

$$\begin{vmatrix} 2 & 3 \\ x & y \end{vmatrix} = 2y - 3x$$

7.6 Find (*a*) the minor of 2 and (*b*) the minor of -3 in the determinant $\begin{vmatrix} 1 & 2 & 3 \\ 0 & 4 & 5 \\ -3 & 0 & -1 \end{vmatrix}$.

(*a*) The element 2 is contained in row 1 and column 2. Delete the elements in row 1 and column 2 to obtain

$\begin{vmatrix} 0 & 5 \\ -3 & -1 \end{vmatrix}$ as the minor of 2.

(*b*) The element -3 is contained in row 3 and column 1. Delete the elements in row 3 and column 1 to obtain

$\begin{vmatrix} 2 & 3 \\ 4 & 5 \end{vmatrix}$ as the minor of -3.

7.7 Evaluate $\begin{vmatrix} 1 & 2 & 3 \\ 0 & 4 & 2 \\ -3 & 0 & -1 \end{vmatrix}$ by expanding by minors of elements in the (*a*) first row; (*b*) first column.

(*a*) Expanding by minors of elements in row 1 we obtain

$$\begin{vmatrix} 1 & 2 & 3 \\ 0 & 4 & 2 \\ -3 & 0 & -1 \end{vmatrix} = 1\begin{vmatrix} 4 & 2 \\ 0 & -1 \end{vmatrix} - 2\begin{vmatrix} 0 & 2 \\ -3 & -1 \end{vmatrix} + 3\begin{vmatrix} 0 & 4 \\ -3 & 0 \end{vmatrix}$$
$$= 1[4(-1) - 2(0)] - 2[0(-1) - 2(-3)] + 3[0(0) - 4(-3)]$$
$$= 1[-4 - 0] - 2[0 + 6] + 3[0 + 12] = 1[-4] - 2[6] + 3[12]$$
$$= -4 - 12 + 36 = 20$$

(*b*) Expanding by minors of elements in column 1 we obtain

$$\begin{vmatrix} 1 & 2 & 3 \\ 0 & 4 & 2 \\ -3 & 0 & -1 \end{vmatrix} = 1\begin{vmatrix} 4 & 2 \\ 0 & -1 \end{vmatrix} - 0\begin{vmatrix} 2 & 3 \\ 0 & -1 \end{vmatrix} + (-3)\begin{vmatrix} 2 & 3 \\ 4 & 2 \end{vmatrix}$$
$$= 1[4(-1) - 2(0)] - 0[2(-1) - 3(0)] - 3[2(2) - 3(4)]$$
$$= 1[-4 - 0] - 0[-2 - 0] - 3[4 - 12] = 1[-4] - 0[-2] - 3[-8]$$
$$= -4 - 0 + 24 = 20$$

Notice the same value was obtained in both cases above. Since the result is unchanged by expansion of minors of elements in any row or column, it is advantageous to choose a row or column that contains the most zeros, if any. Concentrate on the appropriate signs as you do the calculations.

7.8 Evaluate.

(*a*) $\begin{vmatrix} 1 & -2 & 0 \\ -1 & 1 & 3 \\ 2 & 4 & -1 \end{vmatrix}$

Since row 1 and column 3 contain a zero, we could expand by minors of elements in either row 1 or column 3. We choose row 1. Refer to the sign array chart for the appropriate signs of the terms in the expansion.

$$\begin{vmatrix} 1 & -2 & 0 \\ -1 & 1 & 3 \\ 2 & 4 & -1 \end{vmatrix} = 1\begin{vmatrix} 1 & 3 \\ 4 & -1 \end{vmatrix} - (-2)\begin{vmatrix} -1 & 3 \\ 2 & -1 \end{vmatrix} + 0\begin{vmatrix} -1 & 1 \\ 2 & 4 \end{vmatrix}$$
$$= 1[1(-1) - 3(4)] + 2[-1(-1) - 3(2)] + 0$$
$$= 1[-1 - 12] + 2[1 - 6] = 1[-13] + 2[-5] = -13 - 10 = -23$$

(*b*) $\begin{vmatrix} 2 & -1 & -3 \\ 2 & 3 & 2 \\ -2 & 4 & -4 \end{vmatrix}$

No row or column contains a zero. We arbitrarily choose to expand by minors of elements in row 2. Refer to the sign array chart for the appropriate signs of the terms.

$$\begin{vmatrix} 2 & -1 & -3 \\ 2 & 3 & 2 \\ -2 & 4 & -4 \end{vmatrix} = -2\begin{vmatrix} -1 & -3 \\ 4 & -4 \end{vmatrix} + 3\begin{vmatrix} 2 & -3 \\ -2 & -4 \end{vmatrix} - 2\begin{vmatrix} 2 & -1 \\ -2 & 4 \end{vmatrix}$$
$$= -2[-1(-4) - (-3)(4)] + 3[2(-4) - (-3)(-2)] - 2[2(4) - (-1)(-2)]$$
$$= -2[4 - (-12)] + 3[-8 - 6] - 2[8 - 2] = -2[16] + 3[-14] - 2[6]$$
$$= -32 - 42 - 12 = -86$$

(*c*) $\begin{vmatrix} -1 & -2 & 1 \\ 2 & 4 & 0 \\ -2 & -4 & 0 \end{vmatrix}$

Expand by minors of elements in column 3.

$$\begin{vmatrix} -1 & -2 & 1 \\ 2 & 4 & 0 \\ -2 & -4 & 0 \end{vmatrix} = 1\begin{vmatrix} 2 & 4 \\ -2 & -4 \end{vmatrix} - 0\begin{vmatrix} -1 & -2 \\ -2 & -4 \end{vmatrix} + 0\begin{vmatrix} -1 & -2 \\ 2 & 4 \end{vmatrix}$$
$$= 1[2(-4) - 4(-2)] - 0 + 0$$
$$= 1[-8 - (-8)]$$
$$= 1[-8 + 8] = 1[0] = 0$$

See supplementary problem 7.5 to practice evaluating determinants.

7.9 Use Cramer's rule to solve the following.

(*a*) $\begin{cases} -x - y = 1 \\ 5x - 3y = 11 \end{cases}$

$$D = \begin{vmatrix} -1 & -1 \\ 5 & -3 \end{vmatrix} = -1(-3) - (-1)5 = 3 + 5 = 8 \neq 0$$

$$D_x = \begin{vmatrix} 1 & -1 \\ 11 & -3 \end{vmatrix} = 1(-3) - (-1)11 = -3 + 11 = 8$$

$$D_y = \begin{vmatrix} -1 & 1 \\ 5 & 11 \end{vmatrix} = -1(11) - 1(5) = -11 - 5 = -16$$

$x = D_x/D = 8/8 = 1$ and $y = D_y/D = -16/8 = -2$. The solution set is $\{(1, -2)\}$. You should verify $(1, -2)$ satisfies both equations.

(*b*) $\begin{cases} 2x + 3y = 9 \\ 5x - y = -3 \end{cases}$

$$D = \begin{vmatrix} 2 & 3 \\ 5 & -1 \end{vmatrix} = 2(-1) - 3(5) = -2 - 15 = -17 \neq 0$$

$$D_x = \begin{vmatrix} 9 & 3 \\ -3 & -1 \end{vmatrix} = 9(-1) - 3(-3) = -9 + 9 = 0$$

$$D_y = \begin{vmatrix} 2 & 9 \\ 5 & -3 \end{vmatrix} = 2(-3) - 9(5) = -6 - 45 = -51$$

$x = D_x/D = 0/(-17) = 0$ and $y = D_y/D = -51/(-17) = 3$. The solution set is $\{(0, 3)\}$. A mental check readily verifies the result.

(*c*) $\begin{cases} x + 2y = 4 \\ -2x + 4y = -4 \end{cases}$

$$D = \begin{vmatrix} 1 & 2 \\ -2 & 4 \end{vmatrix} = 1(4) - 2(-2) = 4 + 4 = 8 \neq 0$$

$$D_x = \begin{vmatrix} 4 & 2 \\ -4 & 4 \end{vmatrix} = 4(-4) - 2(-4) = 16 + 8 = 24$$

$$D_y = \begin{vmatrix} 1 & 4 \\ -2 & -4 \end{vmatrix} = 1(-4) - 4(-2) = -4 + 8 = 4$$

$x = D_x/D = 24/8 = 3$ and $y = D_y/D = 4/8 = 1/2$. The solution set is $\{(3, 1/2)\}$. A check verifies the stated result.

7.10 Use Cramer's rule to solve the following.

(*a*) $\begin{cases} x + y + z = -1 \\ 2x - y + 2z = 4 \\ -x + 2y - 4z = -5 \end{cases}$

$$D = \begin{vmatrix} 1 & 1 & 1 \\ 2 & -1 & 2 \\ -1 & 2 & -4 \end{vmatrix} = 1\begin{vmatrix} -1 & 2 \\ 2 & -4 \end{vmatrix} - 1\begin{vmatrix} 2 & 2 \\ -1 & -4 \end{vmatrix} + 1\begin{vmatrix} 2 & -1 \\ -1 & 2 \end{vmatrix} = 9 \neq 0$$

$$D_x = \begin{vmatrix} -1 & 1 & 1 \\ 4 & -1 & 2 \\ -5 & 2 & -4 \end{vmatrix} = -1\begin{vmatrix} -1 & 2 \\ 2 & -4 \end{vmatrix} - 1\begin{vmatrix} 4 & 2 \\ -5 & -4 \end{vmatrix} + 1\begin{vmatrix} 4 & -1 \\ -5 & 2 \end{vmatrix} = 9$$

$$D_y = \begin{vmatrix} 1 & -1 & 1 \\ 2 & 4 & 2 \\ -1 & -5 & -4 \end{vmatrix} = 1\begin{vmatrix} 4 & 2 \\ -5 & -4 \end{vmatrix} - (-1)\begin{vmatrix} 2 & 2 \\ -1 & -4 \end{vmatrix} + 1\begin{vmatrix} 2 & 4 \\ -1 & -5 \end{vmatrix} = -18$$

$$D_z = \begin{vmatrix} 1 & 1 & -1 \\ 2 & -1 & 4 \\ -1 & 2 & -5 \end{vmatrix} = 1\begin{vmatrix} -1 & 4 \\ 2 & -5 \end{vmatrix} - 1\begin{vmatrix} 2 & 4 \\ -1 & -5 \end{vmatrix} + (-1)\begin{vmatrix} 2 & -1 \\ -1 & 2 \end{vmatrix} = 0$$

$x = D_x/D = 9/9 = 1, y = D_y/D = -18/9 = -2$ and $z = D_z/D = 0/9 = 0$. The solution set is $\{(1, -2, 0)\}$. Check: $(1, -2, 0)$ satisfies each equation. The details are left to the reader.

(b) $$\begin{cases} 2x - 3y + z = -17 \\ 4y - z = 15 \\ x + 4z = 1 \end{cases}$$

$$D = \begin{vmatrix} 2 & -3 & 1 \\ 0 & 4 & -1 \\ 1 & 0 & 4 \end{vmatrix} = 2\begin{vmatrix} 4 & -1 \\ 0 & 4 \end{vmatrix} - 0\begin{vmatrix} -3 & 1 \\ 0 & 4 \end{vmatrix} + 1\begin{vmatrix} -3 & 1 \\ 4 & -1 \end{vmatrix} = 31 \neq 0$$

$$D_x = \begin{vmatrix} -17 & -3 & 1 \\ 15 & 4 & -1 \\ 1 & 0 & 4 \end{vmatrix} = 1\begin{vmatrix} -3 & 1 \\ 4 & -1 \end{vmatrix} - 0\begin{vmatrix} -17 & 1 \\ 15 & -1 \end{vmatrix} + 4\begin{vmatrix} -17 & -3 \\ 15 & 4 \end{vmatrix} = -93$$

$$D_y = \begin{vmatrix} 2 & -17 & 1 \\ 0 & 15 & -1 \\ 1 & 1 & 4 \end{vmatrix} = 2\begin{vmatrix} 15 & -1 \\ 1 & 4 \end{vmatrix} - 0\begin{vmatrix} -17 & 1 \\ 1 & 4 \end{vmatrix} + 1\begin{vmatrix} -17 & 1 \\ 15 & -1 \end{vmatrix} = 124$$

$$D_z = \begin{vmatrix} 2 & -3 & -17 \\ 0 & 4 & 15 \\ 1 & 0 & 1 \end{vmatrix} = 1\begin{vmatrix} -3 & -17 \\ 4 & 15 \end{vmatrix} - 0\begin{vmatrix} -17 & 1 \\ 1 & 4 \end{vmatrix} + 1\begin{vmatrix} 2 & -3 \\ 0 & 4 \end{vmatrix} = 31$$

$x = D_x/D = -93/31 = -3, y = D_y/D = 124/31 = 4$ and $z = D_z/D = 31/31 = 1$. The solution set is $\{(-3,4,1)\}$. Check: $(-3,4,1)$ satisfies each equation. The details are left to the reader.

7.11 Write the coefficient, constant, and augmented matrices for the following systems.

(a) $$\begin{cases} 4x - 7y = 3 \\ x + 5y = -1 \end{cases}$$

The coefficient matrix is $\begin{bmatrix} 4 & -7 \\ 1 & 5 \end{bmatrix}$. The constant matrix is $\begin{bmatrix} 3 \\ -1 \end{bmatrix}$. The augmented matrix is $\left[\begin{array}{cc|c} 4 & -7 & 3 \\ 1 & 5 & -1 \end{array}\right]$.

(b) $$\begin{cases} 4a - 5b + c = 2 \\ 2a - 4c = 5 \\ 7b - 5c = -1 \end{cases}$$

The coefficient matrix is $\begin{bmatrix} 4 & -5 & 1 \\ 2 & 0 & -4 \\ 0 & 7 & -5 \end{bmatrix}$. The constant matrix is $\begin{bmatrix} 2 \\ 5 \\ -1 \end{bmatrix}$. The augmented matrix is

$$\left[\begin{array}{ccc|c} 4 & -5 & 1 & 2 \\ 2 & 0 & -4 & 5 \\ 0 & 7 & -5 & -1 \end{array}\right].$$

Observe that a missing variable in an equation is simply a variable with coefficient zero. We also point out that the relative positions of the column entries in the coefficient matrix are associated with a particular variable. The first column in part (*b*) above consists of the coefficients of the *a* variable; the second column consists of the coefficients of the *b* variable; and the third column, the coefficients of the *c* variable. It is always necessary to align the variables columnwise before the coefficients matrix is formed.

7.12 Perform the elementary row operations on the matrix $\begin{bmatrix} 1 & 2 & -1 \\ 3 & 7 & 4 \end{bmatrix}$.

$\begin{bmatrix} 1 & 2 & -1 \\ 3 & 7 & 4 \end{bmatrix}$ $R_1 \leftrightarrow R_2$ Interchange rows 1 and 2. $\begin{bmatrix} 3 & 7 & 4 \\ 1 & 2 & -1 \end{bmatrix}$

$\begin{bmatrix} 1 & 2 & -1 \\ 3 & 7 & 4 \end{bmatrix}$ $-3R_1 \rightarrow R_1$ Multiply each element of row 1 by -3 to obtain a new row 1. $\begin{bmatrix} -3 & -6 & 3 \\ 3 & 7 & 4 \end{bmatrix}$

$\begin{bmatrix} 1 & 2 & -1 \\ 3 & 7 & 4 \end{bmatrix}$ $-3R_1 + R_2 \rightarrow R_2$ Multiply each element of row 1 by -3 and add the resulting elements to the corresponding elements of row 2 to obtain a new row 2. $\begin{bmatrix} 1 & 2 & -1 \\ 0 & 1 & 7 \end{bmatrix}$

The elementary row operations stated above are equivalent to the operations previously employed to obtain equivalent systems of equations. See supplementary problem 7.7.

7.13 Transform the following matrices to row echelon form.

(*a*) $\begin{bmatrix} 1 & 3 & 5 \\ 2 & 9 & 16 \end{bmatrix}$

$$\begin{bmatrix} 1 & 3 & 5 \\ 2 & 9 & 16 \end{bmatrix} -2R_1 + R_2 \rightarrow R_2 \begin{bmatrix} 1 & 3 & 5 \\ 0 & 3 & 6 \end{bmatrix} \frac{1}{3}R_2 \rightarrow R_2 \begin{bmatrix} 1 & 3 & 5 \\ 0 & 1 & 2 \end{bmatrix}$$

(*b*) $\begin{bmatrix} -2 & 6 & 4 \\ 1 & -2 & 4 \end{bmatrix}$

$$\begin{bmatrix} -2 & 6 & 4 \\ 1 & -2 & 4 \end{bmatrix} R_1 \leftrightarrow R_2 \begin{bmatrix} 1 & -2 & 4 \\ -2 & 6 & 4 \end{bmatrix} 2R_1 + R_2 \rightarrow R_2 \begin{bmatrix} 1 & -2 & 4 \\ 0 & 2 & 12 \end{bmatrix}$$

$$\frac{1}{2}R_2 \rightarrow R_2 \begin{bmatrix} 1 & -2 & 4 \\ 0 & 1 & 6 \end{bmatrix}$$

(c) $\begin{bmatrix} 1 & -1 & 1 & 1 \\ 1 & -2 & -1 & 2 \\ 0 & 4 & 1 & -4 \end{bmatrix}$

$$\begin{bmatrix} 1 & -1 & 1 & 1 \\ 1 & -2 & -1 & 2 \\ 0 & 4 & 1 & -4 \end{bmatrix} -1R_1 + R_2 \to R_2 \begin{bmatrix} 1 & -1 & 1 & 1 \\ 0 & -1 & -2 & 1 \\ 0 & 4 & 1 & -4 \end{bmatrix}$$

$$-1R_1 \to R_2 \begin{bmatrix} 1 & -1 & 1 & 1 \\ 0 & 1 & 2 & -1 \\ 0 & 4 & 1 & -4 \end{bmatrix} -4R_2 + R_3 \to R_3 \begin{bmatrix} 1 & -1 & 1 & 1 \\ 0 & 1 & 2 & -1 \\ 0 & 0 & -7 & 0 \end{bmatrix}$$

$$\frac{-1}{7}R_3 \to R_3 \begin{bmatrix} 1 & -1 & 1 & 1 \\ 0 & 1 & 2 & -1 \\ 0 & 0 & 1 & 0 \end{bmatrix}$$

(d) $\begin{bmatrix} 2 & 6 & 4 & 8 \\ 4 & 14 & 10 & 20 \\ 3 & 9 & 6 & 12 \end{bmatrix}$

$$\begin{bmatrix} 2 & 6 & 4 & 8 \\ 4 & 14 & 10 & 20 \\ 3 & 9 & 6 & 12 \end{bmatrix} \frac{1}{2}R_1 \to R_1 \begin{bmatrix} 1 & 3 & 2 & 4 \\ 4 & 14 & 10 & 20 \\ 3 & 9 & 6 & 12 \end{bmatrix} -4R_1 + R_2 \to R_2$$

$$\begin{bmatrix} 1 & 3 & 2 & 4 \\ 0 & 2 & 2 & 4 \\ 3 & 9 & 6 & 12 \end{bmatrix} -3R_1 + R_3 \to R_3 \begin{bmatrix} 1 & 3 & 2 & 4 \\ 0 & 2 & 2 & 4 \\ 0 & 0 & 0 & 0 \end{bmatrix} \frac{1}{2}R_2 \to R_2 \begin{bmatrix} 1 & 3 & 2 & 4 \\ 0 & 1 & 1 & 2 \\ 0 & 0 & 0 & 0 \end{bmatrix}$$

See supplementary problem 7.8.

7.14 Use Gaussian elimination to solve the following systems.

(a) $\begin{cases} s + 2t = 7 \\ 2s - \ \ t = 4 \end{cases}$

1. The augmented matrix is $\left[\begin{array}{cc|c} 1 & 2 & 7 \\ 2 & -1 & 4 \end{array}\right]$.

2. $\left[\begin{array}{cc|c} 1 & 2 & 7 \\ 2 & -1 & 4 \end{array}\right] -2R_1 + R_2 \to R_2 \left[\begin{array}{cc|c} 1 & 2 & 7 \\ 0 & -5 & -10 \end{array}\right] \frac{-1}{5}R_2 \to R_2$

   $\left[\begin{array}{cc|c} 1 & 2 & 7 \\ 0 & 1 & 2 \end{array}\right]$

3. The corresponding system is $\begin{cases} s + 2t = 7 \\ t = 2 \end{cases}$

4. $s + 2t = 7$ and $t = 2$ so

$$s + 2(2) = 7$$
$$s + 4 = 7$$
$$s = 3$$

The solution set is $\{(3, 2)\}$.

5. $s + 2t = 7$ and $2s - t = 4$

$3 + 2(2) \stackrel{?}{=} 7$ $\quad$ $2(3) - 2 \stackrel{?}{=} 4$

$7 = 7$ $\quad$ $4 = 4$

(*b*) $\begin{cases} 2x + 3y = 11 \\ x - 2y = -5 \end{cases}$

1. The augmented matrix is $\left[\begin{array}{cc|c} 2 & 3 & 11 \\ 1 & -2 & -5 \end{array}\right]$.

2. $\left[\begin{array}{cc|c} 2 & 3 & 11 \\ 1 & -2 & -5 \end{array}\right] R_1 \leftrightarrow R_2 \left[\begin{array}{cc|c} 1 & -2 & -5 \\ 2 & 3 & 11 \end{array}\right] -2R_1 + R_2 \to R_2$

$\left[\begin{array}{cc|c} 1 & -2 & -5 \\ 0 & 7 & 21 \end{array}\right] \frac{1}{7}R_2 \to R_2 \left[\begin{array}{cc|c} 1 & -2 & -5 \\ 0 & 1 & 3 \end{array}\right]$

3. The corresponding system is $\begin{cases} x - 2y = -5 \\ y = 3 \end{cases}$

4. $x - 2y = -5$ and $y = 3$ so

$x - 2(3) = -5$

$x - 6 = -5$

$x = 1$

The solution set is $\{(1,3)\}$.

5. $2x + 3y = 11$ and $x - 2y = -5$

$2(1) + 3(3) \stackrel{?}{=} 11$ $\quad$ $1 - 2(3) \stackrel{?}{=} -5$

$2 + 9 \stackrel{?}{=} 11$ $\quad$ $1 - 6 \stackrel{?}{=} -5$

$11 = 11$ $\quad$ $-5 = -5$

(*c*) $\begin{cases} x - 2y - 2z = 4 \\ 2x + y - 3z = 7 \\ x - y - z = 3 \end{cases}$

1. The augmented matrix is $\left[\begin{array}{ccc|c} 1 & -2 & -2 & 4 \\ 2 & 1 & -3 & 7 \\ 1 & -1 & -1 & 3 \end{array}\right]$.

2. $\left[\begin{array}{ccc|c} 1 & -2 & -2 & 4 \\ 2 & 1 & -3 & 7 \\ 1 & -1 & -1 & 3 \end{array}\right] \begin{array}{l} -2R_1 + R_2 \to R_2 \\ -R_1 + R_3 \to R_3 \end{array} \left[\begin{array}{ccc|c} 1 & -2 & -2 & 4 \\ 0 & 5 & 1 & -1 \\ 0 & 1 & 1 & -1 \end{array}\right] R_2 \leftrightarrow R_3$

$\left[\begin{array}{ccc|c} 1 & -2 & -2 & 4 \\ 0 & 1 & 1 & -1 \\ 0 & 5 & 1 & -1 \end{array}\right] -5R_2 + R_3 \to R_3 \left[\begin{array}{ccc|c} 1 & -2 & -2 & 4 \\ 0 & 1 & 1 & -1 \\ 0 & 0 & -4 & 4 \end{array}\right] \frac{-1}{4}R_3 \to R_3$

$\left[\begin{array}{ccc|c} 1 & -2 & -2 & 4 \\ 0 & 1 & 1 & -1 \\ 0 & 0 & 1 & -1 \end{array}\right]$

3. The corresponding system is $\begin{cases} x - 2y - 2z = 4 \\ y + z = -1 \\ z = -1 \end{cases}$

4. $y + z = -1$ and $x - 2y - 2z = 4$

$y - 1 = -1$ $\quad x - 2(0) - 2(-1) = 4$

$y = 0$ $\quad x - 0 + 2 = 4$

$x = 2$

$\{(2, 0, -1)\}$ is the solution set.

5. $x - 2y - 2z = 4$ $\quad 2x + y - 3z = 7$ $\quad x - y - z = 3$

$2 - 2(0) - 2(-1) \stackrel{?}{=} 4$ $\quad 2(2) + 0 - 3(-1) \stackrel{?}{=} 7$ $\quad 2 - 0 - (-1) \stackrel{?}{=} 3$

$2 - 0 + 2 \stackrel{?}{=} 4$ $\quad 4 + 0 + 3 \stackrel{?}{=} 7$ $\quad 2 + 1 \stackrel{?}{=} 3$

$4 = 4$ $\quad 7 = 7$ $\quad 3 = 3$

$$(d)\quad \begin{cases} 2a - b - c = -4 \\ a + b + c = -5 \\ a + 3b - 4c = 12 \end{cases}$$

1. The augmented matrix is $\left[\begin{array}{ccc|c} 2 & -1 & -1 & -4 \\ 1 & 1 & 1 & -5 \\ 1 & 3 & -4 & 12 \end{array}\right]$.

2. $\left[\begin{array}{ccc|c} 2 & -1 & -1 & -4 \\ 1 & 1 & 1 & -5 \\ 1 & 3 & -4 & 12 \end{array}\right] R_1 \leftrightarrow R_2 \left[\begin{array}{ccc|c} 1 & 1 & 1 & -5 \\ 2 & -1 & -1 & -4 \\ 1 & 3 & -4 & 12 \end{array}\right] -2R_1 + R_2 \rightarrow R_2$

$\left[\begin{array}{ccc|c} 1 & 1 & 1 & -5 \\ 0 & -3 & -3 & 6 \\ 1 & 3 & -4 & 12 \end{array}\right] -1R_1 + R_3 \rightarrow R_3 \left[\begin{array}{ccc|c} 1 & 1 & 1 & -5 \\ 0 & -3 & -3 & 6 \\ 0 & 2 & -5 & 17 \end{array}\right] \frac{-1}{3}R_2 \rightarrow R_2$

$\left[\begin{array}{ccc|c} 1 & 1 & 1 & -5 \\ 0 & 1 & 1 & -2 \\ 0 & 2 & -5 & 17 \end{array}\right] -2R_2 + R_3 \rightarrow R_3 \left[\begin{array}{ccc|c} 1 & 1 & 1 & -5 \\ 0 & 1 & 1 & -2 \\ 0 & 0 & -7 & 21 \end{array}\right] \frac{-1}{7}R_3 \rightarrow R_3$

$\left[\begin{array}{ccc|c} 1 & 1 & 1 & -5 \\ 0 & 1 & 1 & -2 \\ 0 & 0 & 1 & -3 \end{array}\right]$

3. The corresponding system is $\begin{cases} a + b + c = -5 \\ b + c = -2. \\ c = -3 \end{cases}$

4. $b + c = -2$ and $c = -3$ so

$b + (-3) = -2$ and $a + b + c = -5$

$b = 1$ $\quad a + 1 + (-3) = -5$

$a + (-2) = -5$

$a = -3$

The solution set is $\{(-3, 1, -3)\}$.

5. $2a - b - c = -4$ $\quad a + b + c = -5$ $\quad a + 3b - 4c = 12$

$2(-3) - 1 - (-3) \stackrel{?}{=} -4$ $\quad -3 + 1 + (-3) \stackrel{?}{=} -5$ $\quad -3 + 3(1) - 4(-3) \stackrel{?}{=} 12$

$-6 - 1 + 3 \stackrel{?}{=} -4$ $\quad 1 - 6 \stackrel{?}{=} -5$ $\quad -3 + 3 + 12 \stackrel{?}{=} 12$

$-4 = -4$ $\quad -5 = -5$ $\quad 12 = 12$

(e) $\begin{cases} x - y - z = 2 \\ -x + y + z = 3 \\ -x - y + z = 4 \end{cases}$

1. The augmented matrix is $\left[\begin{array}{ccc|c} 1 & -1 & -1 & 2 \\ -1 & 1 & 1 & 3 \\ -1 & -1 & 1 & 4 \end{array}\right]$.

2. $\left[\begin{array}{ccc|c} 1 & -1 & -1 & 2 \\ -1 & 1 & 1 & 3 \\ -1 & -1 & 1 & 4 \end{array}\right] \begin{array}{l} R_1 + R_2 \to R_2 \\ R_1 + R_3 \to R_3 \end{array} \left[\begin{array}{ccc|c} 1 & -1 & -1 & 2 \\ 0 & 0 & 0 & 5 \\ 0 & -2 & 0 & 6 \end{array}\right] R_2 \leftrightarrow R_3$

$\left[\begin{array}{ccc|c} 1 & -1 & -1 & 2 \\ 0 & -2 & 0 & 6 \\ 0 & 0 & 0 & 5 \end{array}\right] \frac{-1}{2} R_2 \to R_2 \left[\begin{array}{ccc|c} 1 & -1 & -1 & 2 \\ 0 & 1 & 0 & -3 \\ 0 & 0 & 0 & 5 \end{array}\right]$

3. The corresponding system is $\begin{cases} x - y - z = 2 \\ \quad y \quad = -3. \\ \quad 0 = 5 \end{cases}$

The equation $0 = 5$ is never true. There is no solution to the system.

There is no solution to a system if it is not possible to transform the augmented matrix to row echelon form. The situation encountered above is discussed more completely in subsequent courses. See supplementary problem 7.9.

7.15 Solve.

(a) $\begin{cases} y = x^2 + 2x + 2 \\ x + y = 2 \end{cases}$

Solve the linear equation for y, then substitute into the first equation and solve the result.

$$x + y = 2 \Rightarrow y = 2 - x$$
$$2 - x = x^2 + 2x + 2$$
$$0 = x^2 + 3x$$
$$0 = x(x + 3) \Rightarrow x = 0 \text{ or } x = -3$$

If $x = 0, y = 2 - x = 2 - 0 = 2$. If $x = -3, y = 2 - x = 2 - (-3) = 5$. The solution set is $\{(0, 2), (-3, 5)\}$. Both ordered pairs satisfy the original equations. The details are left to the reader.

(b) $\begin{cases} x^2 + y^2 = 26 \\ x + y = 6 \end{cases}$

Solve the linear equation for either variable. We arbitrarily choose y. Then substitute into $x^2 + y^2 = 26$ and solve.

$$x + y = 6 \Rightarrow y = 6 - x$$
$$x^2 + (6 - x)^2 = 26$$
$$x^2 + (36 - 12x + x^2) = 26$$
$$2x^2 - 12x + 36 = 26$$
$$2x^2 - 12x + 10 = 0$$
$$x^2 - 6x + 5 = 0$$
$$(x - 1)(x - 5) = 0 \Rightarrow x = 1 \text{ or } x = 5$$

If $x = 1, y = 6 - x = 6 - 1 = 5$. If $x = 5, y = 6 - x = 6 - 5 = 1$. The solution set is $\{(1, 5), (5, 1)\}$. Both ordered pairs satisfy the original equations. The details are left to the reader.

(c) $\begin{cases} x^2 + y^2 = 1 \\ x - y = 5 \end{cases}$

The linear equation can most readily be solved for x. Hence solve $x - y = 5$ for x and substitute.

$$x - y = 5 \Rightarrow x = y + 5$$
$$(y+5)^2 + y^2 = 1$$
$$(y^2 + 10y + 25) + y^2 = 1$$
$$2y^2 + 10y + 24 = 0$$
$$y^2 + 5y + 12 = 0$$
$$y = \frac{-b \pm \sqrt{b^2 - 4ac}}{2a} = \frac{-5 \pm \sqrt{5^2 - 4(1)(12)}}{2(1)}$$
$$= \frac{-5 \pm \sqrt{25 - 48}}{2} = \frac{-5 \pm \sqrt{-23}}{2} = \frac{-5 \pm i\sqrt{23}}{2}$$

If $y = \frac{-5 + i\sqrt{23}}{2}, x = y + 5 = \frac{-5 + i\sqrt{23}}{2} + \frac{10}{2} = \frac{5 + i\sqrt{23}}{2}$

If $y = \frac{-5 - i\sqrt{23}}{2}, x = y + 5 = \frac{-5 - i\sqrt{23}}{2} + \frac{10}{2} = \frac{5 - i\sqrt{23}}{2}$

The solution set is $\left\{\left(\frac{-5 + i\sqrt{23}}{2}, \frac{-5 + i\sqrt{23}}{2}\right), \left(\frac{5 - i\sqrt{23}}{2}, \frac{-5 - i\sqrt{23}}{2}\right)\right\}$. The check is cumbersome, but it can be accomplished.

(d) $\begin{cases} a^2 - b^2 = 35 \\ ab = 6 \end{cases}$

The second equation is not linear, but either a or b can readily be isolated. We arbitrarily choose to isolate b. Finally substitute and solve.

$$ab = 6 \Rightarrow b = \frac{6}{a}$$
$$a^2 - b^2 = 35$$
$$a^2 - \left(\frac{6}{a}\right)^2 = 35$$
$$a^2 - \frac{36}{a^2} = 35$$
$$a^4 - 36 = 35a^2$$
$$a^4 - 35a^2 - 36 = 0$$

Let $u = a^2$ and solve for u.

$$u^2 - 35u - 36 = 0$$
$$(u - 36)(u + 1) = 0$$
$$u = 36 \text{ or } u = -1$$

If $u = 36, a^2 = 36 \Rightarrow a = \pm\sqrt{36} = \pm 6$.

If $u = -1, a^2 = -1 \Rightarrow a = \pm\sqrt{-1} = \pm i$.

If $a = 6, b = \frac{6}{a} = \frac{6}{6} = 1$ and if $a = -6, b = \frac{6}{a} = \frac{6}{-6} = -1$.

If $a = i, b = \frac{6}{a} = \frac{6}{i} = \frac{6}{i} \cdot \frac{i}{i} = \frac{6i}{i^2} = \frac{6i}{-1} = -6i$

and if $a = -i, b = \frac{6}{a} = \frac{6}{-i} = \frac{6}{-i} \cdot \frac{i}{i} = \frac{6i}{-i^2} = \frac{6i}{-(-1)} = \frac{6i}{1} = 6i.$

The solution set is $\{(6,1), (-6,-1), (i,-6i), (-i,6i)\}$. The check is left to the reader.

7.16 Solve.

(*a*) $\begin{cases} s^2 - t^2 = -5 \\ 4s^2 + t^2 = 25 \end{cases}$

An equation in one unknown is obtained if the equations are added. We obtain

$$\left.\begin{array}{l} s^2 - t^2 = -5 \\ 4s^2 + t^2 = 25 \\ \hline 5s^2 \quad = 20 \end{array}\right\} \Rightarrow s^2 = 4 \text{ so } s = \pm\sqrt{4} = \pm 2.$$

Now use back substitution to find t. We shall substitute into $4s^2 + t^2 = 25$.

If $s = 2$,

$$4s^2 + t^2 = 25$$
$$4(2)^2 + t^2 = 25$$
$$16 + t^2 = 25$$
$$t^2 = 9$$
$$t = \pm\sqrt{9} = \pm 3$$

If $s = -2$,

$$4s^2 + t^2 = 25$$
$$4(-2)^2 + t^2 = 25$$
$$16 + t^2 = 25$$
$$t^2 = 9$$
$$t = \pm\sqrt{9} = \pm 3$$

The solution set is $\{(2, 3), (2, -3), (-2, 3), (-2, -3)\}$. It can be verified that all four ordered pairs satisfy both equations.

(*b*) $\begin{cases} 5a^2 + b^2 = 1 \\ a^2 + 5b^2 = 1 \end{cases}$

Multiply the first equation by -5 and proceed as in part (*a*).

$$\left.\begin{array}{l} -25a^2 - 5b^2 = -5 \\ \quad a^2 + 5b^2 = 1 \\ \hline -24a^2 \quad = -4 \end{array}\right\} \Rightarrow a^2 = \frac{-4}{-24} = \frac{1}{6} \text{ so}$$

$$a = \pm\sqrt{\frac{1}{6}} = \pm\frac{\sqrt{1}}{\sqrt{6}} = \pm\frac{\sqrt{1}}{\sqrt{6}} \cdot \frac{\sqrt{6}}{\sqrt{6}} = \pm\frac{\sqrt{6}}{6}$$

If $a = \frac{\sqrt{6}}{6}$,

$$5a^2 + b^2 = 1$$
$$5\left(\frac{\sqrt{6}}{6}\right)^2 + b^2 = 1$$
$$5\left(\frac{6}{36}\right) + b^2 = 1$$
$$\frac{30}{36} + b^2 = 1$$
$$b^2 = 1 - \frac{30}{36} = \frac{36-30}{36} = \frac{6}{36}$$
$$b = \pm\sqrt{\frac{6}{36}} = \pm\frac{\sqrt{6}}{6}$$

If $a = \frac{-\sqrt{6}}{6}$,

$$5a^2 + b^2 = 1$$
$$5\left(\frac{-\sqrt{6}}{6}\right)^2 + b^2 = 1$$
$$5\left(\frac{6}{36}\right) + b^2 = 1$$
$$b = \pm\frac{\sqrt{6}}{6} \text{ again.}$$

The solution set is $\left\{\left(\frac{\sqrt{6}}{6}, \frac{\sqrt{6}}{6}\right), \left(\frac{\sqrt{6}}{6}, \frac{-\sqrt{6}}{6}\right), \left(\frac{-\sqrt{6}}{6}, \frac{\sqrt{6}}{6}\right), \left(\frac{-\sqrt{6}}{6}, \frac{-\sqrt{6}}{6}\right)\right\}$. The check is left to the reader.

(c) $\begin{cases} 4k^2 + l^2 = 11 \\ l = 4k^2 - 9 \end{cases}$

The second equation is not second degree in both variables. However, if we rewrite the second equation and add, a quadratic equation that we can solve is obtained.

$$\begin{aligned} 4k^2 + l^2 &= 11 \\ -4k^2 + l &= -9 \\ \hline l^2 + l &= 2 \end{aligned}$$

$$l^2 + l - 2 = 0$$
$$(l+2)(l-1) = 0$$
$$l = -2 \text{ or } 1$$

Now use back substitution to find k.

If $l = -2$,

$$4k^2 + l^2 = 11$$
$$4k^2 + (-2)^2 = 11$$
$$4k^2 + 4 = 11$$
$$4k^2 = 7$$
$$k^2 = \frac{7}{4}$$
$$k = \pm\sqrt{\frac{7}{4}} = \pm\frac{\sqrt{7}}{\sqrt{4}} = \pm\frac{\sqrt{7}}{2}$$

If $l = 1$

$$\begin{aligned}4k^2 + l^2 &= 11\\4k^2 + (1)^2 &= 11\\4k^2 + 1 &= 11\\4k^2 &= 10\\k^2 &= \frac{10}{4}\\k &= \pm\sqrt{\frac{10}{4}} = \pm\frac{\sqrt{10}}{\sqrt{4}} = \pm\frac{\sqrt{10}}{2}.\end{aligned}$$

The solution set is $\left\{\left(\frac{\sqrt{7}}{2}, -2\right), \left(\frac{-\sqrt{7}}{2}, -2\right), \left(\frac{\sqrt{10}}{2}, 1\right), \left(\frac{-\sqrt{10}}{2}, 1\right)\right\}$

7.17 Solve the following systems graphically.

(*a*) $\begin{cases} y = \sqrt{x+3} \\ 2x + y = 5 \end{cases}$

Use your knowledge concerning the shape and general position of each graph to pick an appropriate graphing window. We decide to use $x \in [-4, 4]$ and $y \in [-5, 6]$. The graphing window may be changed if the selected window is unsatisfactory. The graphs are shown in Figure 7.4.

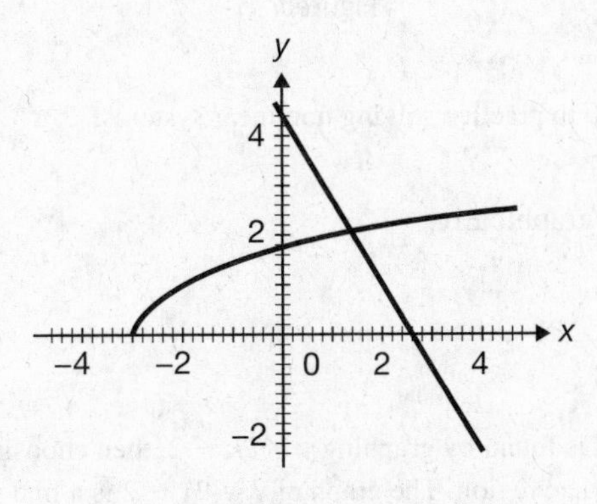

Figure 7.4

Use the trace feature to find the approximate coordinates of the point of intersection of the graphs. We obtain approximately (1.45, 2.11). The solution set is {(1.45, 2.11)}. The zoom feature or built-in function may be employed to obtain a more accurate solution, if your calculator has one. The use of a smaller graphing window that contains the point of intersection is another means of improving our estimate. We should also point out that either the substitution or addition method could be used to find the exact solution to the system.

(*b*) $\begin{cases} 2x^2 + 5y^2 = 20 \\ x^2 + y = 5 \end{cases}$

We must solve for y in both equations in order to graph them.

$$\begin{aligned}2x^2 + 5y^2 &= 20 \\ 5y^2 &= 20 - 2x^2 \\ y^2 &= \frac{20 - 2x^2}{5} \\ y &= \pm\sqrt{\frac{20 - 2x^2}{5}}\end{aligned} \qquad \text{and} \qquad \begin{aligned}x^2 + y &= 5 \\ y &= 5 - x^2\end{aligned}$$

We must graph $y = \sqrt{(20 - 2x^2)/5}$ and $y = -\sqrt{(20 - 2x^2)/5}$ individually in order to obtain the complete graph of $2x^2 + 5y^2 = 20$. We choose a graphing window of $x \in [-4, 4]$ and $y \in [-3,6]$. See Figure 7.5. The graph of $2x^2 + 5y^2 = 20$ is a curve called an *ellipse*, while the graph of $x^2 + y = 5$ is the familiar parabola. Trace the parabola to find the approximate points of intersection of the curves. The points are approximately $(-2.496, -1.228)$, $(-1.836, 1.628)$, $(1.836, 1.628)$, and $(2.496, -1.228)$. The solution set is $\{(-2.5, -1.1)$, $(-2.496, -1.228)$, $(-1.836, 1.628)$, $(1.836, 1.628)$, $(2.496, -1.228)\}$. The remarks made in part (*a*) are applicable here also.

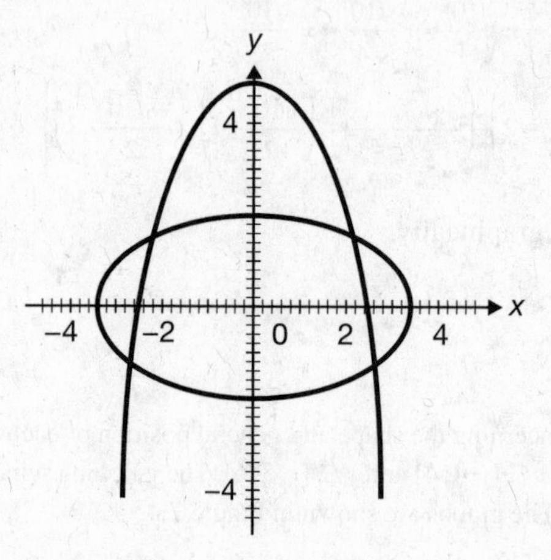

Figure 7.5

See supplementary problem 7.10 to practice solving nonlinear systems.

7.18 Solve the following systems graphically.

(*a*) $\begin{cases} y \geq 2x - 2 \\ y \leq -3x + 2 \end{cases}$

The graph of $y \geq 2x - 2$ is found by graphing $y = 2x - 2$, then choosing a test point, such as $(0, 0)$, and finally shading the appropriate region. The graph of $y = 2x - 2$ is a line with slope 2 and y-intercept -2 (see Figure 7.6 (*a*)). The graph of $y \leq -3x + 2$ is done similarly (see Figure 7.6 (*b*)). Determine the intersection as shown in Figure 7.6 (*c*).

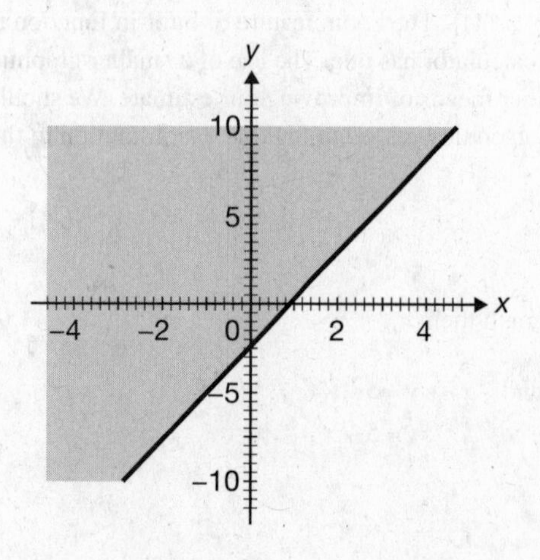

Figure 7.6 (*a*) $y \geq 2x - 2$

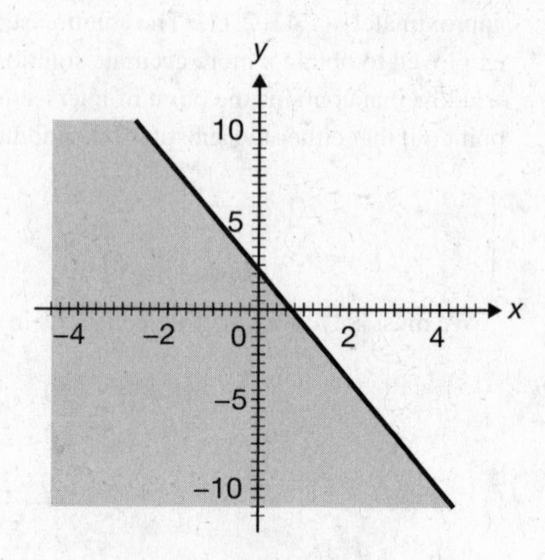

Figure 7.6 (*b*) $y \leq -3x + 2$

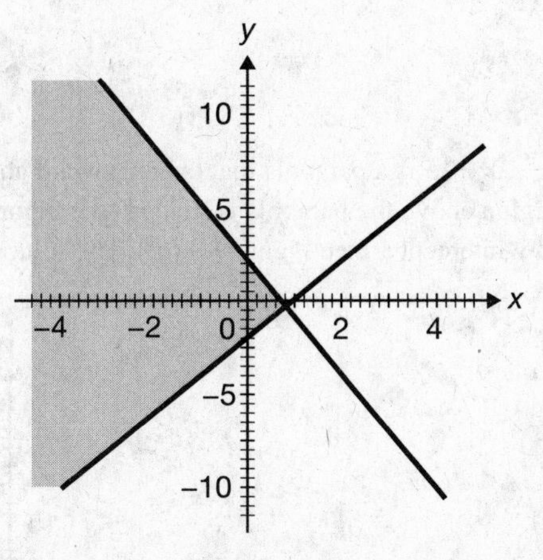

Figure 7.6 (c)

We choose an arbitrary point in the shaded region, such as $(-2, 0)$, to verify that it satisfies each inequality in the system.

(b) $\begin{cases} x - y < 2 \\ y \leq 3 \end{cases}$

Proceed as in part (*a*) above. The inequality $x - y < 2$ is strict, so its boundary is dashed (see Figure 7.7 (*a*)). The boundary of $y \leq 3$ is the solid horizontal line 3 units above the *x*-axis (see Figure 7.7 (*b*)). The solution set is displayed in Figure 7.7 (*c*).

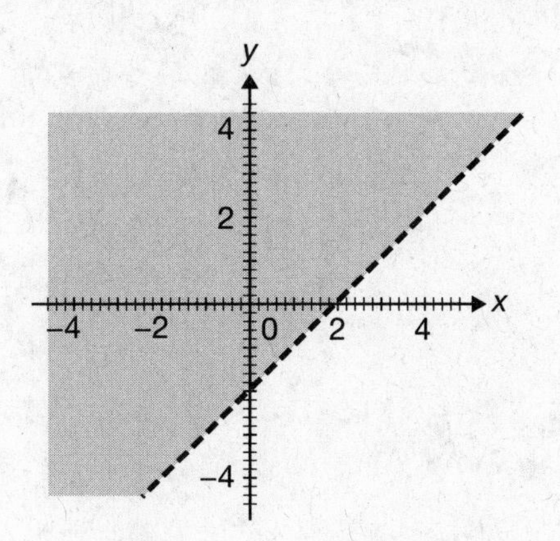

Figure 7.7 (a) $x - y < 2$

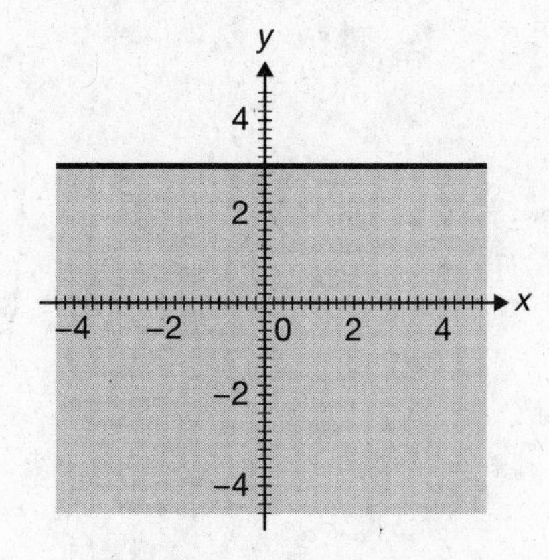

Figure 7.7 (b) $y \leq 3$

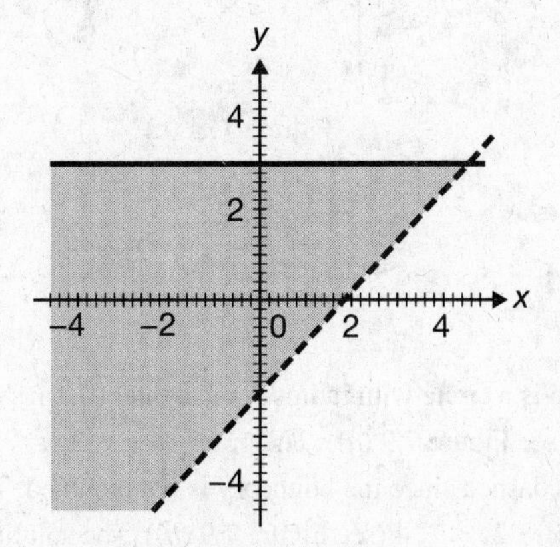

Figure 7.7 (c)

(c) $\begin{cases} y > (x-2)^2 - 5 \\ y \le 1 - x \end{cases}$

The boundary of $y \ge (x-2)^2 - 5$ is a parabola that opens upward and has vertex $(2, -5)$. $(2, 0)$ satisfies the inequality; hence the region above the parabola is shaded (see Figure 7.8 (*a*)). The boundary of $y \le 1 - x$ is a line with slope -1 and y-intercept 1 (see Figure 7.8 (*b*)). The solution set is displayed in Figure 7.8 (*c*).

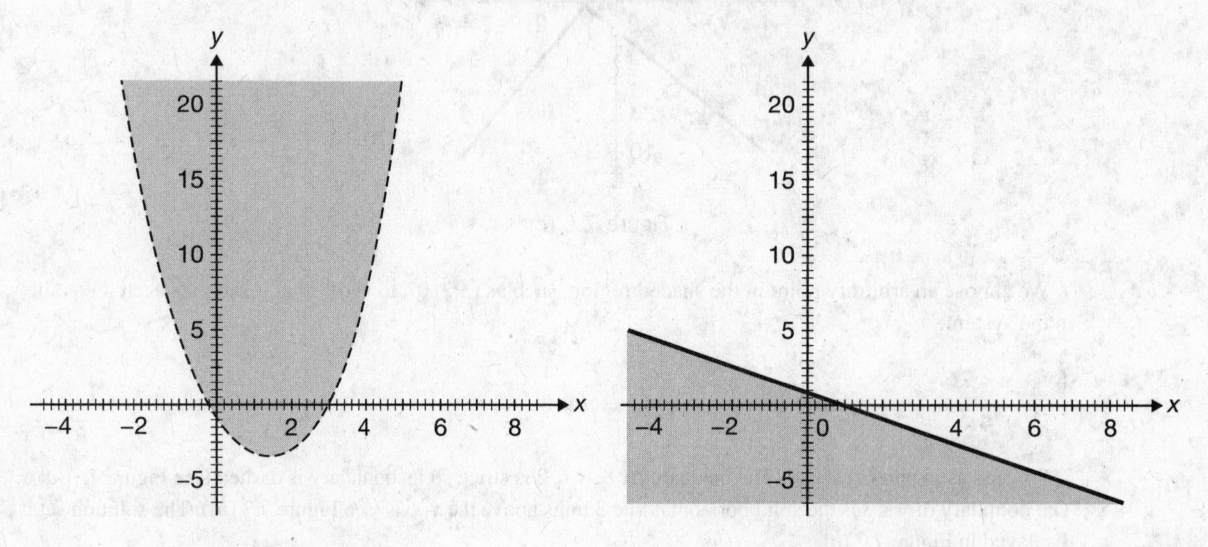

Figure 7.8 (*a*) $y > (x-2)^2 - 5$

Figure 7.8 (*b*) $y \le 1 - x$

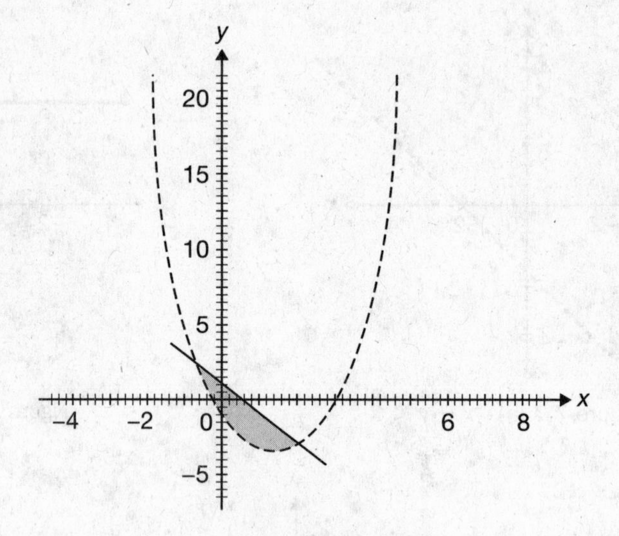

Figure 7.8 (*c*)

(d) $\begin{cases} x^2 + y^2 \le 25 \\ x - 2y < -4 \end{cases}$

The graph of $x^2 + y^2 = 25$ is a circle with radius 5 and center $(0, 0)$. The interior of the circle is shaded, since $(0, 0)$ satisfies $x^2 + y^2 \le 25$ (see Figure 7.9 (*a*)). The graph of $x - 2y = -4$ or $y = \frac{1}{2}x + 2$ is a line with slope $\frac{1}{2}$ and y-intercept 2. The line is dashed since the boundary is not included. The region above the line is shaded, since $(0, 0)$ does not satisfy $x - 2y < -4$ (see Figure 7.9 (*b*)). The solution set is shown in Figure 7.9 (*c*).

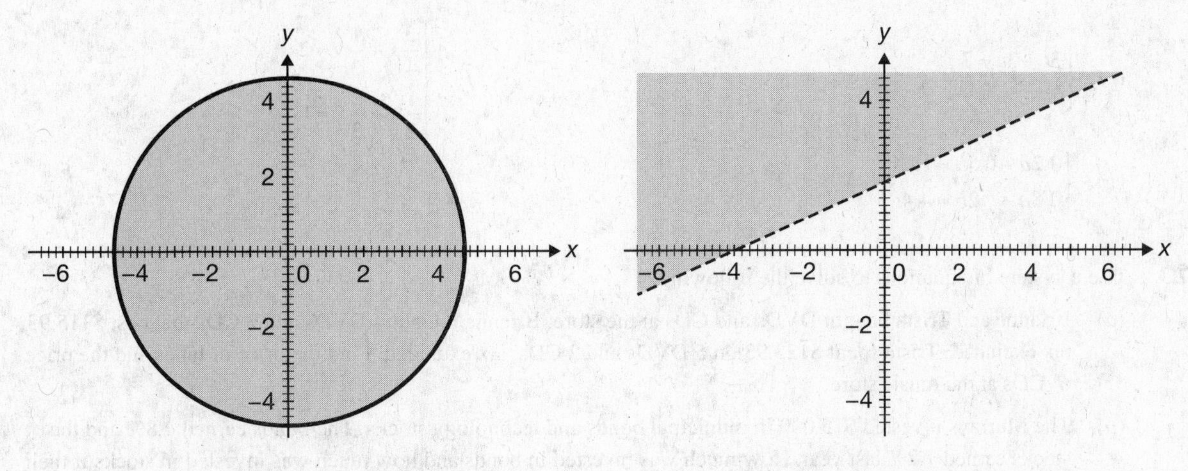

Figure 7.9 (*a*) $x^2 + y^2 \leq 25$ Figure 7.9 (*b*) $x - 2y < -4$

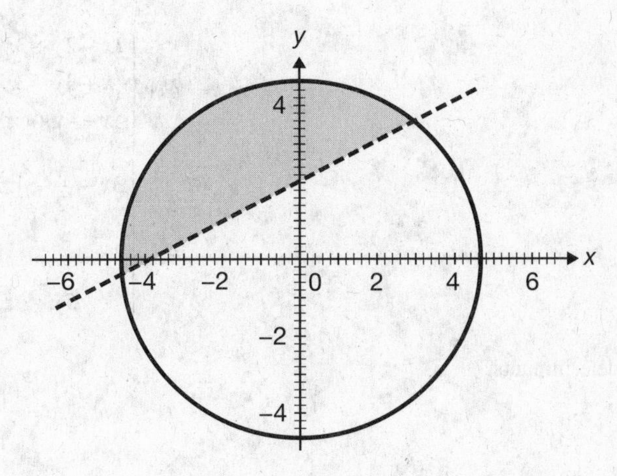

Figure 7.9 (*c*)

SUPPLEMENTARY PROBLEMS

7.1. Use the addition method to solve the following systems.

(*a*) $\begin{cases} x + 3y = 4 \\ 4x + 5y = 2 \end{cases}$ (*b*) $\begin{cases} 7s + 5t = -8 \\ 9s + 2t = 3 \end{cases}$

(*c*) $\begin{cases} \dfrac{7}{10}a + \dfrac{2}{5}b = 0 \\ \dfrac{1}{2}a - \dfrac{4}{5}b = \dfrac{19}{10} \end{cases}$ (*d*) $\begin{cases} -s + \dfrac{4}{5}t = \dfrac{3}{5} \\ 15s - 12t = 7 \end{cases}$

7.2. Use any method to solve the following systems.

(*a*) $\begin{cases} -2a + 3b = 6 \\ a - 2b = -5 \end{cases}$ (*b*) $\begin{cases} -3s + 7t = 14 \\ 2s - t = -13 \end{cases}$

(*c*) $\begin{cases} 3x + 4y = 18 \\ y = -2x + 2 \end{cases}$ (*d*) $\begin{cases} 3s - 4t = 3 \\ 4s + 3t = 14 \end{cases}$

(*e*) $\begin{cases} 3t - 3u = 7 \\ 5t - 5u = 9 \end{cases}$

(*f*) $\begin{cases} \frac{3}{4}x + \frac{1}{3}y = 2 \\ \frac{1}{2}x + \frac{2}{3}y = 1 \end{cases}$

(*g*) $\begin{cases} 0.2a + 0.5b = 1.1 \\ 0.8a + \quad 2b = 4.4 \end{cases}$

7.3. Use a system of equations to solve the following.

(*a*) Brianne and Trista bought DVDs and CDs at the store. Brianne bought 4 DVDs and 3 CDs that cost $115.93, tax excluded. Trista spent $123.93 on 5 DVDs and 2 CDs, tax excluded. Find the price of tapes and the price of CDs at the music store.

(*b*) The Murrays invested $15,000 in municipal bonds and technology stocks. The bonds earned 4.8% and the stocks earned 7.2% last year. How much was invested in bonds and how much was invested in stocks if their total investment earned $856.80 last year?

7.4. Solve the following systems.

(*a*) $\begin{cases} x + 2y - 3z = 3 \\ 2x - \quad y + z = -4 \\ x + \ y - 4z = 1 \end{cases}$

(*b*) $\begin{cases} 2x - 2y + 3z = 1 \\ -x + 3y - 2z = 2 \\ x - \ y + \ z = -1 \end{cases}$

(*c*) $\begin{cases} 2s + \ t - 3u = 1 \\ -s - 2t + \ u = 3 \\ -4s - 2t + 6u = 5 \end{cases}$

(*d*) $\begin{cases} 3x - \frac{y}{2} - 4z = \frac{-5}{2} \\ 6x \qquad\qquad = -4y \\ \qquad 5y + \ z = 5 \end{cases}$

7.5. Evaluate the following determinants.

(*a*) $\begin{vmatrix} 4 & -1 \\ 3 & 2 \end{vmatrix}$

(*b*) $\begin{vmatrix} 0 & 2 \\ 3 & 5 \end{vmatrix}$

(*c*) $\begin{vmatrix} 6 & 1 & -1 \\ 0 & -1 & 4 \\ 2 & -2 & 3 \end{vmatrix}$

(*d*) $\begin{vmatrix} 2 & 3 & -1 \\ 2 & -3 & 4 \\ -2 & -3 & 1 \end{vmatrix}$

(*e*) $\begin{vmatrix} 1 & -2 & 0 \\ 2 & -1 & 3 \\ -3 & 4 & 1 \end{vmatrix}$

(*f*) $\begin{vmatrix} 2 & -1 \\ 3 & 4 \end{vmatrix}$

(*g*) $\begin{vmatrix} 7 & -5 \\ 2 & -2 \end{vmatrix}$

(*h*) $\begin{vmatrix} 0 & 2 & 3 \\ -1 & 3 & -4 \\ -5 & -5 & 2 \end{vmatrix}$

(*i*) $\begin{vmatrix} 4 & 0 & 2 \\ 1 & 2 & 3 \\ -3 & 0 & -8 \end{vmatrix}$

7.6. Use Cramer's rule to solve the following.

(*a*) $\begin{cases} 4x + 5y = 18 \\ 2x - 7y = -48 \end{cases}$

(*b*) $\begin{cases} -2x + y + 5z = 9 \\ 2x + 2y - 3z = -9 \\ x - 4y + \ z = 13 \end{cases}$

(*c*) $\begin{cases} 3x - \ y = 9 \\ -x + 5y = -17 \end{cases}$

(*d*) $\begin{cases} x + \ y - 4z = 0 \\ 4x + 2y - \ z = 6 \\ \qquad y + 3z = -3 \end{cases}$

(*e*) $\begin{cases} 5x - 2y = -3 \\ 4x + 3y = \frac{25}{3} \end{cases}$

(*f*) $\begin{cases} 2x - 4y + 3z = 4 \\ -x + 2y - 2z = -5 \\ -3x + 6y + z = -2 \end{cases}$

7.7. Perform the indicated elementary row operations on the matrix $\begin{bmatrix} 1 & -2 & 3 & 7 \\ 2 & 3 & -1 & 0 \\ 1 & 1 & 1 & 1 \end{bmatrix}$.

(*a*) $R_1 \leftrightarrow R_3$ (*b*) $3R_1 \rightarrow R_1$ (*c*) $-2R_1 + R_2 \rightarrow R_2$

7.8. Transform $\begin{bmatrix} 1 & -2 & 3 & 7 \\ 2 & 3 & -1 & 0 \\ 1 & 1 & 1 & 1 \end{bmatrix}$ to row echelon form.

7.9. Use matrix methods (Gaussian elimination) to solve the following systems.

(*a*) $\begin{cases} x - 2y = 3 \\ -2x + 3y = -6 \end{cases}$ (*b*) $\begin{cases} r + s + t = 2 \\ 2r - 3s + t = 7 \\ 3r + 3s - t = -2 \end{cases}$ (*c*) $\begin{cases} 4l - 8w = 40 \\ 2l + 3w = 6 \end{cases}$ (*d*) $\begin{cases} b - 2c = -1 \\ -3a + 4b - c = -2 \\ a - 2b + 3c = 4 \end{cases}$

7.10. Use an appropriate method to solve the following systems.

(*a*) $\begin{cases} t = s^2 + 2s + 1 \\ t - s = 3 \end{cases}$ (*b*) $\begin{cases} l^2 - lw - 2w^2 = 4 \\ l - w = 2 \end{cases}$ (*c*) $\begin{cases} 3a^2 + b^2 = 15 \\ 11a^2 - 2b^2 = 4 \end{cases}$

(*d*) $\begin{cases} 9x^2 + 4y^2 = 36 \\ x^2 + 4y^2 = 100 \end{cases}$ (*e*) $\begin{cases} y = \sqrt{x+2} + 1 \\ y = x^2 + 2x - 2 \end{cases}$ (*f*) $\begin{cases} 2x^2 + y^2 = 8 \\ 2\sqrt{x+4} + y = 4 \end{cases}$

7.11. Solve the following systems graphically.

(*a*) $\begin{cases} y < 2x + 3 \\ 2y < -x + 4 \end{cases}$ (*b*) $\begin{cases} y \geq (x-2)^2 - 4 \\ y > 2 \end{cases}$ (*c*) $\begin{cases} -3x + y \leq -2 \\ x - 2y \leq 0 \end{cases}$ (*d*) $\begin{cases} y \leq -(x+3)^2 + 2 \\ x^2 + y^2 \leq 9 \end{cases}$

ANSWERS TO SUPPLEMENTARY PROBLEMS

7.1 (*a*) $\{(-2, 2)\}$ (*b*) $\{(1, -3)\}$ (*c*) $\left\{\left(1, \frac{-7}{4}\right)\right\}$ (*d*) Ø or { }

7.2 (*a*) $\{(3, 4)\}$ (*b*) $\{(-7, -1)\}$ (*c*) $\{(-2, 6)\}$

(*d*) $\left\{\left(\frac{13}{5}, \frac{6}{5}\right)\right\}$ (*e*) Ø or { } (*f*) $\left\{\left(3, \frac{-3}{4}\right)\right\}$

(*g*) Infinitely many solutions.
The system is dependent.

7.3 (*a*) The price of DVDs was \$19.99 and the price of CDs was \$11.99 each.

(*b*) \$9,300 were invested in bonds and \$5,700 were invested in stocks last year.

7.4 (*a*) $\{(-1, 2.0)\}$ (*b*) $\{(-2, 2, 3)\}$ (*c*) Ø or { }, inconsistent (*d*) $\left\{\left(\frac{-2}{3}, 1, 0\right)\right\}$

7.5 (*a*) 11 (*b*) −6 (*c*) 36 (*d*) 0 (*e*) 9

(*f*) 11 (*g*) −4 (*h*) 104 (*i*) −52

7.6 (*a*) $\{(-3, 6)\}$ (*b*) $\{(2, -2, 3)\}$ (*c*) $\{(2, -3)\}$

(*d*) $\{(3, -3, 0)\}$ (*e*) $\left\{\left(\frac{1}{3}, \frac{7}{3}\right)\right\}$ (*f*) Ø or { }, no solution. $D = D_z = 0$

7.7 (*a*) $\begin{bmatrix} 1 & 1 & 1 & 1 \\ 2 & 3 & -1 & 0 \\ 1 & -2 & 3 & 7 \end{bmatrix}$ (*b*) $\begin{bmatrix} 3 & -6 & 9 & 21 \\ 2 & 3 & -1 & 0 \\ 1 & 1 & 1 & 1 \end{bmatrix}$ (*c*) $\begin{bmatrix} 1 & -2 & 3 & 7 \\ 0 & 7 & -7 & -14 \\ 1 & 1 & 1 & 1 \end{bmatrix}$

7.8 $\begin{bmatrix} 1 & -2 & 3 & 7 \\ 0 & 1 & -1 & -2 \\ 0 & 0 & 1 & 0 \end{bmatrix}$

7.9 (*a*) $\{(3, 0)\}$ (*b*) $\{(1, -1, 2)\}$ (*c*) $\{(6, -2)\}$ (*d*) $\{(4, 3, 2)\}$

7.10 (*a*) $\{(-2, 1), (1, 4)\}$ (*b*) $\{(3, 1), (2, 0)\}$ (*c*) $\{(\sqrt{2}, 3), (\sqrt{2}, -3), (-\sqrt{2}, 3), (-\sqrt{2}, -3)\}$

(*d*) $\{(2i\sqrt{2}, 3\sqrt{3}), (2i\sqrt{2}, -3\sqrt{3}), (-2i\sqrt{2}, 3\sqrt{3}), (-2i\sqrt{2}, -3\sqrt{3})\}$

(*e*) {Approximately (1.4, 2.8)}

(*f*) {Approximately (−1.9, 1), (1.9, −0.8)}

7.11 (*a*) See Figure 7.10. (*b*) See Figure 7.11. (*c*) See Figure 7.12. (*d*) See Figure 7.13.

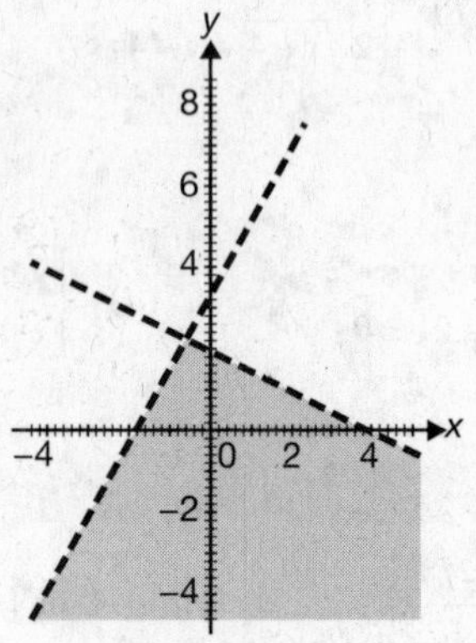

Figure 7.10

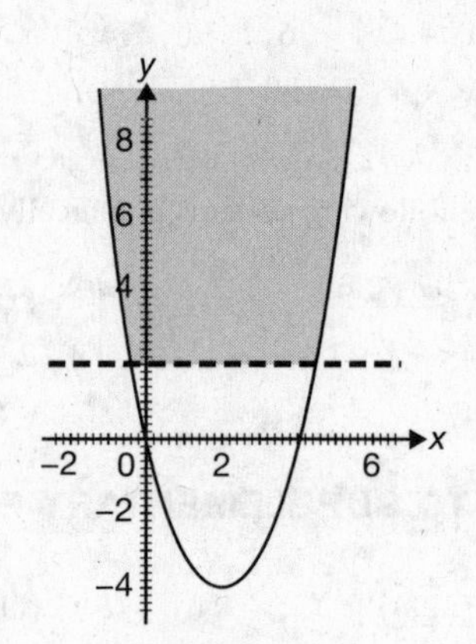

Figure 7.11

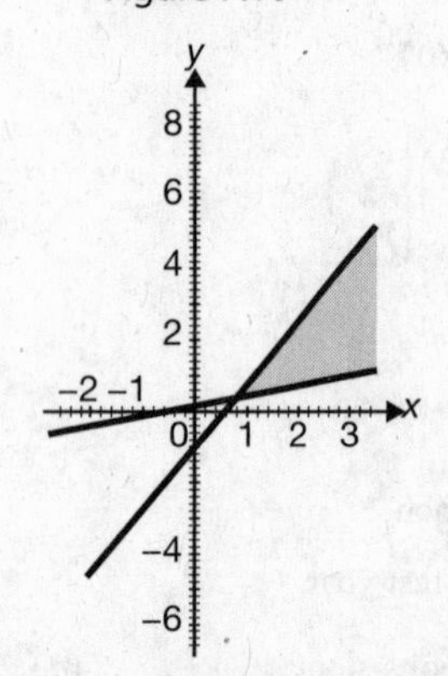

Figure 7.12

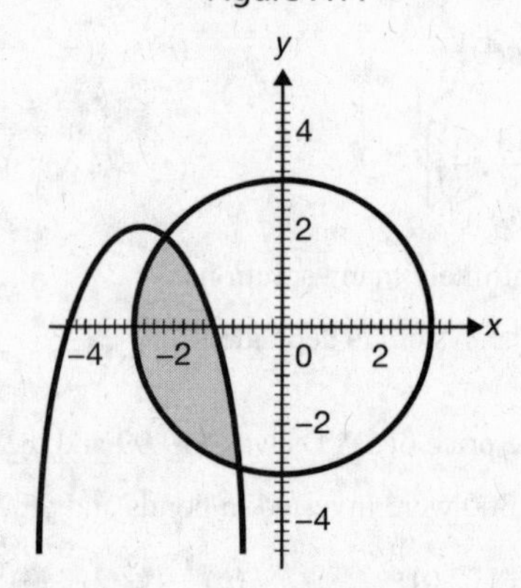

Figure 7.13

Relations and Functions

8.1 Basic Concepts

The *Cartesian Product* is basic to the mathematical concepts of relations and functions.

Definition 1. $A \times B = \{(a, b) \mid a \in A \text{ and } b \in B\}$ is called the *Cartesian Product*.

The *Cartesian Product* of two sets is the set of all ordered pairs for which the first element is a member of the first set and the second element is a member of the second set.

See solved problem 8.1

Definition 2. A *relation* is any set of ordered pairs. The set of first elements is the *domain* and the set of second elements is the *range*.

A *relation* of real numbers is a subset of $R \times R$ (where R symbolizes the set of real numbers). A *graph* is a relation displayed on a coordinate system.

EXAMPLE 1. Relations (in different formats):

(*a*)

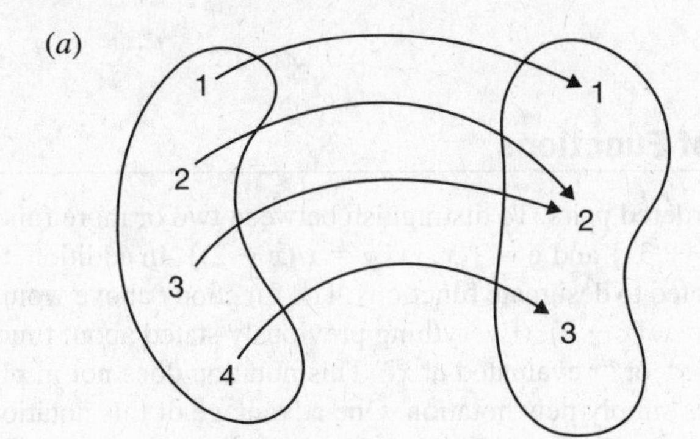

Figure 8.1

The figure illustrates the set

$\{(1, 1), (2, 2), (3, 2), (4, 3)\}$

Domain: $\{1, 2, 3, 4\}$

Range: $\{1, 2, 3\}$

(*b*)

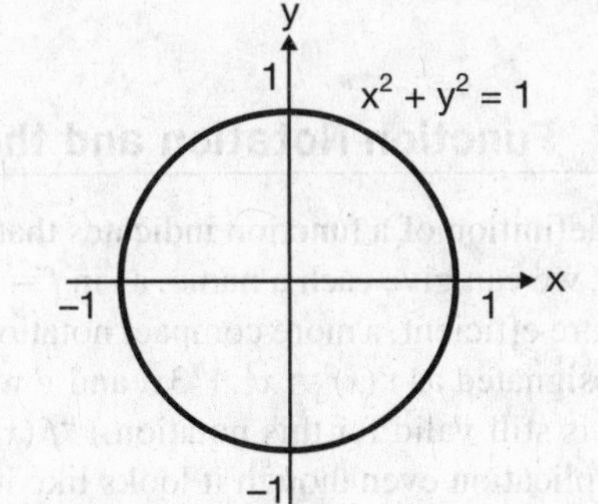

Figure 8.2

The figure illustrates the set

$\{(x, y) \mid x^2 + y^2 = 1\}$

Domain: $\{x \mid -1 \leq x \leq 1\}$

Range: $\{y \mid -1 \leq y \leq 1\}$

(*c*) $\{(1, 3), (2, 5), (1, 7)\}$

Domain: $\{1, 2\}$

Range: $\{3, 5, 7\}$

Definition 3. A *function* is a set of ordered pairs, or A relation, for which no two distinct ordered pairs have the same first element.

Definition 3. (Alternate) A *function* from a set A to a set B is a mapping, rule, or correspondence that assigns to each element x in set A exactly one element y in set B. (Each x is mapped to a unique y.)

A function is a relation with an added condition. The added condition is that for a function, no x value can have more than one y value associated with it.

EXAMPLE 2. In Example 1 above, (*a*) is a function, and (*b*) and (*c*) are not functions.

See solved problem 8.2.

To determine whether a relation in graphical format is a function, as in solved problem 8.2 (*c*) and (*d*), we can use the *Vertical Line Test*.

VERTICAL LINE TEST: A graph is the graph of a function if no vertical line intersects the graph in more than one point.

If any vertical line intersects a graph in two or more points, then the graph is not the graph of a function. The reason is this. Two points on the same vertical line indicate two different ordered pairs that have the same first element. Thus, it can't be a function.

A function is frequently described using set builder notation as in $\{(x, y) \mid y = 3x - 5\}$. In this form, the first variable in the ordered pair is called the *independent variable* and the second variable is the *dependent variable* (since its value depends on the value of the first variable).

See solved problem 8.3.

The concepts of domain and range apply to functions since all functions are also relations though not all relations are functions. When the function is described using an equation and the domain isn't stated (given explicitly), then we need to find the *implicit domain*. The implicit domain is the set of all values for the independent variable that result in real number values for the dependent variable. For the functions we will consider in this chapter, the implicit domain is found by excluding (from the set of real numbers) any values that result in (1) division by zero or (2) square roots of negative numbers.

See solved problem 8.4.

8.2 Function Notation and the Algebra of Functions

The definition of a function indicates that it is a set of ordered pairs. To distinguish between two or more functions, we can give each a name, as in $f = \{(x, y) \mid y = x^2 + 3x\}$ and $g = \{(x, y) \mid y = x/(x - 2)\}$. In addition, to be more efficient, a more compact notation has been created to designate functions. The function f above would be designated as $f(x) = x^2 + 3x$, and g would be $g(x) = x/(x - 2)$. (Everything previously stated about functions is still valid for this notation.) "$f(x)$" is read "f of x" or "f evaluated at x." This notation does not imply multiplication even though it looks like it should. This is simply new notation. One advantage of this notation is its simplicity: e.g., $f(x)$ above could be used anywhere $x^2 + 3x$ would be used. The independent variable x in this notation is just a placeholder for a number or expression. For this function, the symbolism "$f(5)$" means to evaluate the expression "$x^2 + 3x$" at $x = 5$. That is, $f(5) = (5)^2 + 3(5)$. Another way of thinking of this function f is as $f(?) = (?)^2 + 3\,(?)$ where each "?" is reserving space for a number or expression. To evaluate the function for a particular number or expression, replace each occurrence of the independent variable (or the "?") by that number or expression. Use parentheses around the number or expression to avoid confusion and to retain the correct order of operations in the result.

EXAMPLE 3. To evaluate $x^2 + 3x$ for $x = 2$, $f(x) = x^2 + 3x$ can be used. Replace each occurrence of x by 2: $f(2) = (2)^2 + 3\,(2) = 10$.

See solved problem 8.5.

To be consistent between this new notation and the rest of mathematics, another definition is required.

Definition 4. For all x common to the domains of f and g, the arithmetic of f and g is given by the following:

1. The *sum* is $(f+g)(x) = f(x) + g(x)$.
2. The *difference* is $(f-g)(x) = f(x) - g(x)$.
3. The *product* is $(fg)(x) = f(x) \cdot g(x)$.
4. The *quotient* is $\left(\frac{f}{g}\right)(x) = \frac{f(x)}{g(x)}$, for $g(x) \neq 0$.

See solved problem 8.6.

Another important mathematical connection with the algebra of functions follows.

Definition 5. For all x for which the expression is defined, the *composition of functions*, symbolized as $(f \circ g)(x)$, is defined to be $f(g(x))$, sometimes written as $f[g(x)]$ to avoid confusion between grouping symbols.

The composition of functions, $(f \circ g)(x) = f[g(x)]$, evaluates the function f at $g(x)$. It requires the value for $g(x)$ to be in the domain of f. We illustrate the process of evaluating and simplifying the composition of functions below.

See solved problem 8.7.

8.3 Distance and Slope Formulas

Distance Between Points in the Plane

Basic to the distance formula is the following theorem which is restated from Chapter 1.

PYTHAGOREAN THEOREM: The sum of the squares of the lengths of the legs in a right triangle is equal to the square of the length of the hypotenuse.

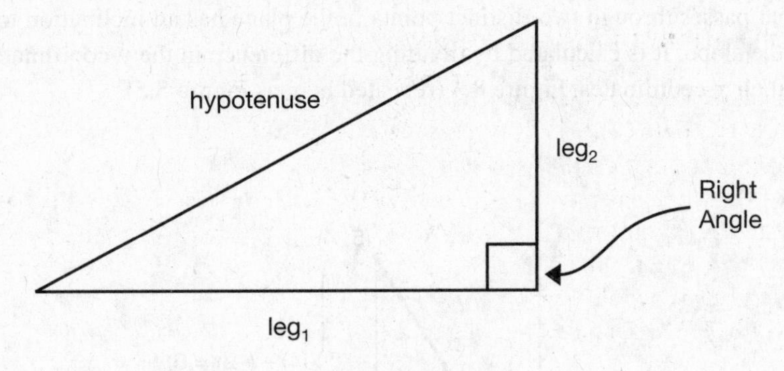

Figure 8.3

$$(leg_1)^2 + (leg_2)^2 = (hypotenuse)^2$$

The distance formula can be better understood using the Pythagorean Theorem and an example. Figure 8.4 below graphically shows the development of the formula for the distance between the points $(1, -2)$ and $(5, 4)$. That is, it shows the distance between the endpoints of the bold line segment d in the figure.

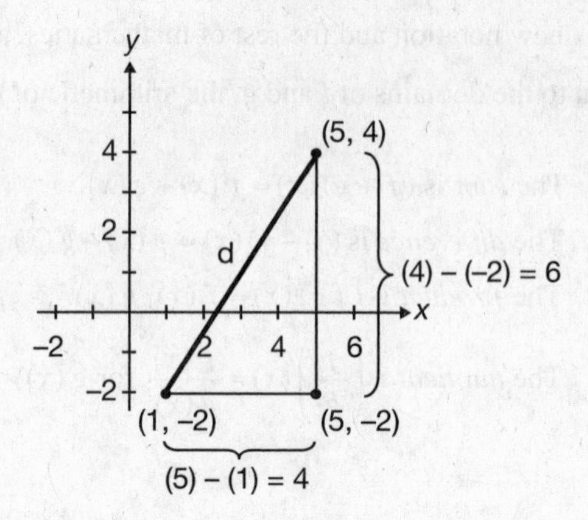

Figure 8.4

By the Pythagorean Theorem, we can now see (with the help of Figure 8.4) that $d^2 = 4^2 + 6^2$ or

$$d = \sqrt{4^2 + 6^2} = \sqrt{16 + 36} = \sqrt{52} = 2\sqrt{13} \approx 7.2111.$$

The distance formula simply generalizes the example above.

DISTANCE FORMULA: The distance d between (x_1, y_1) and (x_2, y_2) is

$$d = \sqrt{(x_2 - x_1)^2 + (y_2 - y_1)^2}.$$

The distance between any two points in the plane is the square root of the sum of the squares of the difference of the x-coordinates and the difference of the y-coordinates. Since the above formula sometimes causes confusion because of all the x's, y's, and subscripts, we will use a simpler looking formula. Namely, $d = \sqrt{(u - a)^2 + (v - b)^2}$, where (a, b) and (u, v) are the two points.

See solved problem 8.8.

Slope of a Line

A nonvertical line that passes through two distinct points in the plane has an inclination to the horizontal. We call that inclination the slope. It is calculated by dividing the difference in the y-coordinates of the two points by the difference in their x-coordinates. Figure 8.4 (repeated here as Figure 8.5)

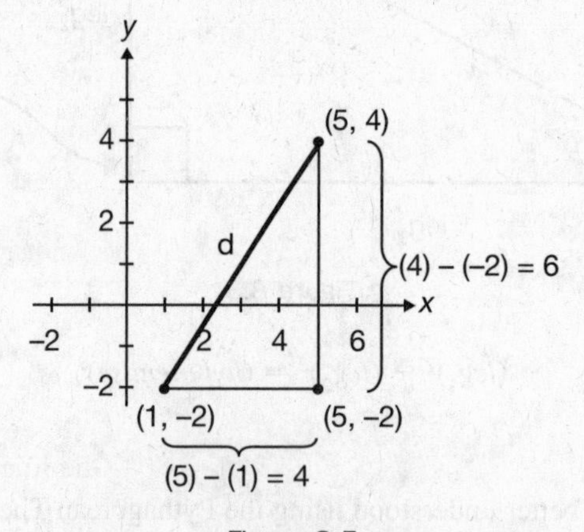

Figure 8.5

indicates the slope of the bold line segment (and thus the line it represents) is: $slope = \frac{6}{4} = 1.5$.

SLOPE FORMULA: The slope of the line passing through the points (x_1, y_1) and (x_2, y_2) is symbolized by the letter m and defined to be

$$m = \frac{y_2 - y_1}{x_2 - x_1} \text{ or } m = \frac{y_1 - y_2}{x_1 - x_2}$$

if the denominator is not zero. If the denominator is zero, the slope is undefined; the line has no slope associated with it. If the slope is undefined, then the line is vertical (and vice versa).

In words, the slope of a line passing through two points is the directed change in y divided by the directed change in x. Defined in terms of coordinates without subscripts we will write $m = (v - b) / (u - a)$ or $m = (b - v) / (a - u)$ for the line passing through the points (a, b) and (u, v).

The order of subtraction of the coordinates of the points is not important as long as there is consistency between the numerator and denominator of the slope formula. A good rule to follow is to make sure that the x-coordinate of a point is directly beneath the y-coordinate of that same point in the formula.

See solved problems 8.9–8.10.

8.4 Linear Equation Forms

From the formula for the slope of a line, we can develop formulas for equations of a line provided we are given enough information about the line. Using Figure 8.6 below, we can find an equation of the line passing through the two points $(-2, 3)$ and $(6, -1)$ by identifying a relation between the coordinates of the general point (x, y).

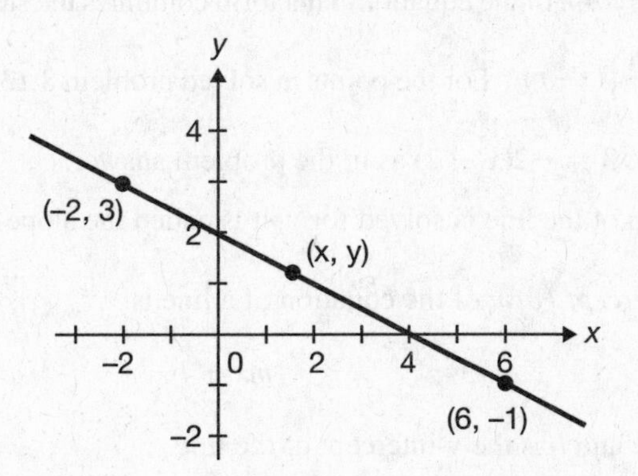

Figure 8.6

In general, to find an equation of a line we need to know two things: (1) its slope and (2) a point on the line. For Figure 8.6 above, the slope of the line passing through the points $(-2, 3)$ and $(6, -1)$ is $m = \frac{v - b}{u - a} = \frac{(-1) - (3)}{(6) - (-2)} = \frac{-1}{2}$. Now using the slope formula again with either one of those points, say $(-2,3)$, and the general point (x, y), we calculate $m = \frac{v - b}{u - a} = \frac{y - (3)}{x - (-2)} = \frac{y - 3}{x + 2}$.

The slope is of the line so it is the same between any pair of points on the line. The two expressions for the slope calculated above must be equal:

$$\frac{-1}{2} = \frac{y - 3}{x + 2}.$$

This equation relates the x and y-coordinates of every point on the line except for the point used in the second calculation, namely $(-2, 3)$ in this case. That is, for $(-2, 3)$, the equation becomes $\frac{(3) - 3}{(-2) + 2} = \frac{-1}{2}$ or $\frac{0}{0} = \frac{-1}{2}$ which is not a valid equation. To avoid this problem, we alter the above equation by clearing the denominator on the side that involves the x:

$$\frac{y-3}{x+2} = \frac{-1}{2}$$

$$\left(\frac{y-3}{x+2}\right)(x+2) = \frac{-1}{2}(x+2)$$

$$y-3 = \frac{-1}{2}(x+2)$$

This last form is used so frequently that it has been given a name: the point-slope form.

Definition 6. For any nonvertical line, the *Point-Slope Form* of the equation of a line is

$$y - y_1 = m\,(x - x_1)$$

where (x_1, y_1) is a known point on the line and m is the slope of the line.

In the point-slope form of the equation of a line, we use the known point and known slope to write an equation that identifies every point on the line. For our calculations, we will use a simpler looking equation without subscripts, namely, $y - s = m\,(x - r)$ for the point (r, s) and slope m.

Note: The slope of a vertical line is undefined. The point-slope form of the equation cannot be written for vertical lines.

See solved problems 8.11–8.13.

To find an equation of the line passing through the points (r, s) and (u, v) as in solved problem 8.13, we can alternately use the *two-point form* of the equation. This form combines the steps shown in the problem answer into one equation: $y - s = \left(\frac{v-r}{u-s}\right)(x-r)$. For the points in solved problem 8.13, we then obtain $y - 3 = \left(\frac{5-3}{1-2}\right)$ $(x - 2)$ which simplifies to $y - 3 = -2(x - 2)$ as in the problem answer.

When the point-slope form of the line is solved for y, it is called the slope-intercept form.

Definition 7. The *Slope-Intercept Form* of the equation of a line is

$$y = mx + b$$

where m is the slope of the line and b is the y-intercept of the line.

The slope-intercept form of a line requires y to be isolated on one side of the equation. In that form, the coefficient of x is the slope and the constant term is the y-intercept. The student should be careful to note that the coefficient of x is the slope of the line in this form only—the coefficient of x is not the slope in the other forms of the equation of the line.

Note: Since the slope of a vertical line is undefined, the slope-intercept form of the equation cannot be written for vertical lines.

See solved problems 8.14–8.17.

Definition 8. The *General Form* of the equation of a line is $ax + by + c = 0$; the *Standard Form* is $ax + by = c$. (Traditionally, a is nonnegative and all fractions are cleared.)

Both the general form and the standard form of the equation of the line require that either a or b is nonzero. Otherwise, the equation becomes $c = 0$ or $0 = c$. If c isn't zero, then this is a contradiction; if c is zero, then it is an identity. Regardless, no possible relationship between the x- and y-coordinates of any point is identified. Additionally, we note that no line has a unique equation in either the general or standard form. As you can see, $x + y + 3 = 0$ and $2x + 2y + 6 = 0$ both represent the same line since they are equivalent equations (multiply both sides of the first equation by 2 to obtain the second equation). Therefore, the solutions in the general and standard forms that you find may differ from the solutions we give, but your solutions should be equivalent to ours.

See solved problem 8.18.

Definition 9. The *Intercept Form* of the equation of a line is $x/r + y/s = 1$ where r is the x-intercept of the line and s is the y-intercept. Neither r nor s may be zero so this form cannot be used when the line passes through the origin.

This form is a variation of the standard form requiring $c = 1$. For the line in solved problem 8.18 (*b*), the equation becomes $x/3 + y/5 = 1$. The student can readily show this to be equivalent to the general form in the problem answer by multiplying both sides of the equation by the LCD $= 15$.

The following are normally proven in College Algebra. We will simply accept them as being true.

Parallel Lines. Nonvertical *parallel* lines have equal slopes. Lines with equal slopes are parallel.

If line l_1 with slope m_1 is parallel to line l_2 with slope m_2, then $m_1 = m_2$. Conversely, if $m_1 = m_2$, then l_1 is parallel to l_2.

Perpendicular Lines. Two *perpendicular* lines, neither of which is vertical, have slopes whose product is -1. (Alternately, nonvertical perpendicular lines have slopes which are negative reciprocals of one another.) Conversely, two lines having slopes whose product is -1 are perpendicular.

If line l_1 with slope m_1 is perpendicular to line l_2 with slope m_2, then $m_1 \cdot m_2 = -1$ or equivalently $m_2 = -1/m_1$. In the reverse perspective, if $m_1 \cdot m_2 = -1$, then l_1 is perpendicular to l_2.

See solved problems 8.19–8.23.

8.5 Types of Functions

The graph of a function f is normally the graph of the equation $y = f(x)$. We will use this equation to identify the graphs of the following special types of functions.

Definition 10. Any function of the form $f(x) = ax + b$ is identified as a *linear function*.

Since $y = f(x) = ax + b$, we see that the graph of a linear function is a straight line with slope $m = a$ and y-intercept b. All linear functions graph into nonvertical lines and all nonvertical lines can be represented by linear functions.

See solved problems 8.24–8.25.

Definition 11. Any function of the form $f(x) = ax^2 + bx + c$ where $a \neq 0$ is identified as a *quadratic function*.

Since $y = f(x) = ax^2 + bx + c = a\left(x + \dfrac{b}{2a}\right)^2 + \dfrac{4ac - b^2}{4a}$, we know that the graph of a quadratic function is a parabola (review Section 6.6). The graph opens upward if $a > 0$, downward if $a < 0$, and has a vertex at $\left(\dfrac{-b}{2a}, f\left(\dfrac{-b}{2a}\right)\right) = \left(\dfrac{-b}{2a}, \dfrac{4ac - b^2}{4a}\right)$. Every quadratic function graphs into a parabola with a vertical axis of symmetry and every parabola with a vertical axis can be represented by a quadratic function.

See solved problem 8.26.

Both linear and quadratic functions are examples of a more general type which we will not analyze in this text. That type is the *polynomial function*. A polynomial function is of the form $f(x) = a_n x^n + a_{n-1} x^{n-1} + \cdots + a_2 x^2 + a_1 x + a_0$ where n is a nonnegative integer and $a_n \neq 0$ (unless $n = 0$, meaning $f(x) = a_0$ and $f(x) = 0$ is valid). An example of a polynomial function that is neither linear nor quadratic is $f(x) = x^3 - x^2 - 6x$. It has x-intercepts at $-2, 0$, and 3 and a y-intercept at 0. See Figure 8.7.

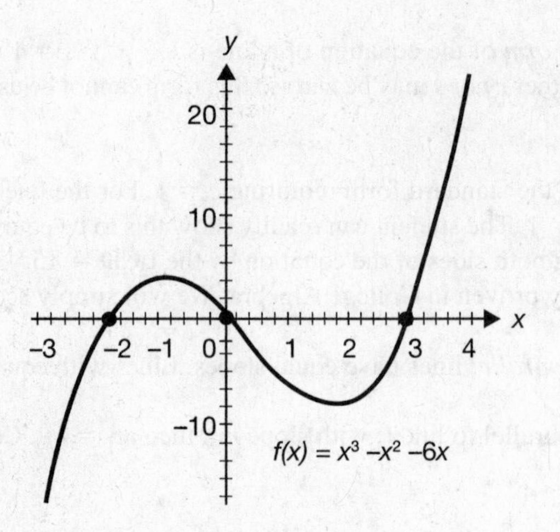

Figure 8.7

Definition 12. The function $f(x) = \sqrt{x}$, where x is a real number and $x \geq 0$, is identified as a *square root function*.

The square root function is the top half of a parabola opening to the right with vertex at the origin. See Figure 8.8.

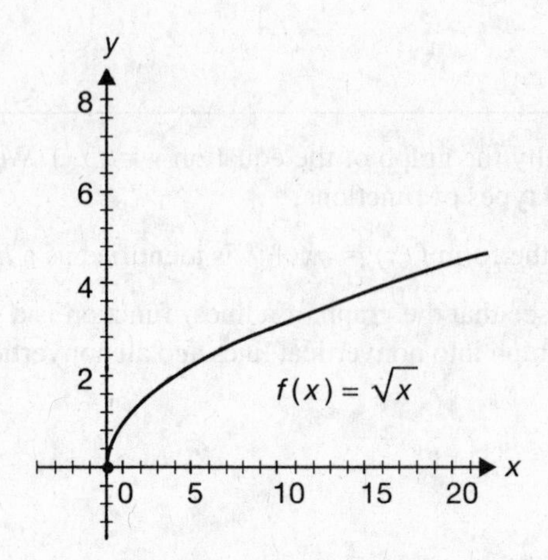

Figure 8.8

The domain of the square root function is $\{x|x \geq 0\}$; the range is $\{y|y \geq 0\}$. Variations of the square root function are of the general form, $g(x) = a + b\sqrt{cx + d}$. For this general form, the "starting point" (parabola vertex) on the graph is $(-d/c, a)$. The graph is the top half of a parabola if $b > 0$ or the bottom half if $b < 0$; it opens right if $c > 0$, left if $c < 0$.

See solved problem 8.27.

Definition 13. The function $f(x) = |x|$ is identified as the *absolute value function*.

The graph of the absolute value function follows from the definition of absolute value:

$$y = |x| = \begin{cases} x, \text{if } x \geq 0 \\ -x, \text{if } x < 0 \end{cases} \quad \text{so} \quad \begin{cases} y = x, \text{if } x \geq 0 \\ y = -x, \text{if } x < 0 \end{cases}$$

See Figure 8.9.

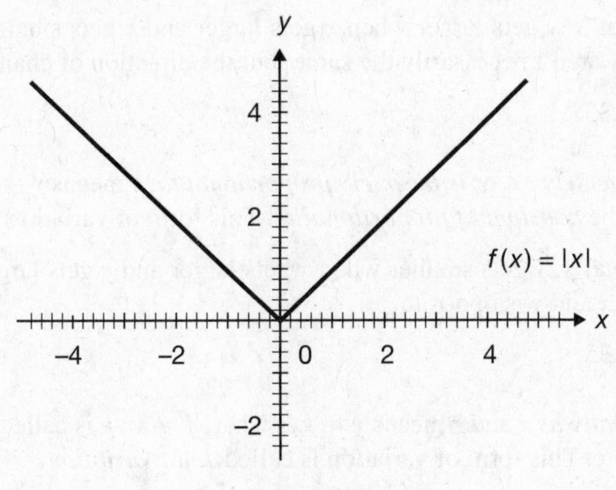

Figure 8.9

At the origin, the graph has a sharp turn or corner, which is called the vertex. The graph resembles the letter "V." As with the square root function, variations exist for the absolute value function and are of the form $f(x) = a|x + b| + c$. The vertex is at the point $(-b, c)$ and the graph opens upward (in a "V") if $a > 0$ or downward (in an inverted "V") if $a < 0$.

See solved problem 8.28.

The absolute value function is an example of a piecewise-defined (or multipart) function. A piecewise-defined function is a function composed of two or more "pieces." Each piece of the function is itself a function for which the domain is explicitly stated. We give examples of piecewise-defined functions next.

See solved problems 8.29–8.30.

The student would do well to memorize the graphs of at least the linear function, the quadratic function, the square root function, and the absolute value function. It is helpful to know the basic form of these graphs and their variations even before plotting points.

Functions can generally be graphed very nicely on graphing calculators. In most cases, the function will need to be set equal to y to identify the equation to graph. The student will need to become familiar with the keyboard of the calculator in order to use some of the special function keys, like $\boxed{\sqrt{\ }}$ (square root) and $\boxed{abs}$ (absolute value). Many of these calculators will even graph piecewise-defined functions. To graph the function in solved problem 8.29, most calculators that are able to do so will use a variation of the following expression:

$$y = (2x + 1) * (x < 1) + (4 - x^2) * (1 \leq x \text{ and } x \leq 3) + (4 - x) * (3 < x).$$

The middle term may require a change to $(4 - x^2) * (1 \leq x) * (x \leq 3)$. With some calculators, the inequalities may need to be replaced by intervals. For $x < 1$ as an example, the interval might be $[-50,1]$ instead of $(-\infty, 1)$, if -50 is outside the window being displayed on the calculator. The best thing to do is to read the manual that came with the calculator and then experiment.

8.6 Variation

A variation is a function involving two or more variables. The type of variation specifies the form of the function. Function notation is rarely used with variations.

Definition 14. *y varies directly as*, or *is proportional to*, x means $y = kx$ where $k \neq 0$; k is called the *constant of variation* or the *constant of proportionality*. This form of variation is called *Direct Variation*.

When y varies directly as x, y gets larger when x gets larger and y gets smaller when x gets smaller. The amounts of change in x and y aren't necessarily the same, but the direction of change is.

See solved problems 8.31–8.33.

Definition 15. *y varies inversely as*, or *is inversely proportional to*, x means $y = k/x$ where $k \neq 0$; k is called the *constant of variation* or the *constant of proportionality*. This form of variation is called *Inverse Variation*.

When y varies inversely as x, y gets smaller when x gets larger and y gets larger when x gets smaller. The direction of change between x and y is opposite.

See solved problems 8.34–8.35.

Definition 16. *y varies jointly as* x and z means $y = kxz$ where $k \neq 0$; k is called the *constant of variation* or the *constant of proportionality*. This form of variation is called *Joint Variation*.

When y varies jointly as x and z, y gets smaller when the product of x and z gets smaller, and y gets larger when the product of x and z gets larger. The direction of change between y and the product of x and z is the same. Be careful to note, however, that the change in y may be opposite that of either x or z. For example, if x changes from 4 to 2 while z changes from 3 to 9, then x becomes smaller but z, the product xz (changing from 12 to 18), and y become larger.

See solved problems 8.36–8.38.

8.7 Inverse Relations and Functions

Definition 17. The *inverse* of a relation or function is the relation that results from interchanging the x- and y-coordinates of all ordered pairs.

See solved problem 8.39.

EXAMPLE 4. The function $y = f(x)$ with a graph above the dotted line $y = x$ (Figure 8.10), is the inverse function of the function $y = g(x)$ whose graph is below the line. That is, f is the inverse of g. Similarly g is the inverse of f.

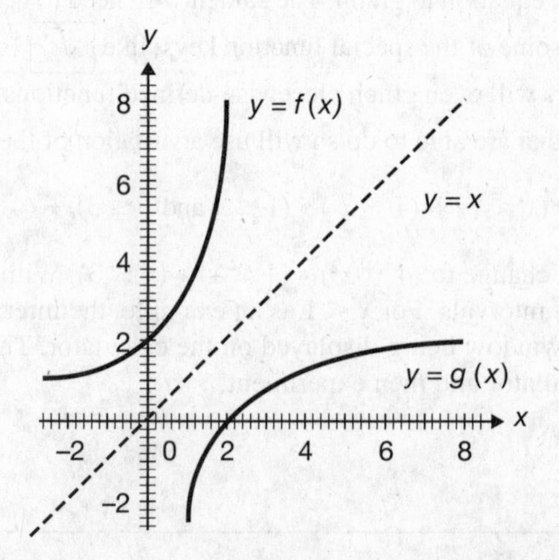

Figure 8.10

In function notation, the inverse of the function f is symbolized by f^{-1}. The superscript is not an exponent. This is just new notation for the concept of the inverse; that is, $f^{-1}(x) \neq 1/f(x)$. In a sense, the inverse function "undoes" what the function does. If $f(3) = 5$, then $f^{-1}(5) = 3$. Using the concept of the composition of functions, the relationship between a function and its inverse can be stated by the equations

$$f[f^{-1}(x)] = f^{-1}[f(x)] = x.$$

See solved problems 8.40–8.41.

The process generally used to find the inverse of a function $f(x)$ uses three steps:

1. Set $y = f(x)$.
2. Interchange x and y. That is, replace each occurrence of x with y and each occurrence of y with x.
3. Solve the resulting equation for y.

See solved problem 8.42.

The inverse of a function is not always itself a function as shown in solved problem 8.42 above. If a function is *one-to-one*, then its inverse is a function; if not, then the inverse is simply a relation.

Definition 18. A function is *one-to-one* if no two different ordered pairs in the function have the same second element.

Recall that for a relation to be a function, no two distinct ordered pairs have the same first element. Hence, for a function to be one-to-one, neither the first elements nor the second elements are the same for distinct ordered pairs.

EXAMPLE 5. (*a*) $\{(1,2),(3,4),(5,2)\}$ is not *one-to-one* since 2 is the second element of two different ordered pairs.

(*b*) $\{(-4,1),(0,2),(3,3),(2,4)\}$ is a *one-to-one* function since no two different ordered pairs have the same first or second element.

(*c*) $p(x) = x^2$ is not *one-to-one* since we can find two different ordered pairs in p which have the same second element, as with $(-2,4)$ and $(2,4)$ (since $p(-2) = 4$ and $p(2) = 4$.)

To determine whether or not a function in graphical format is *one-to-one*, we can use the *Horizontal Line Test*.

HORIZONTAL LINE TEST: The inverse of a function is itself a function if no horizontal line intersects the graph of the function in more than one point. That is, no two points on the graph have the same y-coordinate.

If two or more points on a graph lie on the same horizontal line, then the graph is not the graph of a one-to-one function.

See solved problems 8.43–8.44.

SOLVED PROBLEMS

8.1 Find the Cartesian Product of $\{r,s\}$ and $\{1,2,3\}$.

$$\begin{aligned}\{r,s\} \times \{1,2,3\} &= \{(x,y) \mid x \in \{r,s\} \text{ and } y \in \{1,2,3\}\} \\ &= \{(r,1),(r,2),(r,3),(s,1),(s,2),(s,3)\}\end{aligned}$$

Refer to supplementary problem 8.1 for practice problems.

8.2 Determine which of the following relations are also functions:

(*a*) $\{(3,4),(-2,1),(3,5),(6,7)\}$

This is not a function. (3, 4) and (3, 5) have the same first element.

(*b*) $\{(2,5),(7,-3),(6,7),(16,5)\}$

This is a function. (2, 5) and (16, 5) have the same second element but none have the same first element.

(*c*)

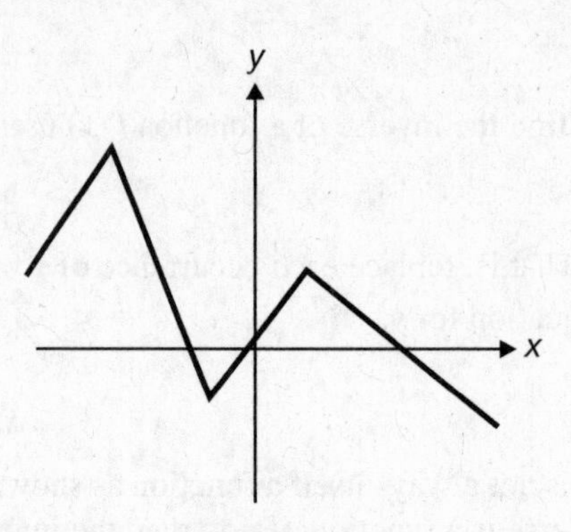

Figure 8.11

This is a function. No ordered pairs or points have the same *x*-coordinate, though there are *y*-coordinates alike.

(*d*)

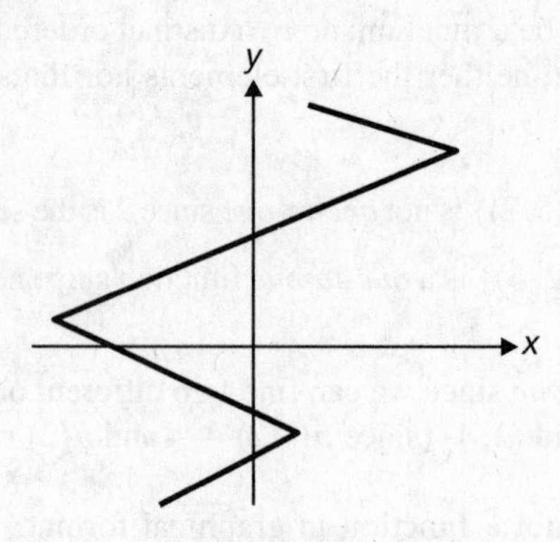

Figure 8.12

This is not a function. Three different points are on the *y*-axis so the *x*-coordinate is zero for each.

Refer to supplementary problem 8.2 for practice problems.

8.3 For $\{(x, y) \mid y = 3x - 5\}$, find the ordered pairs for $x = 1, 2,$ and -5.

First find the *y* values from the stated equation and then put in ordered pair form:

x	y	(x, y)
1	$3(1) - 5 = -2$	$(1, -2)$
2	$3(2) - 5 = 1$	$(2, 1)$
-5	$3(-5) - 5 = -20$	$(-5, -20)$

8.4 Determine the domains of the following functions:

(*a*) $\left\{(x, y) \mid y = \dfrac{3}{x+7}\right\}$

Setting the denominator, $x + 7$, equal to zero and solving for x we get $x = -7$. Therefore, -7 must be excluded from the domain. Domain: $\{x \mid x \neq -7\} = (-\infty, -7) \cup (-7, \infty)$.

(*b*) $\left\{(x, y) \mid y = \sqrt{5 - 3x}\right\}$

We must exclude all values of x for which $5 - 3x < 0$. In other words, we must include all values of x for which $5 - 3x \geq 0$. Solving the latter inequality gives $x \leq \frac{5}{3}$.

$$\text{Domain}: \left\{x \mid x \leq \frac{5}{3}\right\} = \left(-\infty, \frac{5}{3}\right].$$

(*c*) $\left\{(x, y) \mid y = \dfrac{\sqrt{x+5}}{x^2 - 4}\right\}$

We must exclude all values for which (1) $x^2 - 4 = 0$ or (2) $x + 5 < 0$. Solving (1) for x: $x^2 = 4$ so $x = \pm\sqrt{4} = \pm 2$. Hence we must exclude $x = \pm 2$, so we must include everything else, that is, $x \neq \pm 2$. Solving (2) for x: $x < -5$. We must exclude $x < -5$, so we must include $x \geq -5$. We combine these for the domain. Domain: $\{x \mid x \geq -5 \text{ and } x \neq \pm 2\} = [-5, -2) \cup (-2, 2) \cup (2, \infty)$.

(*d*) $\left\{(x, y) \mid y = \sqrt{x^2 - 9}\right\}$

If we must exclude all values for which $x^2 - 9 < 0$, then we must include all values for which $x^2 - 9 \geq 0$. To solve this quadratic inequality, we first factor the polynomial side: $(x - 3)(x + 3) \geq 0$. Then we use the number line method, as we did in Section 6.7, to find that all numbers in the intervals $(-\infty, -3]$ and $[3, \infty)$ satisfy the inequality, and all numbers in the interval $(-3, 3)$ do not. Domain: $\{x \mid x \leq -3$ or $x \geq 3\} = (-\infty, -3] \cup [3, \infty)$.

Refer to supplementary problem 8.3 for practice problems.

8.5 For $f(x) = x^2 + 3x$ and $g(x) = x/(x - 2)$, find the following:

(*a*) $f(-3)$

$$f(-3) = (-3)^2 + 3(-3) = 0$$

(*b*) $g(8)$

$$g(8) = \frac{(8)}{(8) - 2} = \frac{8}{6} = \frac{4}{3}$$

(*c*) $f(r + s)$

$$f(r + s) = (r + s)^2 + 3(r + s) = r^2 + 2rs + s^2 + 3r + 3s$$

(*d*) $g\left(\frac{1}{x}\right)$

This might be a good place for us to use $g(?) = (?)/[(?) - 2]$ for the function to keep from confusing the x's. Now we simply replace each occurrence of the "?" with $1/x$ and simplify the expression:

$$g\left(\frac{1}{x}\right) = \frac{\left(\frac{1}{x}\right)}{\left(\frac{1}{x}\right) - 2} = \frac{\frac{1}{x}}{\frac{1 - 2x}{x}} = \left(\frac{1}{x}\right)\left(\frac{x}{1 - 2x}\right) = \frac{1}{1 - 2x}$$

Refer to supplementary problem 8.4 for more function evaluation problems.

8.6 For $f(x) = 2x^2 - 3$ and $g(x) = 5x + 1$, evaluate the following:

(*a*) $(f + g)(-2)$

$$(f + g)(-2) = f(-2) + g(-2) = \left[2(-2)^2 - 3\right] + [5(-2) + 1] = (5) + (-9) = -4$$

(*b*) $(f - g)(5)$

$$(f-g)(5) = f(5) - g(5) = \left[2(5)^2 - 3\right] - [5(5)+1] = (47) - (26) = 21$$

(*c*) $(fg)(1)$

$$(fg)(1) = f(1)\cdot g(1) = \left[2(1)^2 - 3\right]\cdot[5(1)+1] = (-1)\cdot(6) = -6$$

(*d*) $\left(\frac{f}{g}\right)(0)$

$$\left(\frac{f}{g}\right)(0) = \frac{f(0)}{g(0)} = \frac{2(0)^2 - 3}{5(0)+1} = \frac{-3}{1} = -3$$

Refer to supplementary problem 8.5 for more problems involving function arithmetic.

8.7 Evaluate each of the following for $f(x) = (x-1)/x$ and $g(x) = \sqrt{x+3}$:

(*a*) $(f \circ g)(6)$

$$(f \circ g)(6) = f[g(6)] = f\left(\sqrt{(6)+3}\right) = f\left(\sqrt{9}\right) = f(3) = \frac{(3)-1}{(3)} = \frac{2}{3}$$

(*b*) $(g \circ f)(6)$

$$(g \circ f)(6) = g[f(6)] = g\left[\frac{6-1}{6}\right] = g\left(\frac{5}{6}\right) = \sqrt{\frac{5}{6}+3} = \sqrt{\frac{5+3\cdot 6}{6}} = \sqrt{\frac{23}{6}}\cdot\sqrt{\frac{6}{6}} = \frac{\sqrt{138}}{6}$$

(*c*) $(g \circ f)(x)$

$$\text{Method 1:}\quad (g \circ f)(x) = g[f(x)] = g\left(\frac{x-1}{x}\right) = \sqrt{\left(\frac{x-1}{x}\right)+3} = \sqrt{\frac{x-1}{x}+\frac{3x}{x}}$$

$$= \sqrt{\frac{4x-1}{x}} = \frac{\sqrt{x(4x-1)}}{|x|}$$

$$\text{Method 2:}\quad (g \circ f)(x) = g[f(x)] = \sqrt{[f(x)]+3} = \sqrt{\left[\frac{x-1}{x}\right]+3} = \sqrt{\frac{4x-1}{x}}$$

$$= \frac{\sqrt{x(4x-1)}}{|x|}$$

Note: The two methods above differ in process from the second to the fourth expressions. In method 1, $f(x)$ is first replaced by $(x-1)/x$ and then, since $g(?) = \sqrt{(?)+3}$, the "?" is replaced by $(x-1)/x$. Method 2 first replaces that "?" with $f(x)$ and then replaces $f(x)$ by $(x-1)/x$. Method 1 will be used for the solved problems in this text but method 2 is just as good. The two methods will always result in the same answer. Use the method that seems most comfortable to you.

By this result in (*c*), we have a formula for solving the problem in part (*b*) above. Using this result instead of the process shown in part (*b*), $(g \circ f)(6) = \frac{\sqrt{(6)[4(6)-1]}}{|(6)|} = \frac{\sqrt{135}}{6}$. If we wanted to evaluate $g \circ f$ for many different values, it would be to our advantage to use this formula for $g \circ f$ rather than processing each using the composition of functions.

(*d*) $(f \circ f)(z)$

$$(f \circ f)(z) = f[f(z)] = f\left[\frac{z-1}{z}\right] = \frac{\left[\frac{z-1}{z}\right]-1}{\left[\frac{z-1}{z}\right]} = \frac{\frac{z-1}{z}-\frac{z}{z}}{\frac{z-1}{z}} = \frac{\frac{-1}{z}}{\frac{z-1}{z}}$$

$$= \left(\frac{-1}{z}\right)\left(\frac{z}{z-1}\right) = \frac{-1}{z-1} = \frac{1}{1-z}$$

(*e*) $(f \circ f \circ g)(x)$

$$(f \circ f \circ g)(x) = f\,(f[g\,(x)]) = f\left(f\left[\sqrt{x+3}\right]\right) = f\left(\frac{\left(\sqrt{x+3}\right)-1}{\left(\sqrt{x+3}\right)}\right)$$

$$= \frac{\left(\dfrac{\left(\sqrt{x+3}\right)-1}{\left(\sqrt{x+3}\right)}\right)-1}{\left(\dfrac{\left(\sqrt{x+3}\right)-1}{\left(\sqrt{x+3}\right)}\right)} = \frac{\dfrac{\sqrt{x+3}-1}{\sqrt{x+3}} - \dfrac{\sqrt{x+3}}{\sqrt{x+3}}}{\dfrac{\sqrt{x+3}-1}{\sqrt{x+3}}} = \frac{\dfrac{-1}{\sqrt{x+3}}}{\dfrac{\sqrt{x+3}-1}{\sqrt{x+3}}}$$

$$= \left(\frac{-1}{\sqrt{x+3}}\right)\left(\frac{\sqrt{x+3}}{\sqrt{x+3}-1}\right) = \frac{-1}{\sqrt{x+3}-1} = \frac{-1}{\sqrt{x+3}-1} \cdot \frac{\sqrt{x+3}+1}{\sqrt{x+3}+1}$$

$$= \frac{-\sqrt{x+3}-1}{(x+3)-1} = \frac{-1-\sqrt{x+3}}{x+2}$$

Refer to supplementary problem 8.6 for practice problems.

8.8 Find the distance between each of the following pairs of points.

(*a*) $(1, 5)$ and $(7, 8)$

For $(a, b) = (1, 5)$ and $(u, v) = (7, 8)$, $d = \sqrt{(u-a)^2 + (v-b)^2} = \sqrt{(7-1)^2 + (8-5)^2} = \sqrt{(6)^2 + (3)^2}$ $= \sqrt{36+9} = \sqrt{45} = 3\sqrt{5} \approx 6.7082$,

or for $(a, b) = (7, 8)$ and $(u, v) = (1, 5)$, $d = \sqrt{(u-a)^2 + (u-b)^2} = \sqrt{(1-7)^2 + (5-8)^2}$ $= \sqrt{(-6)^2 + (-3)^2} = \sqrt{36+9} = \sqrt{45} = \sqrt[3]{5} \approx 6.7082$.

Note that the order in which the points are chosen doesn't matter for the formula.

(*b*) $(-2, 3)$ and $(3, 6)$

$$d = \sqrt{(-2-3)^2 + (3-6)^2} = \sqrt{(-5)^2 + (-3)^2} = \sqrt{25+9} = \sqrt{34} \approx 5.8310$$

(*c*) $(3, 8)$ and $(7, 2)$

$$d = \sqrt{(7-3)^2 + (2-8)^2} = \sqrt{16+36} = \sqrt{52} = 2\sqrt{13} \approx 7.2111$$

(*d*) $(-3, 5)$ and $(6, -2)$

$$d = \sqrt{(-3-6)^2 + (5-(-2))^2} = \sqrt{(-9)^2 + (7)^2} = \sqrt{81+49} = \sqrt{130} \approx 11.4018$$

Be sure to use parentheses in these calculations, otherwise $\sqrt{(-9)^2 + (7)^2}$ in problem 8.8 (*d*) above might become $\sqrt{-9^2 + 7^2} = \sqrt{-81 + 49} = \sqrt{-32}$. Since $\sqrt{-32}$ isn't a real number, it can't be a distance. Refer to supplementary problem 8.7 for more practice.

8.9 Find the slope of the line passing through each of the following pairs of points:

(*a*) $(2,4)$ and $(5,10)$

$$m = \frac{v-b}{u-a} = \frac{10-4}{5-2} = \frac{6}{3} = 2 \quad \text{or} \quad m = \frac{b-v}{a-u} = \frac{4-10}{2-5} = \frac{-6}{-3} = 2$$

(*b*) $(1,-2)$ and $(-3,-5)$

$$m = \frac{v-b}{u-a} = \frac{(-5)-(-2)}{(-3)-(1)} = \frac{-3}{-4} = \frac{3}{4}$$

(*c*) $(4,-3)$ and $(-5,8)$

$$m = \frac{v-b}{u-a} = \frac{(8)-(-3)}{(-5)-(4)} = \frac{11}{-9} = \frac{-11}{9}$$

(*d*) (2, −1) and (5, −1)

$$m = \frac{v-b}{u-a} = \frac{(-1)-(-1)}{(5)-(2)} = \frac{0}{3} = 0$$

(*e*) (−3, 4) and (−3, 8)

$$m = \frac{v-b}{u-a} = \frac{(8)-(4)}{(-3)-(-3)} = \frac{4}{0}$$ is undefined. The line has no slope associated with it.

Note: The line passing through the points in problem 8.9(*d*) above is horizontal; the slope of a horizontal line is zero. The line passing through the points in problem 8.9(*e*) is vertical; the slope of a vertical line is undefined.

Refer to supplementary problem 8.8 for practice problems.

8.10 Find the slope of the line given by $3x + 2y = 4$.

First we fin.d two points on the line by choosing a value for one variable and solving for the associated value of the other variable. For example, choosing 0 for x requires y to be 2 and choosing −2 for x requires y to be 5. This means (0, 2) and (−2, 5) are on the line. The slope is

$$m = \frac{(5)-(2)}{(-2)-(0)} = \frac{3}{-2} = \frac{-3}{2}.$$

Any two other points on the line would have given the same result. The student should now find two other points on the line to show this to be true.

Refer to supplementary problem 8.9 for practice problems. Solved problem 8.14 gives an alternate method for finding the slope of a line from the equation.

8.11 Find the point-slope form of the equation of the line with slope 5 and passing through the point (2, 3).

With $(r, s) = (2, 3)$ and $m = 5$, we write $y - 3 = 5\,(x - 2)$.

8.12 Find the point-slope form of the equation of the line with slope −2 and passing through the point (5, −4).

In this case, $(r, s) = (5, -4)$ and $m = -2$, so $y - (-4) = -2\,(x - 5)$, or simplified $y + 4 = -2\,(x - 5)$.

8.13 Find the point-slope form of the equation of the line passing through the two points (2, 3) and (1, 5).

The point-slope formula requires (1) a point and (2) the slope. We have more than enough information for a point, but we need to find the slope. Using the points and the slope formula,

$$m = \frac{v-b}{u-a} = \frac{5-3}{1-2} = -2.$$

The point-slope form then becomes either

$$y - 3 = -2\,(x - 2) \quad \text{or} \quad y - 5 = -2\,(x - 1)$$

depending upon which point is chosen. Either of these equations correctly identifies the line passing through the two given points.

Refer to supplementary problem 8.10 for practice problems.

8.14 Determine the slope and y-intercept of the line given by the equation $3x - 5y = 15$.

We first solve the equation for y:

$$3x - 5y = 15$$
$$-5y = -3x + 15$$
$$y = \frac{-3x + 15}{-5}$$
$$y = \frac{3}{5}x - 3.$$

Now the slope can be read from the coefficient of x as $\frac{3}{5}$ and the y-intercept is read from the constant as -3.

Refer to supplementary problem 8.11 for practice problems.

8.15 Find the slope-intercept form of the equation of the line with $m = 5$ and passing through $(2, 3)$.

From the point-slope form

$$y - 3 = 5(x - 2)$$
$$y - 3 = 5x - 10$$
$$y = 5x - 7$$

Alternate Method: With the slope given, we know that $y = 5x + b$. Since we know that every point on the line satisfies this equation, we know specifically that the given point $(2, 3)$ does. Therefore, we can replace x by 2 and y by 3 and then solve for b.

$$y = 5x + b$$
$$3 = 5(2) + b$$
$$3 = 10 + b$$
$$3 - 10 = b$$
$$-7 = b$$

Now with both m and b identified the equation follows: $y = 5x + (-7)$ or $y = 5x - 7$.

8.16 Find the slope-intercept form of the equation of the line with $m = -3$ and passing through $(0, 4)$.

Since $(0, 4)$ gives us 4 as the y-intercept, the slope-intercept form is $y = -3x + 4$.

Refer to supplementary problem 8.12 for practice problems.

8.17 Find the slope-intercept form of the equation of the line passing through $(3, 7)$ and $(-2, 1)$.

First: $m = \dfrac{v - b}{u - a} = \dfrac{1 - 7}{(-2) - (3)} = \dfrac{6}{5}$. Then in the point-slope form using $(3, 7)$

$$y - 7 = \frac{6}{5}(x - 3) = \frac{6}{5}x - \frac{18}{5}$$
$$y = \frac{6}{5}x - \frac{18}{5} + 7$$
$$y = \frac{6}{5}x + \frac{17}{5},$$

or using (-2, 1)

$$y - 1 = \frac{6}{5}(x - (-2)) = \frac{6}{5}x + \frac{12}{5}$$
$$y = \frac{6}{5}x + \frac{12}{5} + 1$$
$$y = \frac{6}{5}x + \frac{17}{5}.$$

Note that it didn't make any difference which point we used in the calculations after the slope was found. Our result was $y = \frac{6}{5}x + \frac{17}{5}$ using either point.

Refer to supplementary problem 8.13 for practice problems.

8.18 Find the general form of the equations of each of the following:

(*a*) The line passing through the points (2, 3) and (2, −4).

We notice here that the slope $m = (-4 - 3)/(2 - 2) = -7/0$ is undefined so neither the point-slope nor the slope-intercept forms of the equation are available to help us. We can, however, use the reason behind their unavailability to find the equation we want; namely, the line is vertical. This means that all x-coordinates must be the same, which, in this case means they must all be 2. We state that in algebraic form as $x = 2$. In the general form then, $x - 2 = 0$.

(*b*) The line crossing the x-axis at 3 and the y-axis at 5.

The points for the x- and y-intercepts are given as (3, 0) and (0, 5), respectively, so $b = 5$ and $m = \frac{v - b}{u - a} = \frac{0 - 5}{3 - 0} = \frac{-5}{3}$. Therefore,

$$y = \frac{-5}{3}x + 5$$
$$3y = -5x + 15$$
$$5x + 3y - 15 = 0.$$

8.19 Find the slope of a line parallel to the line given by $3x - 2y = 6$.

First, we determine the slope of the given line by solving the equation for y:

$$3x - 2y = 6$$
$$3x - 6 = 2y$$
$$\frac{3}{2}x - 3 = y.$$

m_1, the slope of the given line, is $\frac{3}{2}$. Therefore, the slope of the parallel line m is also $\frac{3}{2}$. That is, $m = m_1 = \frac{3}{2}$.

Refer to supplementary problem 8.14 for practice problems.

8.20 Find the slope of a line perpendicular to the line given by $3x - 2y = 6$.

We first determine the slope of the given line as in problem 8.19 above and find $m_1 = \frac{3}{2}$. Therefore, the slope of the perpendicular line m is

$$m = \frac{-1}{m_1} = \frac{-1}{\frac{3}{2}} = \frac{-2}{3}$$

which is the negative reciprocal of the first slope.

Refer to supplementary problem 8.15 for practice problems.

8.21 Determine the standard form of the equation of the line passing through the point $(2, -5)$ and parallel to the line given by $4x + 3y = 12$.

We start by finding the slope of the given line: $4x + 3y = 12$ or $y = \frac{-4}{3}x + 4$ so the slope is $m_1 = \frac{-4}{3}$. The slope of the parallel line m is also $\frac{-4}{3}$. Using the point-slope form of the line and then converting to standard form, we obtain

$$y-(-5)=\frac{-4}{3}(x-2)$$
$$y+5=\frac{-4}{3}x+\frac{8}{3}$$
$$3y+15=-4x+8$$
$$4x+3y=-7.$$

Notice that for the standard form of the equations of parallel lines, the coefficients of x are the same and the coefficients of y are the same.

8.22 Determine the standard form of the equation of the line passing through the point $(-1, 3)$ and perpendicular to the line given by $2x - 5y + 10 = 0$.

First, we find the slope of the given line. $2x - 5y + 10 = 0$ or $y = \frac{2}{5}x + 2$ so $m_1 = = \frac{2}{5}$. Next, we identify that the slope of the perpendicular line m is $\frac{-5}{2}$. Finally, we use the point-slope form of the line and convert to the standard form to get

$$y-3=\frac{-5}{2}(x-[-1])$$
$$y-3=\frac{-5}{2}x-\frac{5}{2}$$
$$2y-6=-5x-5$$
$$5x+2y=1.$$

Notice that for the standard form of the equations of perpendicular lines, the coefficients of x and the coefficients of y are interchanged and the operation between the terms is the opposite.

8.23 Using what was noticed about the coefficients in problems 8.21 and 8.22 above, determine the standard form of the equation for each of the following:

(*a*) The line passing through $(4, 7)$ and parallel to the line given by $3x + 4y = 10$.

We use the coefficients in the equation of the given line and the coordinates of the point: $3x + 4y = 3(4) + 4(7) = 40$, thus,

$$3x + 4y = 40.$$

(*b*) The line passing through $(4, 7)$ and perpendicular to the line given by $3x + 4y = 10$.

We interchange the coefficients, change the operation between the terms, and use the coordinates of the point: $4x - 3y = 4(4) - 3(7) = -5$, thus,

$$4x - 3y = -5.$$

Note: This method is a useful shortcut when the only objective is to find the equation of the line.

Refer to supplementary problems 8.16 and 8.17 for practice in finding equations of lines parallel or perpendicular to other lines.

8.24 Determine the linear function for which $f(2) = 3$ and $f(5) = 1$.

With $f(x) = y$ identifying the point (x, y), the two given equations identify the points $(2, 3)$ and $(5, 1)$. The slope of the line for the graph is $m = (1 - 3)/(5 - 2) = -2/3$. In point-slope form the equation of the line using the point (2,3) becomes $y - 3 = \frac{-2}{3}(x - 2)$. Changing to slope-intercept form we have:

$$y = \frac{-2}{3}x + \frac{13}{3}. \text{ Thus, } y = f(x) = \frac{-2}{3}x + \frac{13}{3}.$$

Refer to supplementary problem 8.18 for practice problems.

8.25 Graph $f(x) = -3x + 2$.

We first create a table of points or ordered pairs to plot using $f(x) = y = -3x + 2$ and then graph the line through the points. See Figure 8.13.

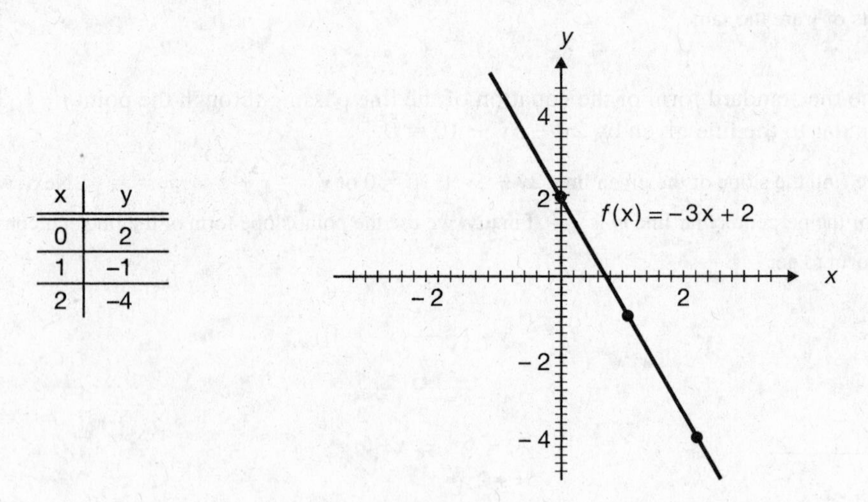

x	y
0	2
1	−1
2	−4

Figure 8.13

8.26 Graph $g(x) = 2x^2 + 4x - 3$.

The vertex is $\left(\frac{-4}{2(2)}, g\left(\frac{-4}{2(2)}\right)\right) = (-1, g(-1)) = (-1, -5)$, so we create a table of points or ordered pairs to plot and graph the parabola through the points. See Figure 8.14.

x	y
−1	−5
0	−3
−2	−3
−5	27
3	27

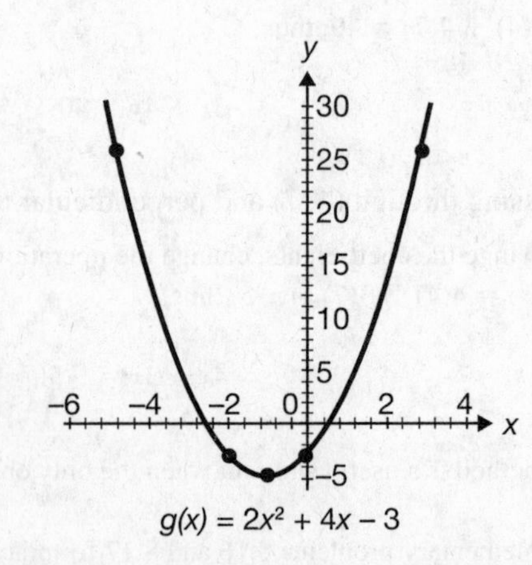

$g(x) = 2x^2 + 4x - 3$

Figure 8.14

8.27 Graph $f(x) = 2 - \sqrt{3 - x}$.

The domain is $\{x | x \leq 3\}$ (since $3 - x \geq 0$). To use the analysis on page 260, we put the function in the appropriate format: $f(x) = 2 + (-1)\sqrt{(-1)x + 3}$. Now the "starting point" is $(-3/(-1), 2) = (3,2)$. The graph is the bottom half of a parabola (since $b < 0$) opening left (since $c < 0$). The graph will continue down to the left from the starting point, passing through (2,1), (−1,0), and (−6, −1) as example points. See Figure 8.15.

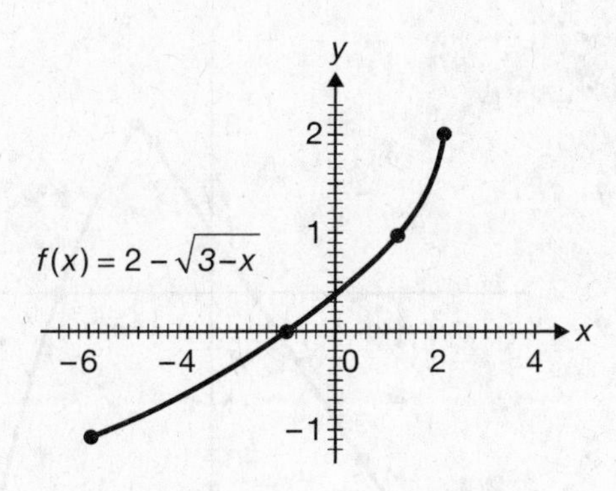

Figure 8.15

8.28 Graph $g(x) = -2|x + 3| - 1$.

The vertex is at (−3, −1) and the graph forms an inverted "V" from there. It passes through (−5, −5), (−4, −3), (−1, −5), and (0, −7). See Figure 8.16.

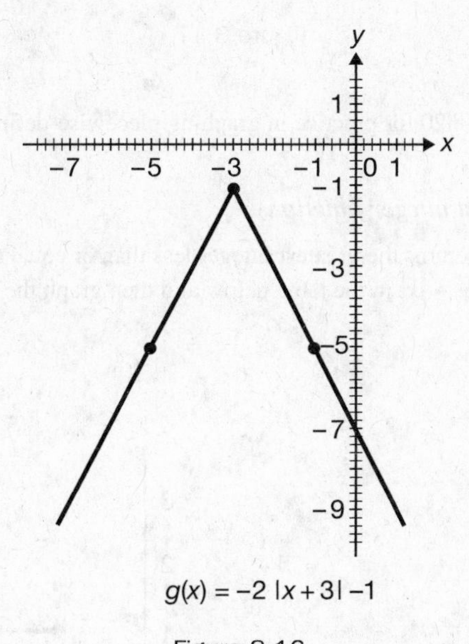

Figure 8.16

Refer to supplementary problem 8.19 for practice in graphing functions.

8.29 Graph $f(x) = \begin{cases} 2x + 1 & \text{if } x < 1 \\ 4 - x^2 & \text{if } 1 \leq x \leq 3 \\ 4 - x & \text{if } x > 3 \end{cases}$

This function is composed of three pieces. The first piece is the function $y = 2x + 1$ and it is defined only for those values of x in the interval $(-\infty, 1)$. The second piece is the function $y = 4 - x^2$ which is defined only for those values of x in the interval $[1, 3]$. The third piece is the function $y = 4 - x$ and is defined only for those values of x in the interval $(3, \infty)$. To graph $f(x)$ calculate the correct value of y for any value x by (1) identifying the interval in which x is found and (2) using the function piece for that interval. See the table below and the graph in Figure 8.17.

x	Interval	Function	y
−2	$(-\infty, 1)$	$2x+1$	−3
0.9	$(-\infty, 1)$	$2x+1$	2.8
1	$[1, 3]$	$4-x^2$	3
2	$[1, 3]$	$4-x^2$	0
3	$[1, 3]$	$4-x^2$	−5
3.1	$(3, \infty)$	$4-x$	0.9
5	$(3, \infty)$	$4-x$	−1

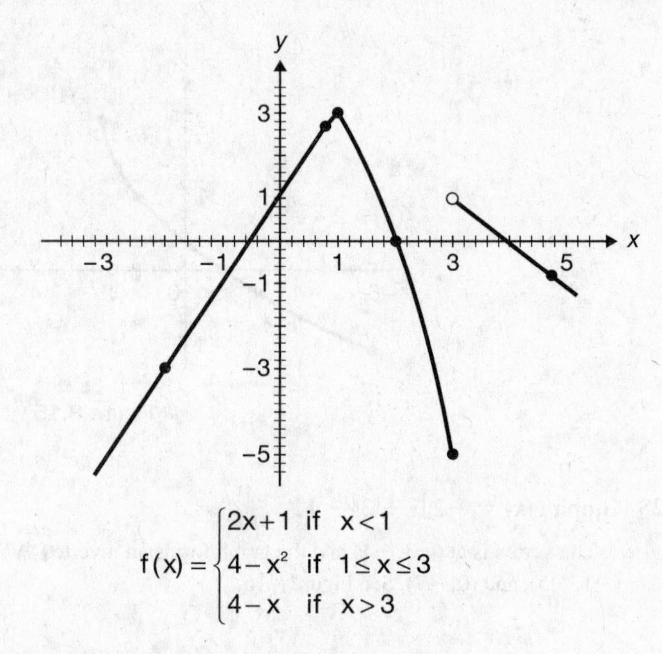

$$f(x) = \begin{cases} 2x+1 & \text{if } x<1 \\ 4-x^2 & \text{if } 1 \le x \le 3 \\ 4-x & \text{if } x>3 \end{cases}$$

Figure 8.17

Refer to supplementary problem 8.20 for practice in graphing piecewise-defined functions.

8.30 Graph $g(x) = [\![x]\!]$ (the *greatest integer function*).

The *greatest integer function* returns the greatest integer less than or equal to the value of the expression between the symbols $[\![$and$]\!]$. We calculate $y = [\![x]\!]$ in the table below and then graph the function. See Figure 8.18.

x	y
−1	−1
−0.5	−1
−0.1	−1
0	0
0.9	0
1	1
1.8	1
2	2

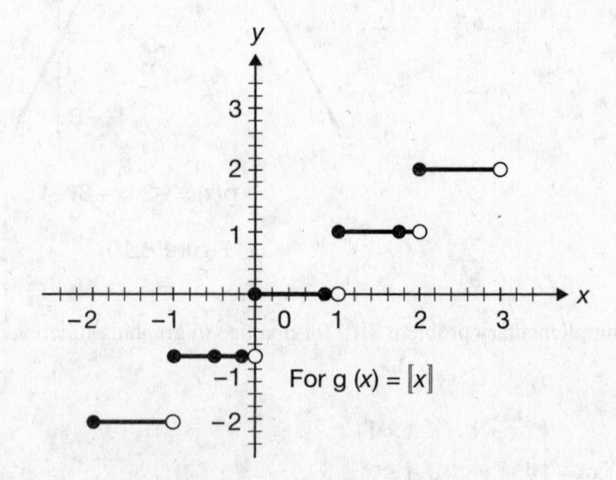

Figure 8.18

The graph will continue in this stair step fashion forever. We've tried to indicate this by including the "steps" for x in the intervals $[-2, -1)$ and $[2, 3)$. This type of function is frequently referred to as a *step function* because of its graph.

8.31 Translate each of the following statements into algebraic form:

(*a*) y varies directly as the square root of x.

$$y = k\sqrt{x}$$

(*b*) Area is proportional to the square of the radius.

$$A = kr^2$$

(*c*) Distance is proportional to time.

$$d = kt$$

(*d*) y varies directly as $x^{3/2}$.

$$y = kx^{3/2}$$

8.32 y varies directly as the cube of x. Find the constant of variation if $y = 6$ when $x = -2$.

$$y = kx^3 \text{ so } 6 = k(-2)^3 = -8k, \text{ thus, } k = 6/(-8) = -3/4.$$

8.33 p is proportional to q^2. If $p = 12$ when $q = 2$, find p when $q = 5$.

$$p = kq^2 \text{ so using the given values, } 12 = k(2)^2 = 4k \text{ or } k = 3. \text{ Thus, } p = 3q^2 = 3(5)^2 = 75.$$

8.34 Translate each of the following statements into algebraic form:

(*a*) y varies inversely as the square of x.

$$y = \frac{k}{x^2}$$

(*b*) Rate is inversely proportional to time.

$$r = \frac{k}{t}$$

8.35 Find y for $x = 9$ when y varies inversely as the square root of x, given that $y = 10$ when $x = 4$.

$$y = k/\sqrt{x} \text{ so using the given values, } 10 = k/\sqrt{4} \text{ or } k = 20. \text{ Thus, } y = 20/\sqrt{9} = 20/3.$$

8.36 Translate each of the following statements into algebraic form:

(*a*) y varies jointly as w, $\sqrt{x}$, and z^3

$$y = kwz^3\sqrt{x}$$

(*b*) Volume varies jointly as the height and the square of the radius.

$$V = khr^2$$

8.37 r varies jointly as s and the cube of t. If $r = 4$ when $s = 5$ and $t = -1$, find r when $s = -3$ and $t = 2$.

$r = kst^3$, so using the given values, $4 = k(5)(-1)^3 = -5k$ or $k = \frac{-4}{5}$.

Thus, $r = \frac{-4}{5}(-3)(2)^3 = \frac{96}{5}$.

Combinations of the above variations can also be identified as we show in the next problem.

8.38 Translate each of the following statements into algebraic form:

(*a*) y varies directly as x and inversely as z.

$$y = \frac{kx}{z}$$

(*b*) r varies jointly as s, t, and u^2 and inversely as $\sqrt{v}$.

$$r = \frac{kstu^2}{\sqrt{v}}$$

Refer to supplementary problems 8.21 through 8.26 for practice with variations.

8.39 Determine the inverse of the relation $\{(2, 3), (3, 9), (7, -1)\}$.

We interchange the order in the ordered pairs and obtain $\{(3, 2), (9, 3), (-1, 7)\}$.

The graphs of a relation or function and its inverse are mirror images of each other in the line $y = x$. In other words, each point (b, a) is the mirror image of each point (a, b) in the line $y = x$.

8.40 Prove that $f(x) = 2x^3 + 5$ and $g(x) = \sqrt[3]{\frac{x-5}{2}}$ are inverses of one another.

$$f[g(x)] = f\left[\sqrt[3]{\frac{x-5}{2}}\right] = 2\left(\sqrt[3]{\frac{x-5}{2}}\right)^3 + 5 = 2\left(\frac{x-5}{2}\right) + 5 = x - 5 + 5 = x \text{ and}$$

$$g[f(x)] = g[2x^3 + 5] = \sqrt[3]{\frac{(2x^3+5)-5}{2}} = \sqrt[3]{\frac{2x^3}{2}} = \sqrt[3]{x^3} = x.$$ Therefore, the functions are inverses of one another; that is, $f^{-1}(x) = g(x)$ and $g^{-1}(x) = f(x)$.

Refer to supplementary problem 8.27 for practice problems.

8.41 Determine the inverse of $f(x) = -2x + 5$.

This is the function, $f = \{(x, y) \mid y = -2x + 5\}$. We can interchange the x- and y-coordinates to obtain the inverse by either switching the order in the ordered pair directly, as (y, x), or by leaving the ordered pair alone and switching its "order" indirectly by interchanging x and y in the equation: $x = -2y + 5$. The latter method is normally used. So, for this function, $f^{-1} = \{(x, y) \mid x = -2y + 5\}$. To help see that this is the inverse, the student should now show that $(1, 3)$ is in f and $(3, 1)$ is in f^{-1}. This form of f^{-1} doesn't easily lend itself to the more common function notation until the embedded equation is solved for y:

$$x = -2y + 5$$

$$2y = -x + 5$$

$$y = \frac{-1}{2}x + \frac{5}{2}.$$

Thus, $f^{-1} = \left\{(x, y) \mid y = \frac{-1}{2}x + \frac{5}{2}\right\}$ and in the more common function notation

$$f^{-1}(x) = \frac{-1}{2}x + \frac{5}{2}.$$

We leave it to the student to prove these functions to be inverses of one another.

8.42 Find the inverse of $g(x) = x^2 + 3$.

Using the three steps listed on page 263,

1. Set $y = g(x) = x^2 + 3$;
2. Interchange x and y: $x = y^2 + 3$; and
3. Solve for y: $y^2 = x - 3$ or $y = \pm\sqrt{x - 3}$.

$y = \pm\sqrt{x - 3}$ identifies a relation but not a function. The inverse of g is not a function, it is just the relation $g^{-1} = \{(x, y) \mid y = \pm\sqrt{x - 3}\}$.

8.43 Determine which of the following functions have inverses that are functions:

(*a*) $f(x) = x^4$

Since $f(-1) = (-1)^4 = 1$ and $f(1) = 1^4 = 1$, $(-1,1)$ and $(1,1)$ are in f. Therefore, f is not *one-to-one* and so its inverse is *not* a function.

(*b*) $g(x) = \sqrt{2x + 1}$

The graph passes the horizontal line test; see Figure 8.19. The inverse is a function.

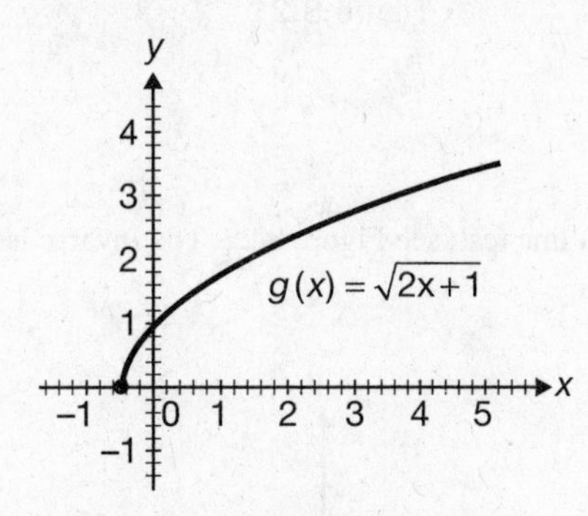

Figure 8.19

(*c*) $y = x^3$

The graph passes the horizontal line test; see Figure 8.20. The inverse is a function.

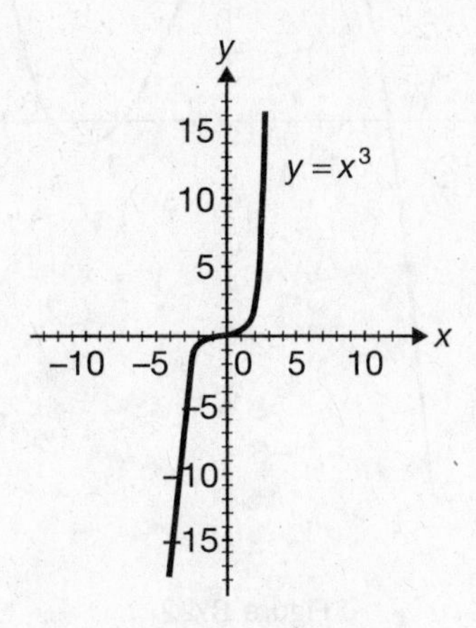

Figure 8.20

(*d*) $y = 4 - x$.

The graph passes the horizontal line test; see Figure 8.21. The inverse is a function.

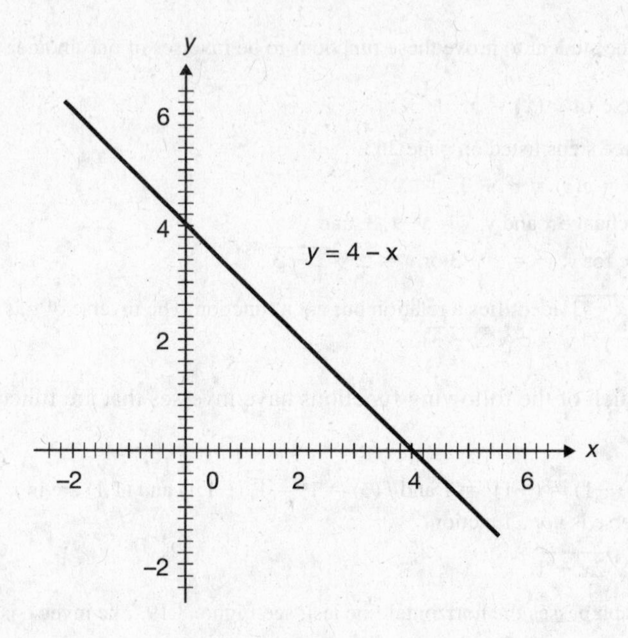

Figure 8.21

(*e*) $y = x^3 - x^2 - 2x + 2$

The graph *fails* the horizontal line test; see Figure 8.22. The inverse is *not* a function.

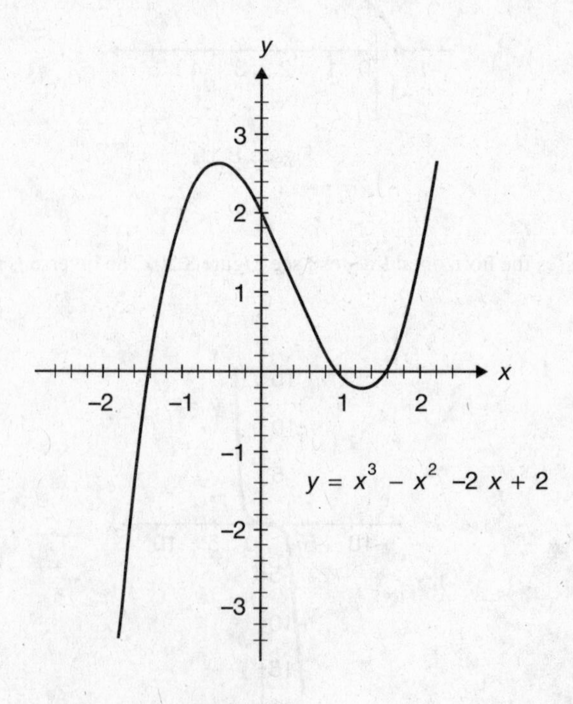

Figure 8.22

(*f*) $f(x) = 3 - \sqrt[2]{x+4}$

The graph passes the horizontal line test; see Figure 8.23. The inverse is a function.

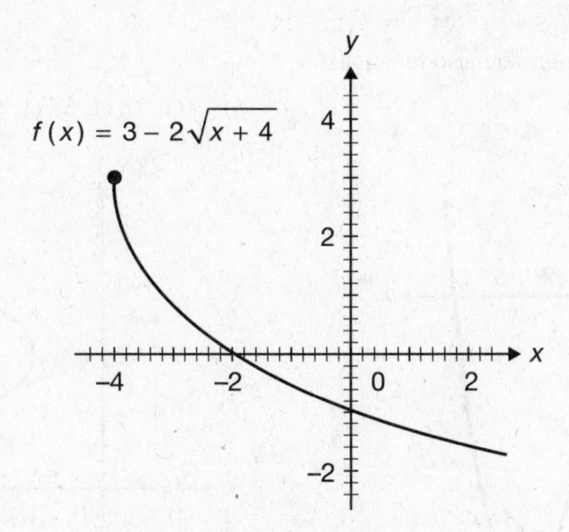

Figure 8.23

Note: The functions in parts (*b*), (*c*), (*d*), and (*f*) above are one-to-one functions. Only one-to-one functions have inverses that are functions.

Refer to supplementary problem 8.28 for practice in determining if inverses are functions or not.

8.44 Determine the inverse of $f(x) = x^2 + 3, x \leq 0$.

1. Set $y = x^2 + 3, x \leq 0$.
2. Replace each occurrence of x with y and vice versa: $x = y^2 + 3, y \leq 0$ (note that $x \leq 0$ changed to $y \leq 0$).
3. Solve the equation for y: $y = \pm\sqrt{x-3}$ and since the new inequality specifies $y \leq 0$, then y must be $-\sqrt{x-3}$. Thus

$$y = f^{-1}(x) = -\sqrt{x-3}.$$

f^{-1} is a function. See Figure 8.24 showing that f^{-1} is the mirror image of f in the line $y = x$.

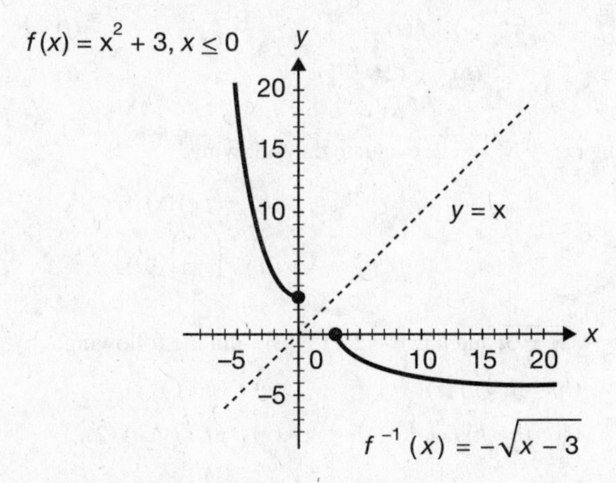

Figure 8.24

Refer to supplementary problem 8.29 for practice in determining inverses of functions.

SUPPLEMENTARY PROBLEMS

8.1 Find the Cartesian Product for the following: (*a*) $\{x, y, z\} \times \{r, s, t\}$ (*b*) $\{1\} \times \{5, 6\}$

8.2 Which of the following relations are also functions?

(*a*) $\{(4, 5), (-2, 5), (7, 5)\}$

(*b*) $\{(1, 3), (1, 5), (1, 6)\}$

(*c*)

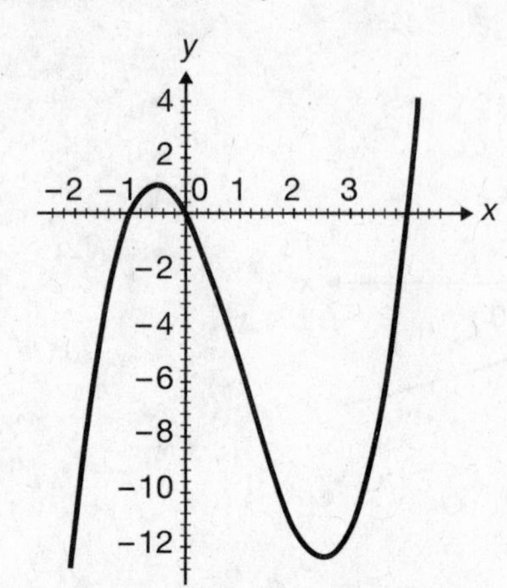

Figure 8.25

(*d*)

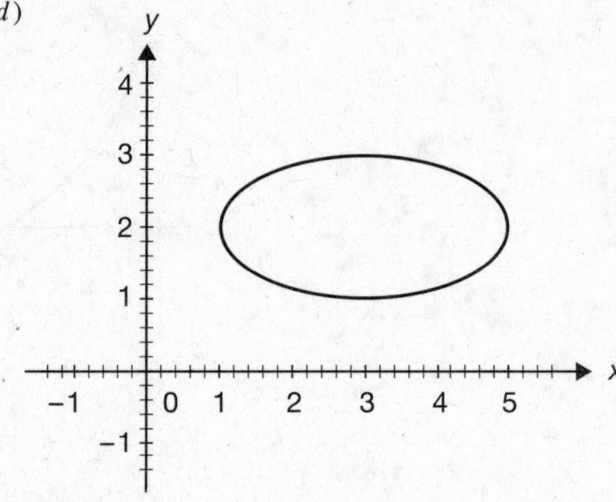

Figure 8.26

8.3 Determine the domain of each of the following functions:

(*a*) $\{(x, y) \mid y = x^2 + 2x - 3\}$

(*b*) $f(x) = \dfrac{5}{x - 3}$

(*c*) $\left\{(x, y) \mid y = \dfrac{1}{x^2 - 3x - 4}\right\}$

(*d*) $\left\{(x, y) \mid y = \sqrt{5x - 15}\right\}$

(*e*) $g(x) = \sqrt{9 - x^2}$

(*f*) $g(x) = \dfrac{\sqrt{x + 2}}{x}$

8.4 Given $f(x) = 2x^2 - x - 3$ and $g(x) = x/(2x - 3)$, evaluate the following:

(*a*) $f(-4)$ (*b*) $g(2)$ (*c*) $g(5)$

(*d*) $f(8)$ (*e*) $f(z)$ (*f*) $g(3u)$

(*g*) $f(a - b)$ (*h*) $g(2x + 1)$ (*i*) $g(x^2 - 2)$

8.5 Given $f(x) = 3x^2 - 5$ and $g(x) = \sqrt{x - 2}$, evaluate the following:

(*a*) $(f + g)(2)$ (*b*) $(fg)(x)$

(*c*) $(g - f)(5)$ (*d*) $\left(\dfrac{f}{g}\right)(6)$

8.6 For $f(x) = 3x + 1$, $g(x) = \sqrt{x + 3}$, and $h(x) = 2/(x - 3)$, find the following:

(*a*) $(f \circ g)(x)$ (*b*) $(h \circ f)(x)$ (*c*) $(g \circ f)(x)$ (*d*) $(g \circ h)(x)$

(*e*) $(g \circ g)(1)$ (*f*) $(h \circ h)(x)$ (*g*) $(f \circ g \circ h)(2)$ (*h*) $(h \circ f \circ h)(x)$

8.7 Find the distance between each of the following pairs of points:

(*a*) $(3, 5)$ and $(2, 8)$ (*b*) $(2, 6)$ and $(2, -1)$ (*c*) $(2, -2)$ and $(-2, 2)$

(*d*) $(0, 0)$ and $(5, 12)$ (*e*) $(3, 0)$ and $(8, 0)$ (*f*) $(-1, -1)$ and $(1, 5)$

8.8 Determine the slope of the line passing through each of the following pairs of points:

(*a*) $(3, 5)$ and $(2, 8)$ (*b*) $(2, 6)$ and $(2, -1)$ (*c*) $(2, -2)$ and $(-2, 2)$

(*d*) $(0, 0)$ and $(5, 12)$ (*e*) $(3, 0)$ and $(8, 0)$ (*f*) $(-1, -1)$ and $(1, 5)$

8.9 Find the slope of the line given by

(*a*) $2x + y = 5$ (*b*) $3x - 4y = 5$ (*c*) $5x + 3y = 8$

8.10 Determine the point-slope form of the equation of

(*a*) The line passing through $(7, 3)$ with slope $m = 2$.

(*b*) The line passing through $(2, -1)$ with $m = -3$.

(*c*) The line passing through the points $(-3, 2)$ and $(2, -3)$.

8.11 Determine the slope and y-intercept of the line given by the following equations:

(*a*) $3x + 2y = 5$ (*b*) $4x - 3y = -6$

8.12 Determine the slope-intercept form of the equation of

(*a*) The line passing through $(2, 8)$ with slope $m = 3$.

(*b*) The line passing through $(1, -3)$ with slope -2.

8.13 Find the slope-intercept form of the equation of

(*a*) The line passing through the points $(1, 3)$ and $(4, 9)$.

(*b*) The line passing through the points $(2, 0)$ and $(4, -4)$.

(*c*) The line passing through the points $(3, -5)$ and $(-2, 2)$.

8.14 Find the slope of the line that is parallel to the line given by

(*a*) $4x - 2y + 6 = 0$ (*b*) $2x + 5y - 7 = 0$

8.15 Find the slope of the line that is perpendicular to the line given by

(*a*) $3x - y - 5 = 0$ (*b*) $2x + 5y - 7 = 0$

8.16 Determine the requested form of the equation of the line indicated by each of the following:

(*a*) The standard form of the equation of the line passing through the point $(7, 3)$ and parallel to the line given by $4x - 5y - 20 = 0$.

(*b*) The slope-intercept form of the equation of the line passing through the point $(-2, 1)$ and parallel to the line given by $y = 2x - 4$.

(*c*) The general form of the equation of the line passing through the point $(5, -3)$ and parallel to the line given by $3x + 2y - 5 = 0$.

(*d*) The slope-intercept form of the equation of the line passing through the point $(5, 0)$ and parallel to the line given by $2x - 4y = 7$.

(*e*) The standard form of the equation of the line passing through the point $(-4, -3)$ and parallel to the line given by $x = 5$.

(*f*) The general form of the equation of the line passing through the point $(2, 7)$ and parallel to the line given by $y = -3$.

8.17 Determine the requested form of the equation of the line indicated by each of the following:

(*a*) The slope-intercept form of the equation of the line passing through the point $(0, -2)$ and perpendicular to the line given by $2x + 5y - 8 = 0$.

(*b*) The standard form of the equation of the line passing through the point $(-5, -4)$ and perpendicular to the line given by $3x - 5y - 5 = 0$.

(*c*) The general form of the equation of the line passing through the point $(4, 2)$ and perpendicular to the line given by $x + 2y = 5$.

(*d*) The slope-intercept form of the equation of the line passing through the point $(1, -3)$ and perpendicular to the line given by $2x - 4y = 7$.

(*e*) The general form of the equation of the line passing through the point $(-3, 5)$ and perpendicular to the line given by $y - 5 = 0$.

(*f*) The standard form of the equation of the line passing through the point $(7, 1)$ and perpendicular to the line given by $x + 6 = 0$.

8.18 Find the linear function g for which

(*a*) $g(2) = -3$ and $g(5) = 2$

(*b*) $g(-5) = 1$ and $g(4) = 0$

8.19 Graph each of the following functions:

(*a*) $f(x) = 2x + 4$

(*b*) $g(x) = \dfrac{2 - 3x}{2}$

(*c*) $h(x) = x^2 - 4x + 3$

(*d*) $p(x) = -3x^2 + 6x$

(*e*) $y = \sqrt{x + 2}$

(*f*) $q(x) = -1 + \sqrt{1 - x}$

(*g*) $r(x) = -2 + |x - 1|$

(*h*) $s(x) = 3 - |x + 3|$

8.20 Graph:

(*a*) $f(x) = \begin{cases} 4 - x^2 & \text{if } x \le 1 \\ 2x + 1 & \text{if } x > 1 \end{cases}$

(*b*) $g(x) = \begin{cases} x + 3 & \text{if } x < -1 \\ 2 & \text{if } -1 \le x < 3 \\ 9 - 2x & \text{if } x \ge 3 \end{cases}$

8.21 Translate each of the following statements into algebraic form:

(*a*) y varies directly as the cube of x.

(*b*) z is proportional to $\sqrt{t}$.

(*c*) Volume V is proportional to depth d.

(*d*) x varies inversely as q^2.

(*e*) p varies jointly as r and $\sqrt[3]{s}$.

(*f*) a is inversely proportional to bc.

(*g*) y varies jointly as w and v^3.

(*h*) Pressure varies as temperature and inversely as volume.

8.22 Find the constant of variation for

(*a*) $y = -2$ and $x = 3$ when y is proportional to x^2.

(*b*) $z = 1$ and $v = 5$ when z varies inversely as v.

(*c*) $t = 2$, $s = 3$, and $r = -1$ when t varies directly as s and inversely as r^3.

8.23 Find z for $p = -3$ when z varies directly as p^2 given that $z = 4$ when $p = -4$.

8.24 Find r for $s = 4$ when r is inversely proportional to $\sqrt{s}$ given that $r = 5$ when $s = 3$.

8.25 Find p when $q = 3$ and $r = 5$ when p varies jointly as q^2 and r given that $p = 2$ when $q = 1$ and $r = 3$.

8.26 Find a when $b = -2$ and $c = -3$ when a varies as b^2 and inversely as c given that $a = 5$ when $b = 3$ and $c = 4$.

8.27 Show that the following pairs of functions are inverses of one another:

(*a*) $f(x) = 5x - 7$ and $g(x) = \dfrac{x+7}{5}$ (*b*) $p(x) = \sqrt[3]{3x-2}$ and $q(x) = \dfrac{1}{3}x^3 + \dfrac{2}{3}$

8.28 Determine which of the following functions have inverses that are functions and justify your answers:

(*a*) $f(x) = 3x^3 - 1$ (*b*) $g(x) = 4 - x^2$ (*c*) $h(x) = -3\sqrt{2-x}$

(*d*)

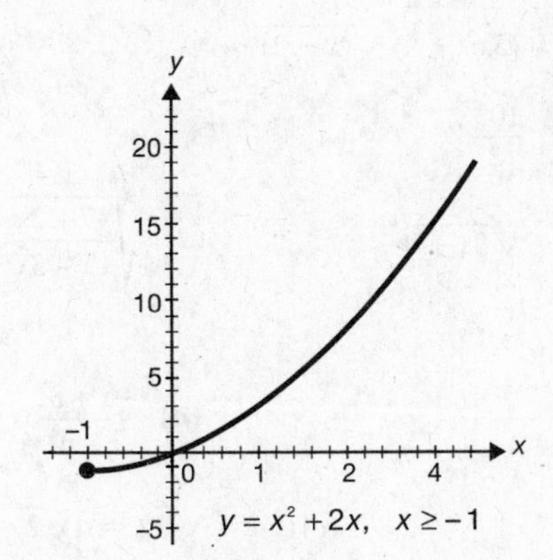

Figure 8.27

(*e*)

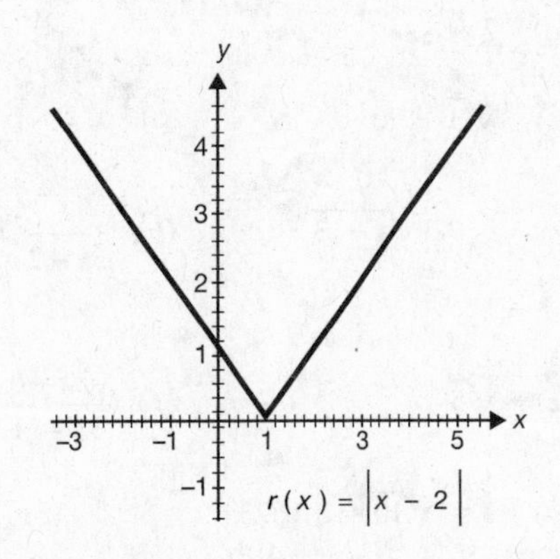

Figure 8.28

8.29 Determine the inverse of each of the following:

(*a*) $\{(5, 3), (-2, 4), (1, 6)\}$ (*b*) $f(x) = \dfrac{2x+5}{4}$

(*c*) $f(x) = 3x - 4$ (*d*) $u(x) = 9 - 2x$

(*e*) $p(x) = 2x^2 - 1, x < 0$ (*f*) $r(x) = \sqrt{3x+5}$

(*g*) $g(x) = (5x)^3$ (*h*) $h(x) = \dfrac{1}{x}$

(*i*) $q(x) = \dfrac{x}{1+x}$ (*j*) $s(x) = \sqrt{x^2 - 4}, x \geq 2$

ANSWERS TO SUPPLEMENTARY PROBLEMS

8.1 (*a*) $\{(x, r), (x, s), (x, t), (y, r), (y, s), (y, t), (z, r), (z, s), (z, t)\}$

(*b*) $\{(1, 5), (1, 6)\}$

8.2 (*a*) A function (*b*) Not a function

(*c*) A function (*d*) Not a function

8.3 (*a*) Domain: All Real Numbers $= (-\infty, \infty)$

(b) Domain: $\{x | x \neq 3\}$ $= (-\infty, 3) \cup (3, \infty)$

(*c*) Domain: $\{x \mid x \neq 4 \text{ and } x \neq -1\}$ $= (-\infty, -1) \cup (-1, 4) \cup (4, \infty)$

(*d*) Domain: $\{x \mid x \geq 3\}$ $= [3, \infty)$

(*e*) Domain: $\{x \mid -3 \leq x \leq 3\}$ $= [-3, 3]$

(*f*) Domain: $\{x \mid x \geq -2 \text{ and } x \neq 0\}$ $= [-2, 0) \cup (0, \infty)$

8.4 (*a*) 33 (*b*) 2 (*c*) $\frac{5}{7}$

(*d*) 117 (*e*) $2z^2 - z - 3$ (*f*) $\frac{u}{2u-1}$

(*g*) $2(a-b)^2 - (a-b) - 3 = 2a^2 - 4ab + 2b^2 - a + b - 3$ (*h*) $\frac{2x+1}{4x-1}$ (*i*) $\frac{x^2-2}{2x^2-7}$

8.5 (*a*) 7 (*b*) $(3x^2 - 5)\sqrt{(x-2)}$

(*c*) $\sqrt{3} - 70$ (*d*) $\frac{103}{2}$

8.6 (*a*) $1 + 3\sqrt{x+3}$ (*b*) $\frac{2}{3x-2}$ (*c*) $\sqrt{3x+4}$ (*d*) $\sqrt{\frac{-7+3x}{x-3}}$

(*e*) $\sqrt{5}$ (*f*) $\frac{-2x+6}{3x-11}$ (*g*) 4 (*h*) $\frac{-x+3}{x-6}$

8.7 (*a*) $d = \sqrt{10} \approx 3.1623$ (*b*) $d = 7$ (*c*) $d = \sqrt{32} = 4\sqrt{2} \approx 5.6569$

(*d*) $d = 13$ (*e*) $d = 5$ (*f*) $d = \sqrt{40} = 2\sqrt{10} \approx 6.3246$

8.8 (*a*) $m = -3$ (*b*) Undefined (*c*) $m = -1$

(*d*) $m = \frac{12}{5}$ (*e*) $m = 0$ (*f*) $m = 3$

8.9 (*a*) $m = -2$ (*b*) $m = \frac{3}{4}$ (*c*) $m = \frac{-5}{3}$

8.10 (*a*) $y - 3 = 2(x - 7)$ (*b*) $y - (-1) = -3(x - 2)$

(*c*) $y - 2 = -1(x - (-3))$ or $y - (-3) = -1(x - 2)$

8.11 (*a*) $m = \frac{-3}{2}$; y-intercept: $\frac{5}{2}$ (*b*) $m = \frac{4}{3}$; y-intercept: 2

8.12 (*a*) $y = 3x + 2$ (*b*) $y = -2x - 1$

8.13 (*a*) $y = 2x + 1$ (*b*) $y = -2x + 4$ (*c*) $y = \frac{-7}{5}x - \frac{4}{5}$

8.14 (*a*) $m = 2$ (*b*) $m = \frac{-2}{5}$

8.15 (*a*) $m = \frac{-1}{3}$ (*b*) $m = \frac{5}{2}$

8.16 (*a*) $4x - 5y = 3$ (*b*) $y = 2x + 5$

(*c*) $3x + 2y - 9 = 0$ (*d*) $y = \frac{1}{2}x - \frac{5}{2}$

(*e*) $x = -4$ (*f*) $y - 7 = 0$

8.17 (*a*) $y = \frac{5}{2}x - 2$ (*b*) $5x + 3y = -37$

(*c*) $2x - y - 6 = 0$ (*d*) $y = -2x - 1$

(*e*) $x + 3 = 0$ (*f*) $y = 1$

8.18 (*a*) $g(x) = \frac{5}{3}x - \frac{19}{3}$ (*b*) $g(x) = \frac{-1}{9}x + \frac{4}{9}$

8.19 (*a*) See Figure 8.29. (*b*) See Figure 8.30. (*c*) See Figure 8.31. (*d*) See Figure8.32.
(*e*) See Figure 8.33. (*f*) See Figure 8.34. (*g*) See Figure 8.35. (*h*) See Figure 8.36.

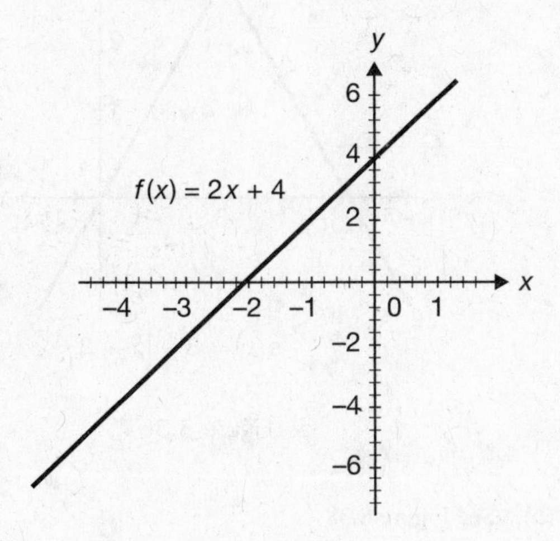

Figure 8.29

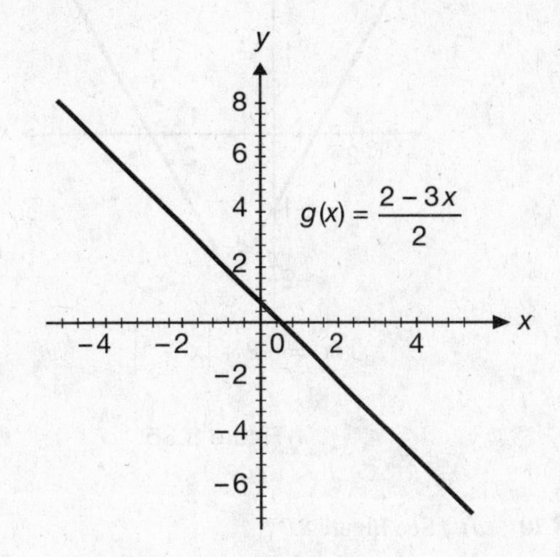

Figure 8.30

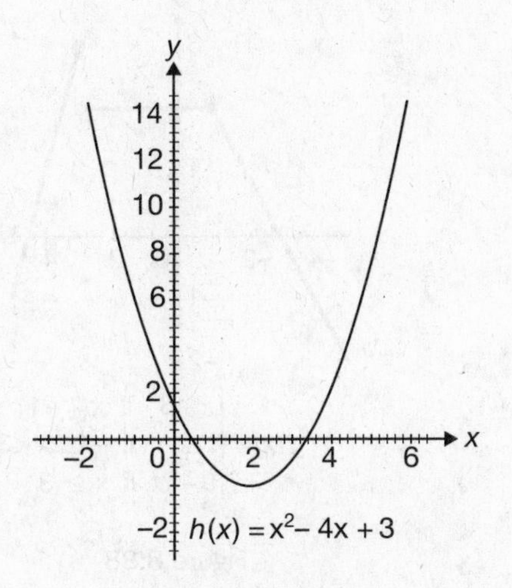

Figure 8.31

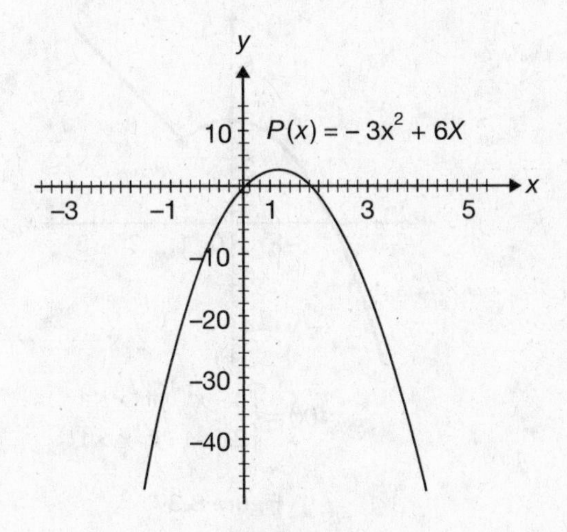

Figure 8.32

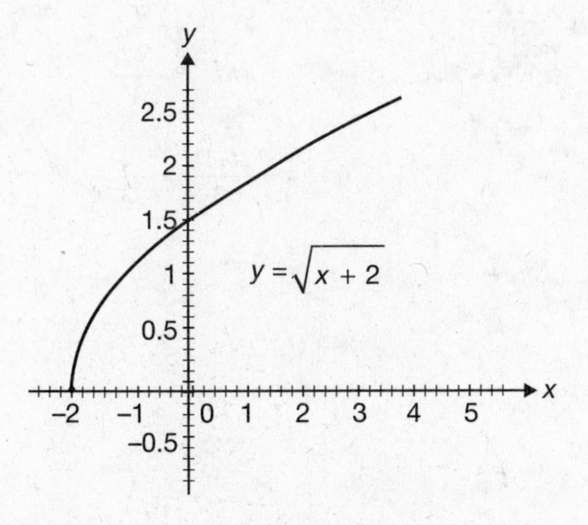

Figure 8.33

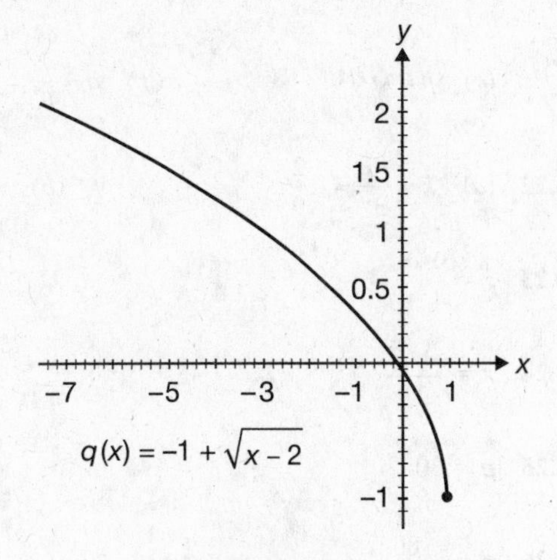

Figure 8.34

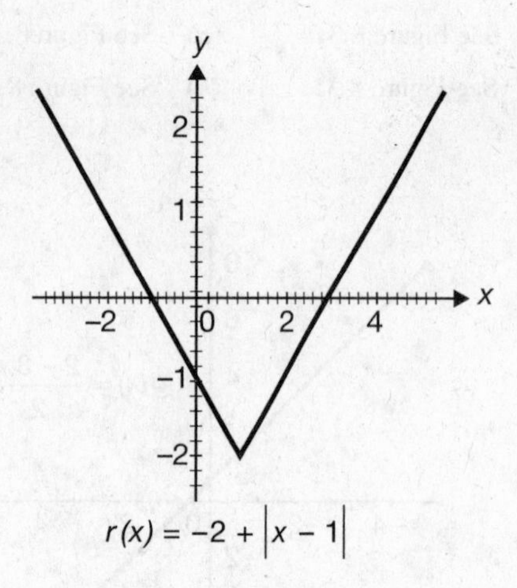

Figure 8.35

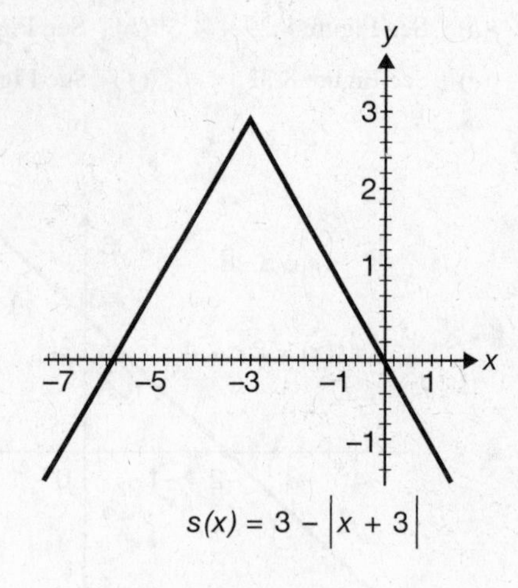

Figure 8.36

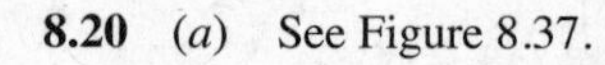

8.20 (*a*) See Figure 8.37. (*b*) See Figure 8.38.

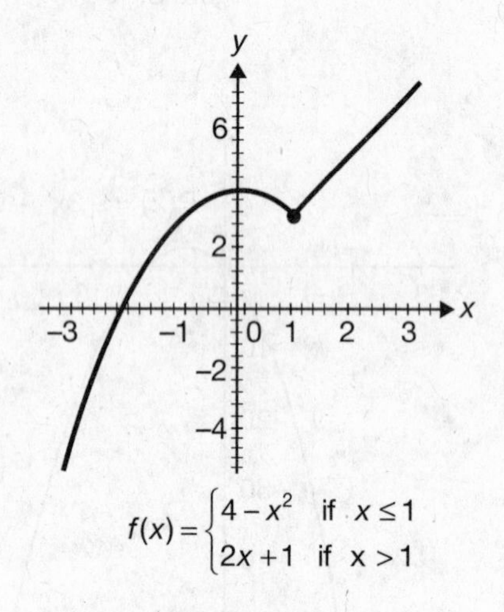

$$f(x) = \begin{cases} 4 - x^2 & \text{if } x \le 1 \\ 2x + 1 & \text{if } x > 1 \end{cases}$$

Figure 8.37

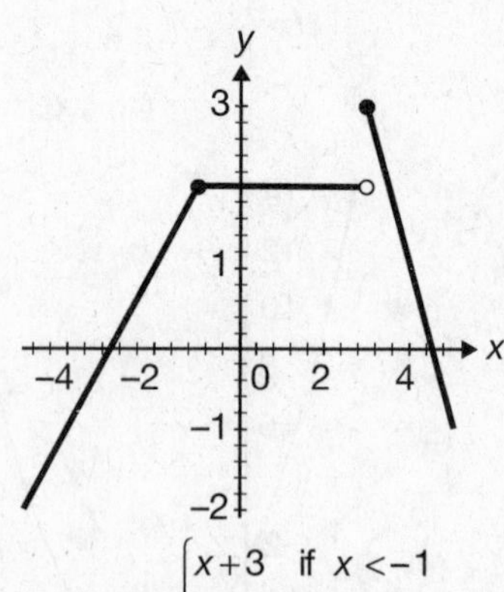

$$g(x) = \begin{cases} x + 3 & \text{if } x < -1 \\ 2 & \text{if } -1 \le x < 3 \\ 9 - 2x & \text{if } x \ge 3 \end{cases}$$

Figure 8.38

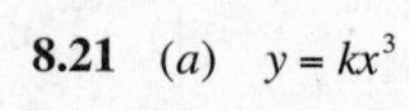

8.21 (*a*) $y = kx^3$ (*b*) $z = k\sqrt{t}$ (*c*) $V = kd$ (*d*) $x = \dfrac{k}{q^2}$

(*e*) $p = kr\sqrt[3]{s}$ (*f*) $a = \dfrac{k}{bc}$ (*g*) $y = kwv^3$ (*h*) $P = \dfrac{kT}{V}$

8.22 (*a*) $k = \dfrac{-2}{3^2} = -\dfrac{2}{9}$ (*b*) $k = (1)\,5 = 5$ (*c*) $k = \dfrac{2(-1)^3}{3} = -\dfrac{2}{3}$

8.23 $z = \dfrac{9}{4}$

8.24 $r = \dfrac{5\sqrt{3}}{2}$

8.25 $p = 30$

8.26 $a = \dfrac{-80}{27}$

8.27 (*a*) $f(g(x)) = f\left(\frac{x+7}{5}\right) = 5\left(\frac{x+7}{5}\right) - 7 = x + 7 - 7 = x$, and

$$g(f(x)) = g(5x-7) = \frac{(5x-7)+7}{5} = \frac{5x}{5} = x$$

so $g(x) = f^{-1}(x)$ and $f(x) = g^{-1}(x)$.

(*b*) $p(q(x)) = p\left(\frac{1}{3}x^3 + \frac{2}{3}\right) = \sqrt[3]{3\left(\frac{1}{3}x^3 + \frac{2}{3}\right) - 2} = \sqrt[3]{x^3 + 2 - 2} = \sqrt[3]{x^3} = x$, and

$$q(p(x)) = q\left(\sqrt[3]{3x-2}\right) = \frac{1}{3}\left(\sqrt[3]{3x-2}\right)^3 + \frac{2}{3} = \frac{1}{3}(3x-2) + \frac{2}{3} = x$$

so $q(x) = p^{-1}(x)$ and $p(x) = q^{-1}(x)$.

8.28 The inverse is

(*a*) A function (by the horizontal line test, see Figure 8.39)

(*b*) Not a function. $(1, 3)$ and $(-1,3) \in g$, i.e., $g(1) = g(-1) = 3$

(*c*) A function (by the horizontal line test see Figure 8.40).

(*d*) A function (by the horizontal line test, as Figure 8.27 shows)

(*e*) Not a function (by the horizontal line test as Figure 8.28 shows).

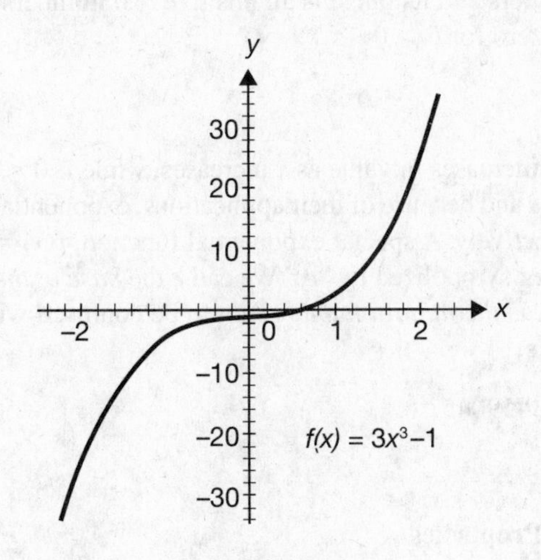

Figure 8.39

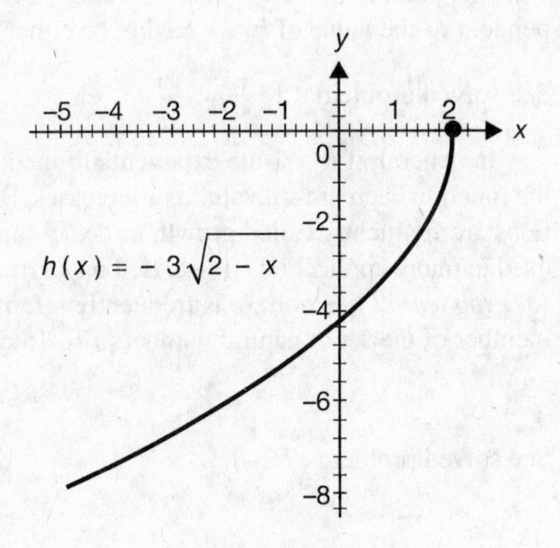

Figure 8.40

8.29 (*a*) $\{(3,5), (4,-2), (6,1)\}$

(*b*) $f^{-1}(x) = 2x - \frac{5}{2}$

(*c*) $f^{-1}(x) = \frac{1}{3}x + \frac{4}{3}$

(*d*) $u^{-1}(x) = \frac{9}{2} - \frac{1}{2}x$

(*e*) $p^{-1}(x) = \frac{-1}{2}\sqrt{(2x+2)}$

(*f*) $r^{-1}(x) = \frac{x^2 - 5}{3}$

(*g*) $g^{-1}(x) = \frac{1}{5}\sqrt[3]{x}$

(*h*) $h^{-1}(x) = \frac{1}{x} = h(x)$

(*i*) $q^{-1}(x) = \frac{x}{1-x}$

(*j*) $s^{-1}(x) = \sqrt{x^2 + 4}$

Exponential and Logarithmic Functions

9.1 Exponential Functions

Definition 1. An *exponential function* is of the form $f(x) = b^x$ where $b > 0, b \neq 1$, and x is any real number.

The domain of the exponential function is all real numbers and its range is all positive real numbers (independent of the value of b); b^x cannot be either negative or zero for $b > 0$.

See solved problem 9.1.

In general, if $b > 1$ the exponential function $f(x) = b^x$ increases in value as x increases, while if $0 < b < 1$ the function decreases in value as x increases. Because of this and because of their applications, exponential functions are frequently called growth or decay functions, respectively. A special exponential function, $f(x) = e^x$, is used in many applications. It is based on an irrational number symbolized by "e." We call e the *base of the natural exponential function*. (e is frequently referred to simply as *the natural number*, not to be confused with any member of the set of natural numbers.) To 15 decimal places,

$$e \approx 2.718281828459045.$$

See solved problems 9.2–9.3.

Exponential Properties

For $a, b > 0$ and $a, b \neq 1$

1. $a^n = b^n$ if and only if $a = b$
2. $a^n = a^m$ if and only if $n = m$

See solved problem 9.4.

9.2 Logarithmic Functions

Since the exponential function is a one-to-one function, we know its inverse is also a function. The inverse of the exponential function is the logarithmic function, or log for short. We attempt to find the inverse of $f(x) = b^x$ by the conventional method: (1) $f(x) = y = b^x$; (2) $x = b^y$; (3) then solve for y. To solve for y, however, requires a new function, namely $\log_b x$. This function is read: "the logarithm to the base b of x." The inverse then is $f^{-1}(x) = y = \log_b x$.

Definition 2. If b and x are positive real numbers with $b \neq 1$, the function

$$f(x) = \log_b x$$

is called the *logarithmic function* to the base b.

The domain of the logarithmic function is all positive real numbers (the range of the exponential function) and its range is all real numbers (the domain of the exponential function).

Because the exponential and logarithmic functions are inverses of one another, the following is identified as the fundamental relationship between them.

Fundamental Relationship

$$b^a = c \text{ is equivalent to } a = \log_b c$$

Note that b is the base in each equation; the base of the power in the exponential equation and the base of the logarithm in the logarithmic equation. The value of a logarithm is, in essence, an exponent. In the equivalence stated above, it is the exponent to which the base b must be raised to obtain the number c. This relationship is the key relationship employed in many problems that involve the exponential and logarithmic functions.

See solved problems 9.5–9.9.

The following identities frequently prove useful:

Identities

$\log_b b^x$	$=$	x	Identity 1
$b^{\log_b x}$	$=$	x	Identity 2
$\log_b 1$	$=$	0	Identity 3

We will utilize these in some application problems (see section 9.5).

Common Logs and Natural Logs

The following abbreviations allow us to represent the indicated logarithms without specifying a base:

Common logarithms	$\log_{10} x = \log x$ (no base on log means base 10)
Natural logarithms	$\log_e x = \ln x$

Calculators with scientific calculation capabilities have a log key for $\log_{10}$ calculations and a ln key for $\log_e$ calculations. The solved problem 9.10 illustrates keystrokes for evaluating logs by some general calculator types.

See solved problems 9.10–9.11.

9.3 Properties of Logarithms

The following five basic properties are used for manipulations involving logarithms.

Basic Logarithmic Properties

1. $\log_b (ac) = \log_b a + \log_b c$
2. $\log_b \left(\dfrac{a}{c}\right) = \log_b a - \log_b c$

3. $\log_b (a^c) = c \log_b a$

4. $\log_b a = \dfrac{\log_c a}{\log_c b}$

5. $\log_b a = \log_b c$ if and only if $a = c$

In all five properties above, $a > 0, b > 0, b \neq 1$ and $c > 0$. In property 4, $c \neq 1$.

Property 4 above is frequently referred to as the "change of base formula." It is used to change the base of a logarithm from a base that cannot be evaluated directly using a calculator into a base that can (using a quotient of logarithms), as in the solved problem 9.12.

See solved problems 9.12–9.16.

Historically, before the advent of calculators and computers, logarithms were used to evaluate expressions involving products, quotients, powers, and roots. Logarithms employ simpler arithmetic procedures in each case: products become sums, quotients become differences, and powers and roots become products.

See solved problem 9.17

9.4 Exponential and Logarithmic Equations

Equations involving exponential functions are called *exponential equations*. Equations involving logarithmic functions are called *logarithmic equations*. Both frequently require the use of the logarithmic properties. When introducing logarithms, generally we want to use either natural logarithms (ln) or common logarithms (log) since they can be entered directly into a calculator. We use natural logarithms in each of the solved problems 9.18 and 9.19, but common logarithms could be used throughout instead. The properties listed are from those given in Section 9.3.

See solved problems 9.18–9.20.

9.5 Applications

Interest

Compound interest occurs when an initial amount of money earns interest at a constant rate at the end of each period for multiple periods. If that interest rate per period is i and the initial amount of money or principal is P, then the accumulated amount at the end of each period A is

$$
\begin{array}{ll}
A = P + Pi = P(1+i) & \text{Period 1} \\
A = P(1+i) + P(1+i)\,i = P(1+i)(1+i) = P(1+i)^2 & \text{Period 2} \\
A = P(1+i)^2 + P(1+i)^2\,i = P(1+i)^2(1+i) = P(1+i)^3 & \text{Period 3} \\
\vdots & \vdots \\
A = P(1+i)^n & \text{Period } n
\end{array}
$$

See solved problems 9.21–9.22.

For a principal to be compounded continuously (instead of periodically), the formula is $A = Pe^{rt}$ where A is the accumulated amount, P is the principal, r is the annual interest rate, and t is the number of years of compounding.

See solved problems 9.23–9.24.

Exponential Growth and Decay

Exponential growth and decay functions follow the equation $A = A_0e^{kt}$ where A is the final amount, A_0 is the initial amount, t is the time span involved, and k is related to the rate of growth or decay. k is positive for growth functions and k is negative for decay functions.

See solved problems 9.25–9.31 for growth and decay applications.
See solved problems 9.32–9.35 for miscellaneous applications.

SOLVED PROBLEMS

9.1 Graph each of the following exponential functions:

(*a*) $f(x) = 2^x$

x	2^x
-3	$\frac{1}{8}$
-2	$\frac{1}{4}$
-1	$\frac{1}{2}$
0	1
1	2
2	4
3	8

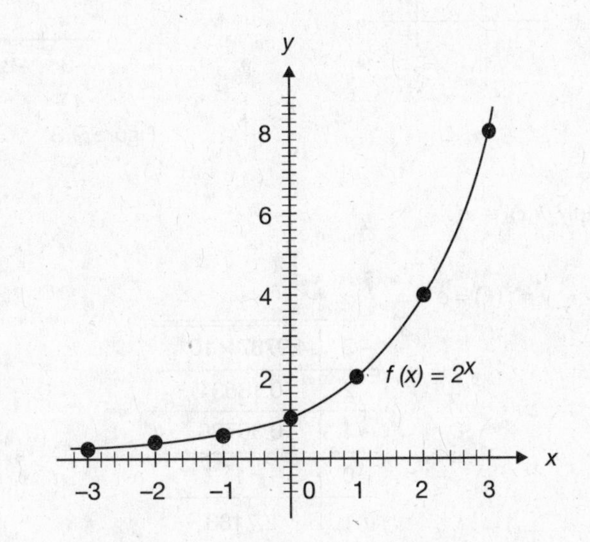

Figure 9.1

(*b*) $g(x) = 10^x$

x	10^x
-2	$\frac{1}{100}$
-1	$\frac{1}{10}$
0	1
1	10
2	100

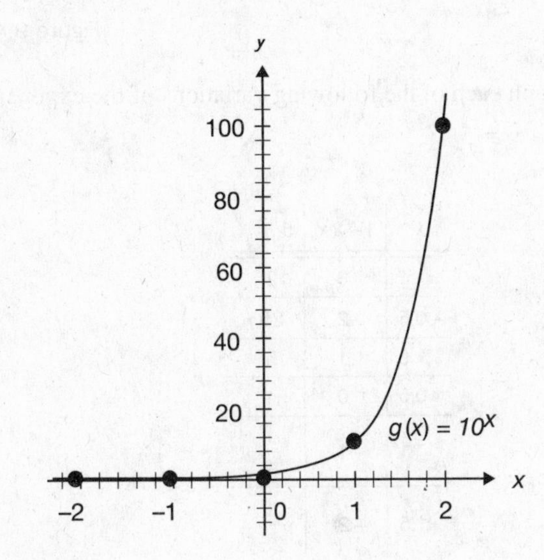

Figure 9.2

(c) $y = \left(\frac{1}{2}\right)^x$

x	$\left(\frac{1}{2}\right)^x = 2^{-x}$
−3	8
−2	4
−1	2
0	1
1	$\frac{1}{2}$
2	$\frac{1}{4}$
3	$\frac{1}{8}$

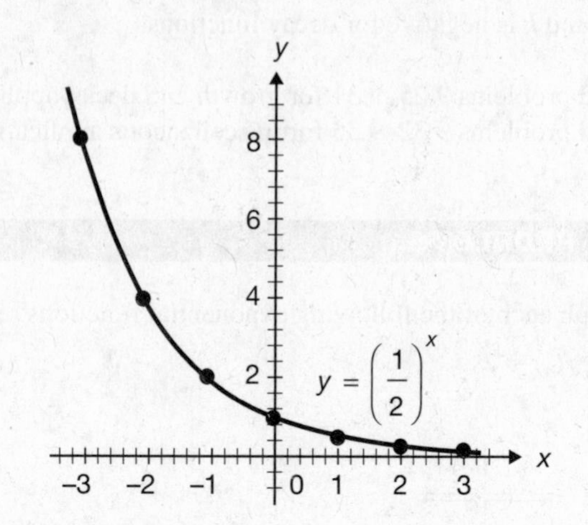

Figure 9.3

9.2 Graph $f(x) = e^x$.

$y = f(x) = e^x \Rightarrow$

x	y
−3	4.9787×10^{-2}
−2	0.13534
−1	0.36788
0	1
1	2.7183
2	7.3891
3	20.086

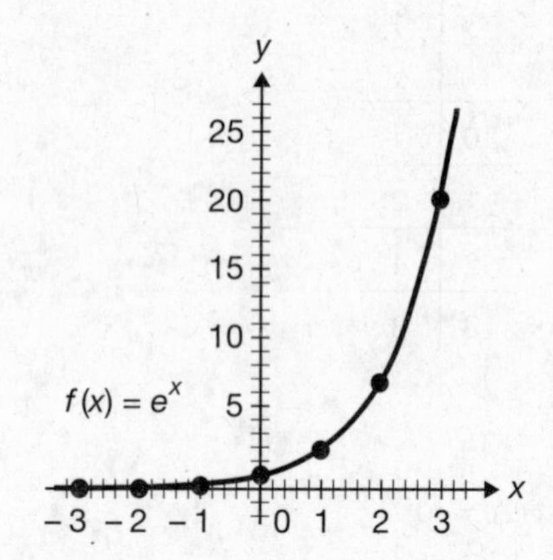

Figure 9.4

9.3 Graph each of the following variations of the exponential function:

(a) $y = 5^{1-2x}$

x	$1-2x$	5^{1-2x}
−1	3	125
−0.5	2	25
0	1	5
0.5	0	1
1	−1	$\frac{1}{5}$
1.5	−2	$\frac{1}{25}$

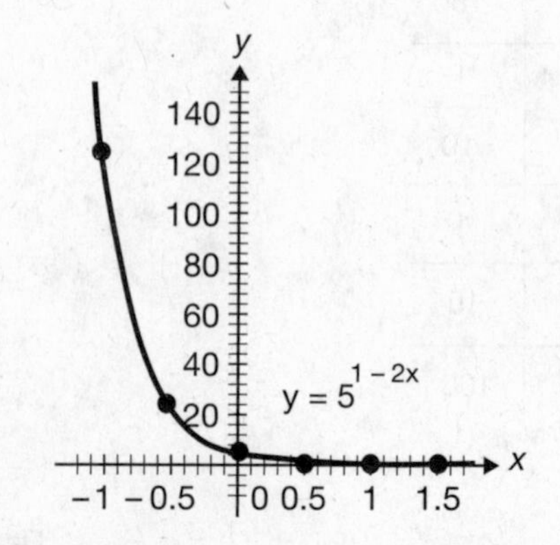

Figure 9.5

(*b*) $f(x) = -2 \cdot 3^{x+1}$

x	3^{x+1}	$-2 \cdot 3^{x+1}$
-3	$\frac{1}{9}$	$\frac{-2}{9}$
-2	$\frac{1}{3}$	$\frac{-2}{3}$
-1	1	-2
0	3	-6
1	9	-18
2	27	-54

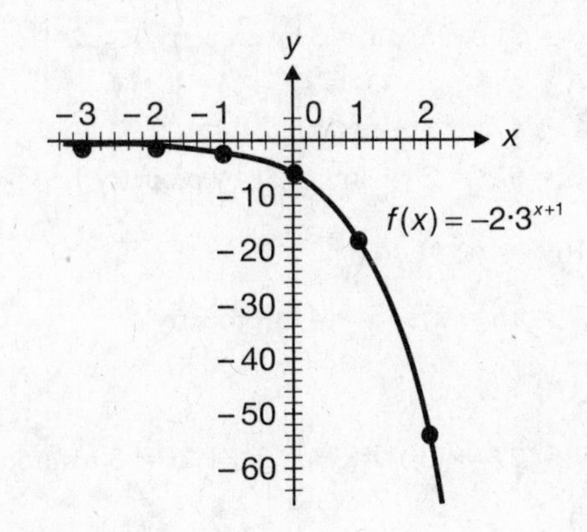

Figure 9.6

(*c*) $g(x) = e^{\sqrt{x}}$

x	$\sqrt{x}$	$e^{\sqrt{x}}$
0	0	1.0
1	1	$e^1 \approx 2.7$
2	$\sqrt{2}$	$e^{\sqrt{2}} \approx 4.1$
4	2	$e^2 \approx 7.4$
9	3	$e^3 \approx 20.1$
16	4	$e^4 \approx 54.6$

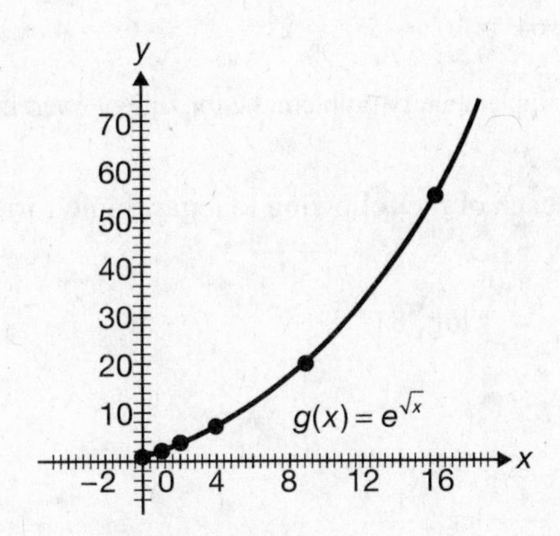

Figure 9.7

(*d*) $h(x) = 2 + 2^x$

x	2^x	$2+2^x$
-3	0.125	2.125
-2	0.25	2.25
-1	0.5	2.5
0	1	3
1	2	4
2	4	6
3	8	10

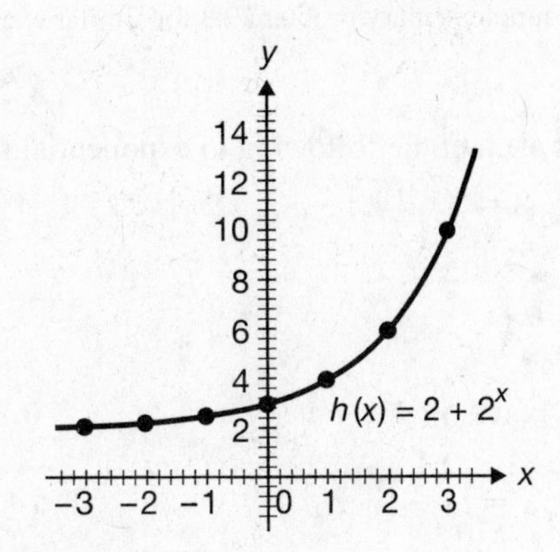

Figure 9.8

Refer to supplementary problem 9.1 for more practice in graphing exponential functions.

9.4 Use the exponential properties stated on page 288 to find the value of x in each of the following:

(*a*) $x^4 = 81$

$x^4 = 81 = 3^4$ so $x = 3$ by property 1.

(*b*) $x^2 = 9k^2$

$x^2 = 9k^2 = (3k)^2$ so $x = 3k$ by property 1.

(*c*) $2^x = 16$

$2^x = 16 = 2^4$ so $x = 4$ by property 2.

(*d*) $9^x = 27$

$9^x = 27 \Rightarrow (3^2)^x = 3^{2x} = 3^3 \Rightarrow 2x = 3$ by property 2 so $x = \frac{3}{2}$.

(*e*) $5^{2x} = \frac{1}{25}$

$5^{2x} = \frac{1}{25} = \frac{1}{5^2} = 5^{-2} \Rightarrow 2x = -2$ by property 2 so $x = -1$.

(*f*) $64^x = 16$

$64^x = 16 \Rightarrow (2^6)^x = 2^{6x} = 2^4 \Rightarrow 6x = 4$ by property 2 so $x = \frac{2}{3}$.

Refer to supplementary problem 9.2 for similar exercises.

9.5 Convert each of the following to logarithmic form:

(*a*) $3^4 = 81$

$4 = \log_3 81$

(*b*) $16^{\frac{-1}{2}} = \frac{1}{4}$

$\frac{-1}{2} = \log_{16} \frac{1}{4}$

(*c*) $p^q = m$

$q = \log_p m$

Refer to supplementary problem 9.3 for similar exercises.

9.6 Convert each of the following to exponential form:

(*a*) $\log_2 8 = 3$

$8 = 2^3$

(*b*) $\log_{0.5} 4 = -2$

$4 = 0.5^{-2}$

(*c*) $\log_r s = t$

$s = r^t$

Refer to supplementary problem 9.4 for similar exercises.

9.7 Graph each of the following:

(*a*) $y = \log_2 x$

Use the fundamental relationship: choose values for y, then evaluate x. Finally plot ordered pairs (x, y) (in that order — x first, y second).

y	$x=2^y$
-2	$\frac{1}{4}$
-1	$\frac{1}{2}$
0	1
1	2
2	4
3	8

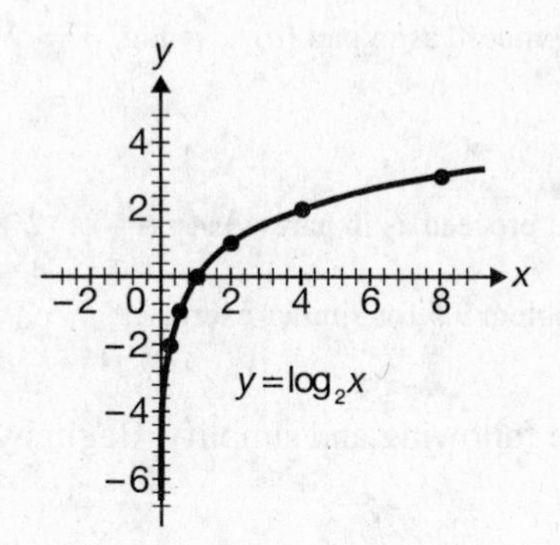

Figure 9.9

(*b*) $f(x) = \log_5 x$

Proceed as in part (*a*).

y	$x=5^y$
-2	$\frac{1}{25}$
-1	$\frac{1}{5}$
0	1
1	5
2	25

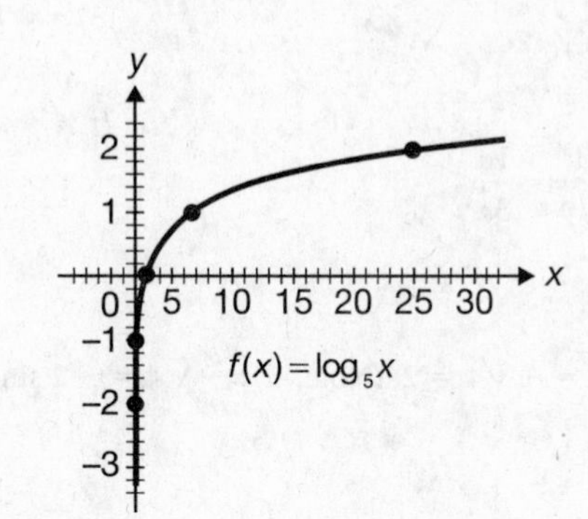

Figure 9.10

(*c*) $g(x) = \log_{0.1} x$

Proceed as in part (*a*).

y	$x=0.1^y$
-2	100
-1	10
0	1
1	0.1
2	0.01

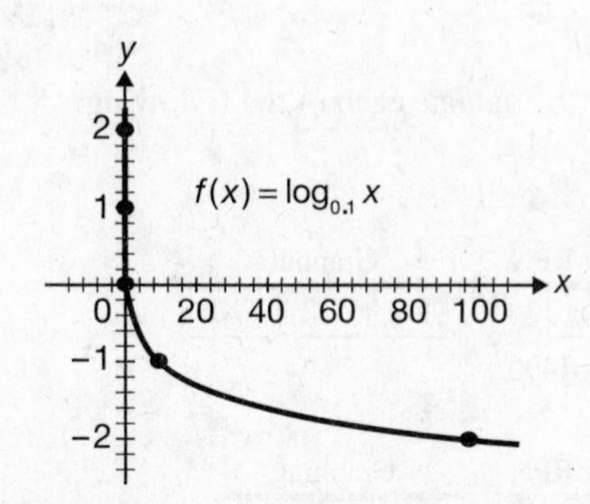

Figure 9.11

9.8 Evaluate each of the following without using a calculator:

(*a*) $\log_2 16$

Let $x = \log_2 16$, then apply the fundamental relationship. Hence $2^x = 16 = 2^4$; so $x = 4$.

(*b*) $\log_9 3$

Let $x = \log_9 3$ and proceed as in part (*a*). $x = \log_9 3 \Rightarrow 9^x = 3$ or $(3^2)^x = 3^{2x} = 3^1$. Hence $2x = 1$ or $x = \frac{1}{2}$.

(*c*) $\log_4 \frac{1}{2}$

Let $x = \log_4 \frac{1}{2}$ and proceed as in part (*a*). $4^x = \frac{1}{2}$ or $(2^2)^x = 2^{2x} = \frac{1}{2} = 2^{-1}$. Hence $2x = -1$ or $x = \frac{-1}{2}$.

Refer to supplementary problem 9.5 for similar exercises.

9.9 Solve for x in each of the following and simplify. Begin by applying the fundamental relationship in each case.

(*a*) $\log_2 x = 3$

$x = 2^3 = 8$

(*b*) $\log_5 x = 0$

$x = 5^0 = 1$

(*c*) $\log_9 x = \frac{-1}{2}$

$x = 9^{\frac{-1}{2}} = \frac{1}{\sqrt{9}} = \frac{1}{3}$

(*d*) $\log_x 4 = 2$

$4 = x^2 \Rightarrow x = +\sqrt{4} = 2$. (Note: $x \neq -\sqrt{4} = -2$ since the base of a logarithm is positive).

(*e*) $\log_x 81 = 4$

$81 = x^4 \Rightarrow x = +\sqrt[4]{81} = 3$

(*f*) $\log_x 5 = 3$

$5 = x^3 \Rightarrow x = \sqrt[3]{5} \approx 1.70998$

Refer to supplementary problem 9.6 for similar exercises.

9.10 Use a calculator to evaluate each of the following:

(*a*) $\log 412$

Algebraic or RPN	Graphing
412 log	log 412 EXE

$\log 412 \approx 2.61490$

(*b*) $\ln 52$

Algebraic or RPN	Graphing
52 ln	ln 52 EXE

$\ln 52 \approx 3.95124$

(*c*) $\log(4.87 + 5.2)$

Algebraic

[( ] [4.87] [+] [5.2] [ )] [log]

RPN

[4.87] [ENTER] [5.2] [+] [log]

Graphing

[log] [( ] [4.87] [+] [5.2] [ )] [EXE]

$\log(4.87 + 5.2) \approx 1.00303$

(*d*) $(5 + \ln 23.4)^3$

Algebraic

[( ] [5] [+] [23.4] [ln] [ )] [y^x] [3] [=]

RPN

[5] [ENTER] [23.4] [ln] [+] [3] [y^x]

Graphing

[( ] [5] [+] [ln] [23.4] [ )] [y^x] [3] [EXE]

$(5 + \ln 23.4)^3 \approx 541.889$

Refer to supplementary problem 9.7 for similar exercises.

9.11 Graph each of the following:

(*a*) $f(x) = \log(5x + 3)$

Domain: $5x + 3 > 0 \Rightarrow x > \dfrac{-3}{5} = -0.6$

x	$y = f(x)$
−0.59	−1.30
−0.05	−0.30
−0.25	0.24
0	0.48
1	0.90
2	1.11
3	1.26
4	1.36

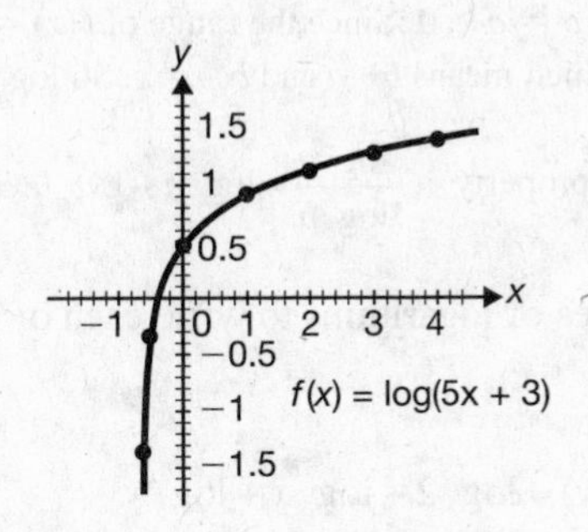

Figure 9.12

(*b*) $g(x) = 2 - \ln(3x)^2$

Domain: $(3x)^2 > 0 \Rightarrow x \neq 0$

x	$\ln((3x)^2)$	$y = g(x)$
±0.1	−2.41	4.41
±1	2.2	−0.20
±2	3.58	−1.58
±3	4.39	−2.39
±4	4.97	−2.97

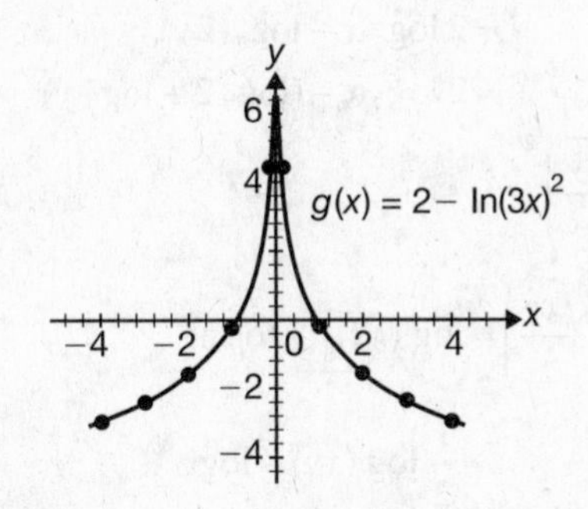

Figure 9.13

Refer to supplementary problem 9.8 for additional practice in graphing logarithmic functions.

9.12 Evaluate $\log_3 15$.

$$\log_3 15 = \frac{\log 15}{\log 3} \approx \frac{1.176091}{0.477121} \approx 2.46497 \text{ or } \log_3 15 = \frac{\ln 15}{\ln 3} \approx \frac{2.708050}{1.098612} \approx 2.46497$$

Note: Either common logarithms or natural logarithms may be used in these types of calculations. The result is identical in either case.

Refer to supplementary problem 9.9 for more change of base calculation problems.

9.13 Prove each of the on pages 289–290 five properties.

1. Let $x = \log_b a$, $y = \log_b c$, and $z = \log_b (ac)$. Then $a = b^x$, $c = b^y$, and $ac = b^z$; so $b^z = ac = b^x b^y = b^{x+y}$. Therefore, $z = x + y$ or $\log_b (ac) = \log_b a + \log_b c$.

2. Let $x = \log_b a$, $y = \log_b c$, and $z = \log_b\left(\frac{a}{c}\right)$. Then $a = b^x$, $c = b^y$, and $\frac{a}{c} = b^z$; so $b^z = \frac{a}{c} = \frac{b^x}{b^y} = b^{x-y}$.

   Therefore, $z = x - y$ or $\log_b\left(\frac{a}{c}\right) = \log_b a - \log_b c$.

3. Let $x = \log_b a$ and $y = \log_b (a^c)$. Then $a = b^x$ and $b^y = a^c = (b^x)^c = b^{cx}$. Therefore, $y = cx$ or $\log_b (a^c) = c \log_b a$.

4. Let $x = \log_b a$, $y = \log_c a$, and $z = \log_c b$. Then $a = b^x = c^y$ and $b = c^z$; so $c^y = a = b^x = (c^z)^x = c^{xz}$.

   Therefore, $y = xz \Rightarrow x = \frac{y}{z}$ or $\log_b a = \frac{\log_c a}{\log_c b}$.

5. (Part *I*) Let $x = \log_b a = \log_b c$. Then $b^x = a = c$. Therefore, if $\log_b a = \log_b c$, then $a = c$.

   (Part *II*) Let $a = c > 0$. Since the range of $f(x) = b^x$ is all positive numbers, there must be some x_1 such that $f(x_1) = a$, which means $b^{x_1}\ a$ and $b^{x_1} = c$. So $\log_b a = x_1 = \log_b c$. Therefore, if $a = c$, then $\log_b a = \log_b c$.

   Note that in property 4, $\frac{\log_c a}{\log_c b} \neq \log_c a - \log_c b$. In other words, property 2 does not simplify property 4.

9.14 Use the properties of logarithms to write each of the following in terms of the logarithms of x, y, and z:

(*a*) $\log_3 (2xy)$

$\log_3 (2xy) = \log_3 2 + \log_3 x + \log_3 y$ — Property 1

(*b*) $\log_7\left(\frac{x^2}{2y}\right)$

$\log_7\left(\frac{x^2}{2y}\right) = \log_7 x^2 - \log_7 (2y)$ — Property 2

$= 2\log_7 x - \log_7 (2y)$ — Property 3

$= 2\log_7 x - (\log_7 2 + \log_7 y)$ — Property 1

(*c*) $\log\left(\frac{\sqrt{xy}}{z}\right)$

$\log\left(\frac{\sqrt{xy}}{z}\right) = \log (xy)^{\frac{1}{2}} - \log z$ — Property 2

$= \frac{1}{2}\log (xy) - \log z$ — Property 3

$= \frac{1}{2}(\log x + \log y) - \log z$ — Property 1

(*d*) $\ln\left(\frac{x^3y^4}{z^2}\right)$

$\ln\left(\frac{x^3y^4}{z^2}\right) = \ln x^3y^4 - \ln z^2$ Property 2

$= \ln x^3 + \ln y^4 - \ln z^2$ Property 1

$= 3\ln x + 4\ln y - 2\ln z$ Property 3

Refer to supplementary problem 9.10 for similar exercises.

9.15 Use the properties of logarithms to write each of the following as a single logarithm with a coefficient of 1:

(*a*) $\log_2 3 - \log_2 x$

$\log_2 3 - \log_2 x = \log_2\left(\frac{3}{x}\right)$ Property 2

(*b*) $4(\log_5 y + \log_5 z)$

$4(\log_5 y + \log_5 z) = 4\log_5(yz)$ Property 1

$= \log_5(yz)^4$ Property 3

(*c*) $\ln x - \ln y - \ln(z-3)$

$\ln x - \ln y - \ln(z-3) = \ln x - (\ln y + \ln(z-3))$

$= \ln x - \ln[y(z-3)]$ Property 1

$= \ln\left(\frac{x}{y(z-3)}\right)$ Property 2

(*d*) $2\log 5 + 3\log x - z\log y$

$2\log 5 + 3\log x - z\log y = \log 5^2 + \log x^3 - \log y^z$ Property 3

$= \log(5^2x^3) - \log y^z$ Property 1

$= \log\left(\frac{5^2x^3}{y^z}\right)$ Property 2

Refer to supplementary problem 9.11 for similar exercises.

9.16 Find the value of each of the following expressions given that $\log_b 2 = 0.7$, $\log_b 3 = 1.1$ and $\log_b 5 = 1.6$:

(*a*) $\log_b 6$

$\log_b 6 = \log_b(2\cdot 3) = \log_b 2 + \log_b 3 = 0.7 + 1.1 = 1.8$

(*b*) $\log_b 75$

$\log_b 75 = \log_b(3\cdot 5^2) = \log_b 3 + 2\log_b 5 = 1.1 + 2(1.6) = 4.3$

(*c*) $\log_b\left(\frac{3}{5}\right)$

$\log_b\left(\frac{3}{5}\right) = \log_b 3 - \log_b 5 = 1.1 - 1.6 = -0.5$

(*d*) $\log_b\left(\frac{12}{25}\right)$

$\log_b\left(\frac{12}{25}\right) = \log_b\frac{2^2\cdot 3}{5^2} = 2\log_b 2 + \log_b 3 - 2\log_b 5 = 2(0.7) + 1.1 - 2(1.6) = -0.7$

Refer to supplementary problem 9.12 for similar exercises.

9.17 Evaluate $\dfrac{(49830)^3\sqrt{987}}{(2348)(1548)^4}$ using common logarithms.

Let $N = \dfrac{(49830)^3\sqrt{987}}{(2348)(1548)^4} = \dfrac{(4.983\times10^4)^3(9.87\times10^2)^{\frac{1}{2}}}{(2.348\times10^3)(1.548\times10^3)^4}$. Then

$$\log N = \log\frac{(4.983\times10^4)^3(9.87\times10^2)^{\frac{1}{2}}}{(2.348\times10^3)(1.548\times10^3)^4}$$

$$= 3\log(4.983\times10^4)+\frac{1}{2}\log(9.87\times10^2)-\log(2.348\times10^3)-4\log(1.548\times10^3)$$

$$= 3(\log(4.983)+\log 10^4)+\frac{1}{2}(\log(9.87)+\log 10^2)-(\log(2.348)+\log(10^3))$$

$$-4(\log(1.548)+\log(10^3))$$

$$\approx 3(0.69749+4)+\frac{1}{2}(0.99432+2)-(0.3707+3)-4(0.18977+3)$$

$$= -0.54015$$

So, $N \approx 10^{-0.54015} \approx 0.288304$ from the fundamental relationship.

9.18 Use the properties of logarithms stated in the previous section to solve each of the following for x to 6 significant digits. (We will use 7 significant digits in the intermediate steps to help ensure 6 digit accuracy in the answers.)

(*a*) $2^x = 15$

$\ln(2^x) = \ln 15$	Property 5
$x\ln 2 = \ln 15$	Property 3
$x = \dfrac{\ln 15}{\ln 2}$	
$x \approx \dfrac{2.708050}{0.6931472}$	
$x \approx 3.90689$	

Check: $2^{3.90689} \stackrel{?}{=} 15$

$15.0 = 15$ True

(*b*) $12^{2-x} = 20$

$\ln(12^{2-x}) = \ln 20$	Property 5
$(2-x)\ln 12 = \ln 20$	Property 3
$2\ln 12 - x\ln 12 = \ln 20$	
$-x\ln 12 = \ln 20 - 2\ln 12$	
$x = \dfrac{\ln 20 - 2\ln 12}{-\ln 12}$	
$x \approx \dfrac{-1.974081}{-2.484907} \approx 0.794429$	

Check: $12^{2-(0.794429)} \stackrel{?}{=} 20$

$19.99998 \approx 20$ True, accounting for round-off error.

(*c*) $5^x = 3^{2x-1}$

$$\ln(5^x) = \ln(3^{2x-1}) \qquad \text{Property 5}$$
$$x\ln 5 = (2x-1)\ln 3 \qquad \text{Property 3}$$
$$x\ln 5 = 2x\ln 3 - \ln 3$$
$$x\ln 5 - 2x\ln 3 = -\ln 3$$
$$x(\ln 5 - 2\ln 3) = -\ln 3$$
$$x = \frac{-\ln 3}{\ln 5 - 2\ln 3}$$
$$x \approx 1.86907$$

Check: $5^{1.86907} \stackrel{?}{=} 3^{2(1.86907)-1}$

$20.2499 \approx 20.2500$ Again true (with round-off).

(*d*) $8^{5x+1} = 18^{2x-3}$

$$\ln(8^{5x+1}) = \ln(18^{2x-3}) \qquad \text{Property 5}$$
$$(5x+1)\ln 8 = (2x-3)\ln 18 \qquad \text{Property 3}$$
$$5x\ln 8 + \ln 8 = 2x\ln 18 - 3\ln 18$$
$$5x\ln 8 - 2x\ln 18 = -\ln 8 - 3\ln 18$$
$$x(5\ln 8 - 2\ln 18) = -\ln 8 - 3\ln 18$$
$$x = \frac{-\ln 8 - 3\ln 18}{5\ln 8 - 2\ln 18}$$
$$x \approx -2.32874$$

Check: $8^{5(-2.32874)+1} \stackrel{?}{=} 18^{2(-2.32874)-3}$

$2.44220 \times 10^{-10} \approx 2.44217 \times 10^{-10}$ True again after accounting for round-off.

9.19 Solve each of the following for x to 6 significant digits:

(*a*) $\log_3 x + \log_3 4 = 2$

$$\log_3 x + \log_3 4 = 2$$
$$\log_3(4x) = 2 \qquad \text{Property 1}$$
$$4x = 3^2 \qquad \text{Fundamental relationship}$$
$$x = \frac{3^2}{4} = \frac{9}{4} = 2.25$$

Check: $\log_3\left(\frac{9}{4}\right) + \log_3 4 \stackrel{?}{=} 2$

$$\frac{\ln\left(\frac{9}{4}\right)}{\ln 3} + \frac{\ln 4}{\ln 3} \stackrel{?}{=} 2 \qquad \text{Property 4}$$
$$\frac{0.8109302}{1.098612} + \frac{1.386294}{1.098612} \stackrel{?}{=} 2$$
$$0.7381405 + 1.261859 \stackrel{?}{=} 2$$
$$2.00000 = 2 \qquad \text{True.}$$

(*b*) $\log_5(2x) - 2\log_5 4 = \log_5 3$

$$\log_5(2x) - 2\log_5 4 = \log_5 3$$

$$\log 5\left(\frac{2x}{4^2}\right) = \log_5 3 \qquad \text{Properties 2 and 3}$$

$$\frac{2x}{4^2} = 3 \qquad \text{Property 5}$$

$$x = \frac{3\cdot 4^2}{2} = 24$$

Check: $\log_5(2[24]) - 2\log_5 4 \overset{?}{=} \log_5 3$

$$\frac{\ln(48)}{\ln 5} - 2\left(\frac{\ln 4}{\ln 5}\right) \overset{?}{=} \frac{\ln 3}{\ln 5} \qquad \text{Property 4}$$

$$\frac{3.871201}{1.609438} - 2\left(\frac{1.386294}{1.609438}\right) \overset{?}{=} \frac{1.098612}{1.609438}$$

$$0.405312 - 2(0.8613531) \overset{?}{=} 0.682606$$

$$0.682606 = 0.682606 \qquad \text{True, accounting for round-off.}$$

(*c*) $3\log_x 6 = 4$

$$3\log_x 6 = 4$$

$$\log_x(6^3) = 4 \qquad \text{Property 3}$$

$$6^3 = x^4 \qquad \text{Fundamental relationship}$$

$$x = \sqrt[4]{6^3} \approx 3.83366$$

Check: $3\log_{(3.83366)} 6 \overset{?}{=} 4$

$$3\left(\frac{\ln 6}{\ln(3.83366)}\right) \overset{?}{=} 4 \qquad \text{Property 4}$$

$$3\left(\frac{1.791759}{1.343820}\right) \overset{?}{=} 4$$

$$3(1.333333) \overset{?}{=} 4$$

$$3.99999 \approx 4 \qquad \text{True, accounting for round-off error.}$$

(*d*) $\log_2 8 + \log_2 9 = \log_x 3$

$$\log_2 8 + \log_2 9 = \log_x 3$$

$$\log_2(8\cdot 9) = \log_x 3 \qquad \text{Property 1}$$

$$\frac{\ln 72}{\ln 2} = \frac{\ln 3}{\ln x} \qquad \text{Property 4}$$

$$\ln x = \frac{(\ln 2)(\ln 3)}{\ln 72} \approx 0.1780593 \qquad \text{Evaluate}$$

$$x \approx e^{0.1780593} \approx 1.19490 \qquad \text{Fundamental relationship}$$

Check: $\log_2 8 + \log_2 9 \overset{?}{=} \log_{1.19490} 3$

$$\frac{\ln 8}{\ln 2} + \frac{\ln 9}{\ln 2} \overset{?}{=} \frac{\ln 3}{\ln(1.19490)}$$

$$\frac{2.079442}{0.6931472} + \frac{2.197225}{0.6931472} \overset{?}{=} \frac{1.098612}{0.1780625}$$

$$3.000000 + 3.169920 \overset{?}{=} 6.169813$$

$$6.16992 \approx 6.16981 \qquad \text{True, accounting for round-off error.}$$

(e) $\log_2 5 - \log_3 7 = \log_5 x$

$$\log_2 5 - \log_3 7 = \log_5 x$$

$$\frac{\ln 5}{\ln 2} - \frac{\ln 7}{\ln 3} = \log_5 x \qquad \text{Property 4}$$

$$\left(\frac{1.609438}{0.6931472} - \frac{1.945910}{1.098612}\right) = \log_5 x$$

$$0.5506843 = \log_5 x$$

$$5^{0.5506843} = x \qquad \text{Fundamental relationship}$$

$$x \approx 2.42611$$

Check: $\log_2 5 - \log_3 7 \stackrel{?}{=} \log_5 (2.42611)$

$$\frac{\ln 5}{\ln 2} - \frac{\ln 7}{\ln 3} \stackrel{?}{=} \frac{\ln 2.42611}{\ln 5} \qquad \text{Property 4}$$

$$\frac{1.609438}{0.6931472} - \frac{1.945910}{1.098612} \stackrel{?}{=} \frac{0.8862892}{1.609438}$$

$$2.321928 - 1.771244 \stackrel{?}{=} 0.5506824$$

$$0.550684 \approx 0.550682 \qquad \text{Again true (with round-off)}$$

(f) $3 - \ln(1 + x) = \ln(1 - x)$

$$3 - \ln(1 + x) = \ln(1 - x)$$

$$3 = \ln(1 - x) + \ln(1 + x) \qquad \text{Algebra}$$

$$3 = \ln[(1 - x)(1 + x)] \qquad \text{Property 1}$$

$$e^3 = (1 - x)(1 + x) \qquad \text{Fundamental relationship}$$

$$e^3 = 1 - x^2$$

$$x^2 = 1 - e^3$$

$$x = \pm\sqrt{1 - e^3} \approx \pm 4.36870i \qquad x \text{ must be real so there is no solution.}$$

Refer to supplementary problem 9.13 for more practice in solving exponential and logarithmic equations.

9.20 Solve for x to 6 significant digits: $x^x = 16$.

This equation doesn't have an algebraic solution without approximation, so we'll use a graphic approach. Let $y = x^x - 16$, graph the equation, and approximate the x-intercept or root. That is, approximate where $y = 0$ since $x^x - 16 = 0$. This process may require using the zoom function on a graphing calculator many times until the accuracy required is achieved.

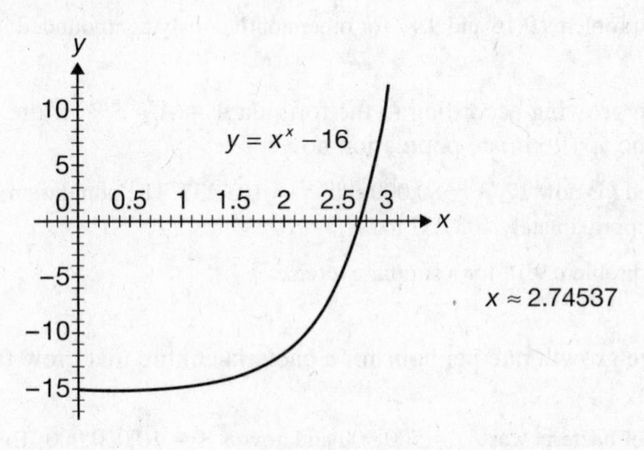

$x \approx 2.74537$

Figure 9.14

Check: $(2.74537)^{(2.74537)} \stackrel{?}{=} 16$

$16.0001 \approx 16$ which is true (accounting for round-off error).

9.21 Calculate the accumulated amount to the nearest cent for a principal of \$5,000 compounded monthly at an annual rate of 4.5% for 10 years.

$P = \$5{,}000$, $i = 0.045/12 = 0.00375$, $n = 10 \cdot 12 = 120$, so $A = \$5{,}000\,(1 + 0.00375)^{120} = \$7{,}834.96$. The amount accumulated after 10 years is \$7,834.96

Refer to supplementary problem 9.14 for a similar exercise.

9.22 Determine the time required for the principal to double in value when it is compounded quarterly at an annual rate of 8%.

$i = 0.08/4$, $n = 4t$ where t is the number of years required. Then

$$A = 2P = P\left(1+\frac{0.08}{4}\right)^{4t}$$

$$2 = \left(1+\frac{0.08}{4}\right)^{4t}$$

$$\ln 2 = \ln\left(1+\frac{0.08}{4}\right)^{4t} = 4t\ln\left(1+\frac{0.08}{4}\right)$$

$$t = \frac{\ln 2}{4\ln\left(1+\frac{0.08}{4}\right)} \approx 8.7507 \text{ years}$$

The principal will double in value in just over 8.75 years.

Refer to supplementary problem 9.15 for a similar exercise.

9.23 How much is accumulated for an initial investment of \$200 compounded continuously at an annual rate of 6% for 5 years?

$P = \$200$, $r = 0.06$, $t = 5$, so $A = \$200e^{0.06(5)} \approx \$269.97$. The original \$200 investment will grow to almost \$270 when compounded continuously at this rate for 5 years.

9.24 Find the value of a piece of property now valued at \$100,000 that appreciates continuously at an annual rate of 3% for 10 years.

$P = \$100{,}000$, $r = 0.03$, $t = 10$, so $A = \$100{,}000e^{0.03(10)} \approx \$134{,}985.88$. In 10 years, the property should see close to a 35% increase in value to be worth almost \$135,000.

Refer to supplementary problems 9.16 and 9.17 for other continuously compounded interest exercises.

9.25 A population has been growing according to the formula $A = A_0e^{0.015t}$. If the population 17 years ago was 80,000, what is the approximate population now?

Since $A_0 = 80{,}000$ and t is now 17, $A = 80{,}000e^{0.015(17)} \approx 103{,}237$. The population should grow from 80,000 seventeen years ago to approximately 103,237 today.

Refer to supplementary problem 9.18 for a similar exercise.

9.26 What is the percentage growth rate per hour for a bacteria culture that grew from 500,000 twelve hours ago to 10,000,000 now?

The original number of bacteria was $A_0 = 500{,}000$ and now is $A = 10{,}000{,}000$. To find k, the constant for the problem, solve the following:

$$10{,}000{,}000 = 500{,}000e^{k(12)}$$
$$\frac{10{,}000{,}000}{500{,}000} = e^{k(12)}$$
$$20 = e^{12k}$$
$$12k = \ln 20$$
$$k = \frac{\ln 20}{12}$$
$$k = 0.249644$$

Now we can use this value for k to determine the number of bacteria present after 1 hour.

$$A = 500{,}000e^{k(1)}$$
$$A = 500{,}000e^{(0.249644)(1)}$$
$$A = 500{,}000\ (1.28357)$$
$$A = 641{,}785$$

By this, we can determine the growth rate for that first hour (and therefore for each hour) to be $\frac{641{,}785 - 500{,}000}{500{,}000} = 0.28357$. (You might notice that this is the fractional part of the number for $e^{k(1)}$ we found above.) The percentage growth rate from 500,000 to 10,000,000 in 12 hours is 28.357% per hour.

Refer to supplementary problem 9.19 for a similar exercise.

9.27 **Bacteria in a certain culture doubles every hour. If there are 1,000,000 bacteria now, how many will there be in 5.5 hours?**

$A_0 = 1{,}000{,}000$ so at the end of one hour, 2,000,000 bacteria will be present. To find k, the constant for the problem, solve the following:

$$2{,}000{,}000 = 1{,}000{,}000e^{k(1)}$$
$$\frac{2{,}000{,}000}{1{,}000{,}000} = e^{k(1)}$$
$$2 = e^{k}$$
$$k = \ln 2$$

Now knowing k and using the property $e^{\ln x} = e^{\log_e x} = x$ from Section 9.2,

$$A = 10^6 e^{t \ln 2}$$
$$= 10^6\,(e^{\ln 2})^t$$
$$= 10^6(2)^t$$
$$= 10^6\,(2)^{(5.5)} \qquad \text{(evaluated at } t = 5.5 \text{ hours)}$$
$$\approx 4.52548 \times 10^7 = 45{,}254{,}800 \text{ bacteria.}$$

In other words, in just $5\frac{1}{2}$ hours the number of bacteria mushrooms from 1 million to over 45 million.

Refer to supplementary problem 9.20 for a similar exercise.

9.28 **Assuming the world population is growing continuously at a rate of 1.7% per year, how long will it take for the population to double at that rate?**

Let A_0 be the population now. For an increase of 1.7%, the population next year will be $1 + 0.017 = 1.017$ times what it is now or $1.017A_0$. At the end of one year $t = 1$ so set the population $A = A_0e^{k(1)}$ equal to $1.017\,A_0$; then solve for k to find that constant for the problem.

$$1.017A_0 = A_0e^{k(1)}$$
$$\frac{1.017A_0}{A_0} = e^k$$
$$1.017 = e^k$$
$$k = \ln 1.017$$

So to find the time for doubling, the population then will be twice what it is now or $A = 2A_0$. Solve for t using the k found above.

$$2A_0 = A_0e^{t \ln 1.017}$$
$$\frac{2A_0}{A_0} = e^{t \ln 1.017}$$
$$2 = e^{t \ln 1.017}$$
$$\ln 2 = t \ln 1.017$$
$$\text{Thus } t = \frac{\ln 2}{\ln 1.017} \approx 41.1190 \text{ years.}$$

Therefore, at the rate of increase of only 1.7% per year, the world population would double in just over 41 years.

9.29 The population for a fictitious county is estimated as $P = 250{,}000e^{0.06t}$ t years after 1990.

(*a*) Estimate the population for this county in the year 2000.

In the year 2000, $t = 10$ so $P = 250{,}000e^{0.06(10)}$ or $P \approx 455{,}530$. Since the population in 1990 would have been 250,000 (set $t = 0$), in a span of 10 years the population would grow from 250,000 to over 455,000.

(*b*) Estimate the population for this county in the year 2015.

In the year 2015, $t = 25$ so $P = 250{,}000e^{0.06(25)}$ or $P \approx 1{,}120{,}422$. In a span of 25 years, this county would expand from a population of a quarter of a million people to well over a million people.

Refer to supplementary problem 9.21 for a similar exercise.

9.30 The half-life of a substance is the time it takes for half of that substance to decay (and half to be left). The half-life of a certain radioactive substance is known to be 5.2 years. What percent of the original amount is left after 12 years?

After 5.2 years $\frac{1}{2}A_0$ will be left, so $A = \frac{1}{2}A_0$. Solve for the constant k.

$$\frac{1}{2}A_0 = A_0e^{k(5.2)}$$
$$\frac{\frac{1}{2}A_0}{A_0} = e^{5.2k}$$
$$\frac{1}{2} = e^{5.2k}$$
$$\ln 0.5 = 5.2k$$
$$\text{so } k = \frac{\ln 0.5}{5.2} \approx -0.133298$$

After 12 years using k above, the amount present will be

$$A = A_0e^{-0.133298(12)}$$
$$\approx 0.201982A_0.$$

Therefore, approximately 20.1982% or just over one-fifth of the original amount is left after 12 years.

Refer to supplementary problem 9.22 for a similar exercise.

9.31 A hypothetical substance decays to 37% of its original amount in 125 years. What is its half-life?

The amount A after $t = 125$ years will be 0.37 of the original amount present or $0.37\,A_0$. Use this to solve for the constant k.

$$A = 0.37A_0$$
$$0.37A_0 = A_0 e^{k(125)}$$
$$\frac{0.37A_0}{A_0} = e^{125k}$$
$$0.37 = e^{125k}$$
$$\ln 0.37 = 125k$$
$$\text{so } k = \frac{\ln 0.37}{125} \approx -0.007954$$

To find its half-life, the amount present at that time t will be one-half of the original amount or $A = \frac{1}{2}A_0$. Use the value of k above and solve for t.

$$\frac{1}{2}A_0 = A_0 e^{-0.007954t}$$
$$\frac{\frac{1}{2}A_0}{A_0} = e^{-0.007954t}$$
$$\frac{1}{2} = e^{-0.007954t}$$
$$\ln 0.5 = -0.007954t$$
$$\text{so } t = \frac{\ln 0.5}{-0.007954} \approx 87.1445 \text{ years.}$$

The half-life of this substance is about 87.1445 years.

Refer to supplementary problem 9.23 for a similar exercise.

9.32 *Newton's Law of Cooling* for a substance can be stated as

$$T = T_R + (T_0 - T_R)\,e^{-kt}$$

where T is the temperature for the substance at time t, T_R is the room (or surrounding medium) temperature, and T_0 is the initial temperature for the substance. After 10 minutes, the temperature of a cup of coffee had dropped from 170° to 140°F in a 70°F room. How long did it take the coffee to reach 90°F?

The given information is $T_R = 70$, $T_0 = 170$ and we know that $T = 140$ when $t = 10$. From that information, we can find the constant k.

$$140 = 70 + (170 - 70)\,e^{-k(10)}$$
$$140 - 70 = (170 - 70)\,e^{-k(10)}$$
$$70 = (100)\,e^{-10k}$$
$$\frac{70}{100} = e^{-10k}$$
$$\ln\left(\frac{70}{100}\right) = -10k$$
$$\text{so } k = \frac{\ln\left(\frac{70}{100}\right)}{-10} \approx \frac{-0.356675}{-10} \approx 0.0356675$$

To determine how long it took the temperature of the coffee to drop to 90°F, use the k found above, set $T = 90$, then solve for t.

$$90 = 70 + (170 - 70)\,e^{-(0.0356675)t}$$
$$90 - 70 = (170 - 70)\,e^{-(0.0356675)t}$$
$$20 = 100e^{-(0.0356675)t}$$
$$\frac{20}{100} = e^{-(0.0356675)t}$$
$$\ln\left(\frac{20}{100}\right) = -(0.0356675)t$$
$$\text{so } t = \frac{\ln\left(\frac{20}{100}\right)}{-(0.0356675)} \approx 45.12 \text{ minutes.}$$

Therefore, in this environment, the temperature of the coffee would drop from 170° to 90°F in just about three-quarters of an hour.

Refer to supplementary problem 9.24 for a similar exercise.

9.33 The *atmospheric pressure P*, in pounds per square inch, can be calculated approximately using the formula $P = 14.7e{-}0.21h$, where h is the altitude in miles above sea level. Find the atmospheric pressure at 6,030 feet.

6,030 feet converts to 6,030/5,280 ≈ 1.14205 miles, so $P = 14.7e^{-0.21(1.14205)} \approx 11.5654$ pounds per square inch. Since the formula indicates the pressure at sea level is 14.7 pounds per square inch (set $h = 0$), the pressure drops about 3.1346 pounds per square inch or over 21% to 11.5654 as we climb to an elevation of 6,030 feet.

Refer to supplementary problem 9.25 for a similar exercise.

9.34 In chemistry, *hydrogen potential* or pH, is a way to describe the acidity or alkalinity of a solution. The pH of distilled water is 7. A pH above 7 indicates the solution is alkaline while a pH below 7 indicates the solution is acidic. The measure is calculated by the equation pH $= -\log\,[H^+]$, where $[H^+]$ is the concentration in moles per liter of the hydrogen ion. Compute the pH of orange juice for which $[H^+]$ is about 5.4×10^{-4} moles per liter.

$$\begin{aligned} \text{pH} &= -\log(5.4 \times 10^{-4}) \\ &= -\log 5.4 - \log 10^{-4} \\ &= -\log 5.4 - (-4) \approx 3.3 \end{aligned}$$

Thus, orange juice is acidic.

Refer to supplementary problem 9.26 for a similar exercise.

9.35 The *Richter scale* is a measure of the magnitude (intensity) of earthquakes; it is generally used to compare magnitudes of earthquakes. The Richter number is given by the formula $R = \log(I/I_0)$ where I is the intensity of the earthquake measured and I_0 is the intensity of the least movement that can be felt and is usually equated to 1. Therefore, $I = 10^R$. How much more powerful was the San Francisco earthquake of 1906 at 8.3 on the Richter scale than an earthquake of 6.5 on the Richter scale?

The comparison of the magnitudes is given by the ratio of their intensities:

$$\begin{aligned} \frac{I_{8.3}}{I_{6.5}} &= \frac{10^{8.3}}{10^{6.5}} \\ &= 10^{8.3-6.5} \\ &= 10^{1.8} \approx 63 \end{aligned}$$

The San Francisco quake was about 63 times more powerful!

Refer to supplementary problem 9.27 for a similar exercise.

SUPPLEMENTARY PROBLEMS

9.1 Graph each of the following functions:

(*a*) $f(x) = 3^x$ (*b*) $y = (0.7)^x$ (*c*) $g(x) = e^{-x}$ (*d*) $h(x) = 1^x$

(*e*) $y = -2(5^{-x})$ (*f*) $f(x) = 3^{x-3}$ (*g*) $g(x) = 2 \cdot 3^{1-x}$ (*h*) $h(x) = 3^{x-3}$

9.2 Use the exponential properties to find x for each of the following without a calculator:

(*a*) $3^x = 27$ (*b*) $4^x = 32$ (*c*) $81^x = 27$

(*d*) $16^{2x} = \frac{1}{8}$ (*e*) $9^{1-x} = 27^{3x}$ (*f*) $\left(\frac{1}{2}\right)^{x-6} = 4$

9.3 Convert each of the following to logarithmic form:

(*a*) $5^2 = 25$ (*b*) $7^x = 8$ (*c*) $10^4 = 10000$ (*d*) $e^3 = x$

9.4 Convert each of the following to exponential form:

(*a*) $\log_2 8 = 3$ (*b*) $\log_x 9 = 4$ (*c*) $\log x = -2$ (*d*) $\ln 5 = x + 1$

9.5 Evaluate each of the following without using a calculator:

(*a*) $\log_2 8$ (*b*) $\log_5 25$ (*c*) $\log_9 27$

(*d*) $\log_8 4$ (*e*) $\log_{100} 0.1$ (*f*) $\log_3\left(\frac{1}{9}\right)$

9.6 Find x for each of the following without using a calculator:

(*a*) $\log_x 16 = 2$ (*b*) $\log_3 9 = x$ (*c*) $\log_s x = \frac{-1}{3}$

(*d*) $2 \log_5 x = 3$ (*e*) $4 \log_x 8 = 2$ (*f*) $\log_2 x + 3\log_2 4 = 5$

(*g*) $\log_x 8 - \log_x 5 = 3$ (*h*) $2 \log_5 (3x) = \log_5 4$ (*i*) $-3 \log(x + 2) + \log 5 = 0$

9.7 Use a calculator to evaluate each of the following correct to 6 significant digits:

(*a*) $\log 2365$ (*b*) $\ln 2365$ (*c*) $2 \log 0.00034$

(*d*) $\ln 32 - \ln 8$ (*e*) $(2 - \log 42)^{1/2}$ (*f*) $\ln (8^3 + 17^2)$

9.8 Graph each of the following functions:

(*a*) $y = \log x$ (*b*) $f(x) = \log_{0.7} x$ (*c*) $g(x) = \ln x$ (*d*) $h(x) = -2 \log_3 x$

(*e*) $y = \log (2x + 3)$ (*f*) $f(x) = \log_5 (5x)^2$ (*g*) $g(x) = 2 - \ln (x - 1)$ (*h*) $h(x) = \log\left(\frac{x-1}{x+1}\right)$

9.9 Evaluate each of the following using a calculator, state answers accurate to 6 significant digits:

(*a*) $\log_3 49$ (*b*) $\log_5 10$ (*c*) $\log_9 4$ (*d*) $\log_5 (8^2)$ (*e*) $(\log_5 8)^2$ (*f*) $2 \log_3 5 + 3 \log_5 2$

9.10 Use the properties of logarithms to write each of the following in terms of the logs of x, y, and z:

(*a*) $\log_2 (2xz)$ (*b*) $\log_5 (x^2y^3)$ (*c*) $\log \sqrt{y^z}$

(*d*) $\log_7\left(\frac{xy^4}{z^6}\right)$ (*e*) $\ln \frac{\sqrt{x}}{y^3 z}$ (*f*) $\log_4 \sqrt[5]{\frac{x^3 y}{x^2}}$

9.11 Use the properties of logarithms to write each of the following as a single logarithm with a coefficient of 1:

(*a*) $2 \log_2 x + \log_2 y$ (*b*) $\log_3 x + 3 \log_3 (y + 1)$ (*c*) $\ln x + \frac{1}{3} \ln y - 2 \ln z$

(*d*) $8(\log_7 x + \log_7 y) - 3 \log_7 z$ (*e*) $\frac{2}{3}(\log x - \log y) + 5 \log z$ (*f*) $2 \log_5 x - 3 (\log_5 y + \log_5 z)$

9.12 Find the value of each of the following expressions given that $\log_a 2 = 0.545$, $\log_a 3 = 0.864$, and $\log_a 5 = 1.266$:

(*a*) $\log_a 8$ (*b*) $\log_a 18$ (*c*) $\log_a \left(\frac{2}{5}\right)$

(*d*) $\log_a \left(\frac{4}{75}\right)$ (*e*) $\log_a \left(\frac{5}{3}\right)^2$ (*f*) $\log_a \left[2^3 \cdot \left(\frac{9}{5}\right)\right]$

9.13 Solve each of the following for x to 6 significant digits:

(*a*) $3^x = 5$ (*b*) $5^x = 100$ (*c*) $8x = 80$

(*d*) $100^x = 50$ (*e*) $4^{2x-1} = 9$ (*f*) $3^{2x} = 7^{x+1}$

(*g*) $10^{x+2} = 20^{x+1}$ (*h*) $x = \log_2 9$ (*i*) $\log_x 15 = 3$

(*j*) $\log_7 x = -1.5$ (*k*) $\log_3 (x+1) + \log_3 4 = 2$ (*l*) $\log_x 5 + 2 \log_x 3 = 1$

(*m*) $\ln x - 2 \ln 3 = -2$ (*n*) $3 \log 2 + \log 7 = \log x$ (*o*) $\log_2 (x-1) + 2 = \log_2 5$

9.14 Calculate the accumulated amount for a principal of \$800 compounded monthly at an annual rate of 4.7% for 5 years.

9.15 How long will it take for an initial amount to triple in value when it is compounded quarterly at 5% annual interest?

9.16 Calculate the accumulated amount for a principal of \$800 compounded continuously at an annual rate of 4.7% for 5 years.

9.17 How long will it take for an initial amount to triple in value when it is compounded continuously at 5% annual interest?

9.18 If the population of a certain city increased from 96,000 five years ago to 100,000 now, what is the expected population 20 years from now (assuming the same rate of growth)?

9.19 What is the percentage growth rate per hour for a bacteria culture that grew from 2,500,000 7 hours ago to 10,000,000 now?

9.20 Bacteria in a certain culture doubles every 2.5 hours. If there are 1,000,000 bacteria now, how many will there be in 12 hours?

9.21 The population for a fictitious country is estimated as $P = 65{,}000{,}000e^{0.06t}$ t years after 1997.

(*a*) Estimate the population for this country in the year 2000.

(*b*) Estimate the population for this country in the year 2027.

9.22 For a substance with a half-life of 20 days.

(*a*) What percent is left after 30 days?

(*b*) How long it takes to have only 15% left?

9.23 Radioactive strontium-90 decays according to the formula $A = A_0 e^{-0.02476t}$. Find the half-life of strontium-90.

9.24 Using Newton's Law of Cooling, find the time it takes for boiling water at 100°C to cool to 40°C in a room kept at 22°C given that it cooled to 96°C in the first minute.

9.25 Atmospheric pressure P in pounds per square inch can be calculated using $P = 14.7e^{-0.21h}$, where h is the altitude in miles above sea level. Find the atmospheric pressure at 4000 feet above sea level.

9.26 Find the pH of a solution for which $[H^+] = 3.25 \times 10^{-8}$ using $pH = -\log [H^+]$.

9.27 Determine the relative magnitude of an earthquake that measured 6.5 on the Richter scale to an earthquake that measured 7.6 on the Richter scale.

ANSWERS TO SUPPLEMENTARY PROBLEMS

9.1 (*a*) See Figure 9.15. (*b*) See Figure 9.16.
(*c*) See Figure 9.17. (*d*) See Figure 9.18.
(*e*) See Figure 9.19. (*f*) See Figure 9.20.
(*g*) See Figure 9.21. (*h*) See Figure 9.22.

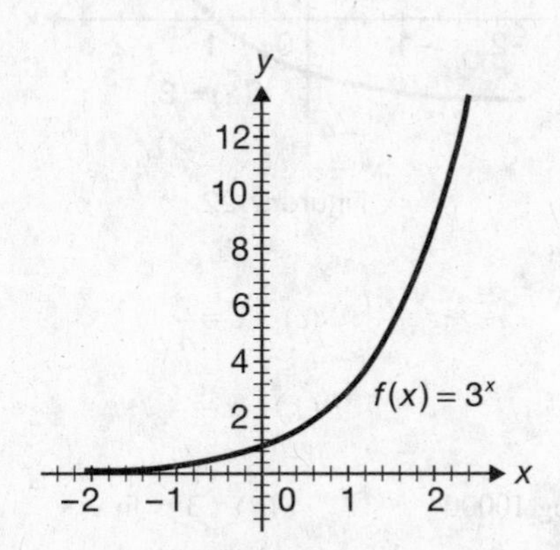

Figure 9.15

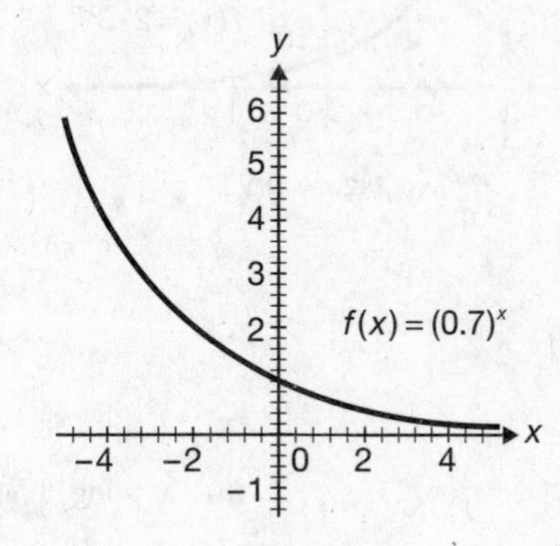

Figure 9.16

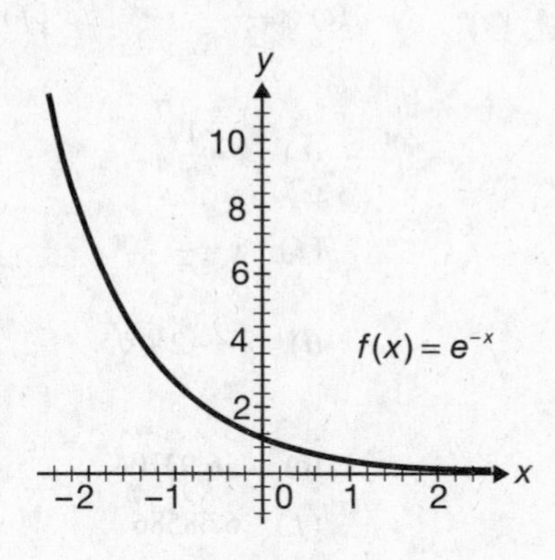

Figure 9.17

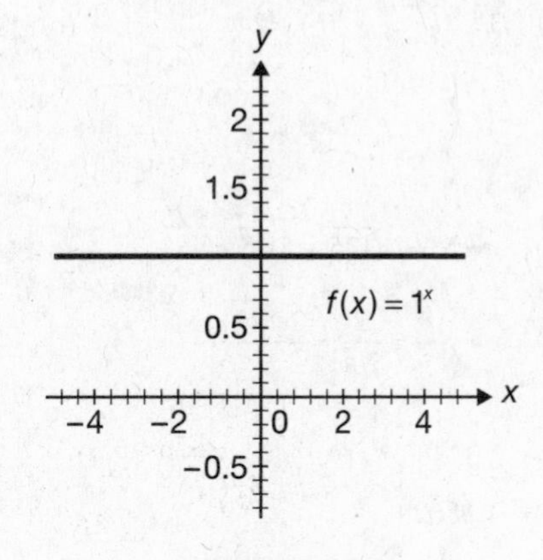

Figure 9.18

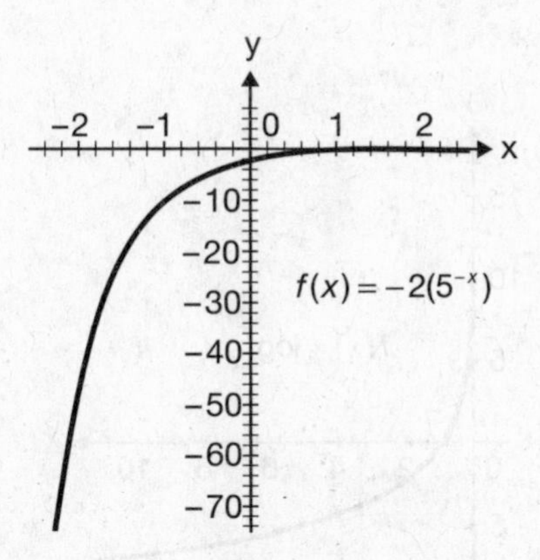

Figure 9.19

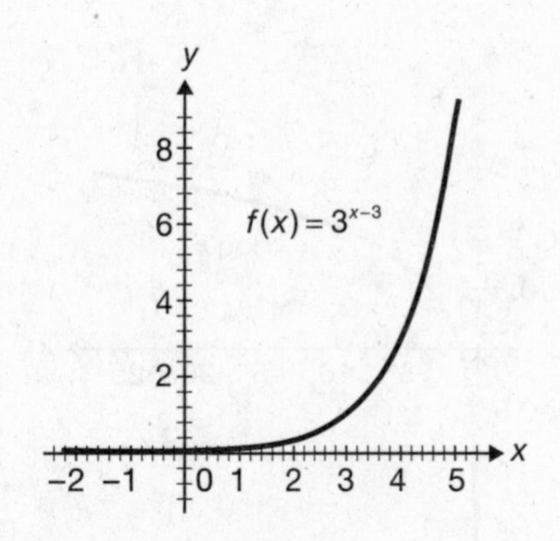

Figure 9.20

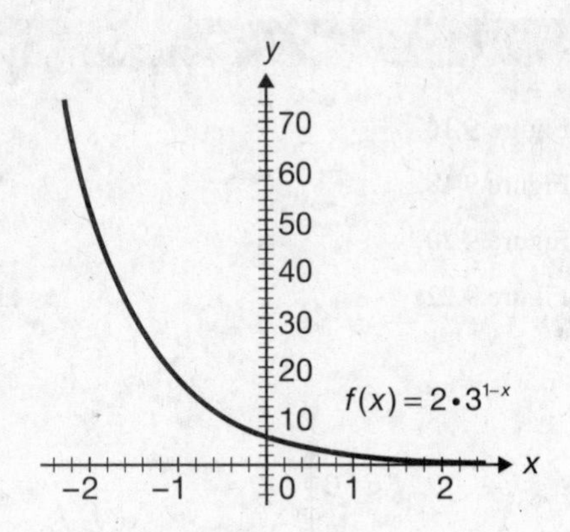

Figure 9.21

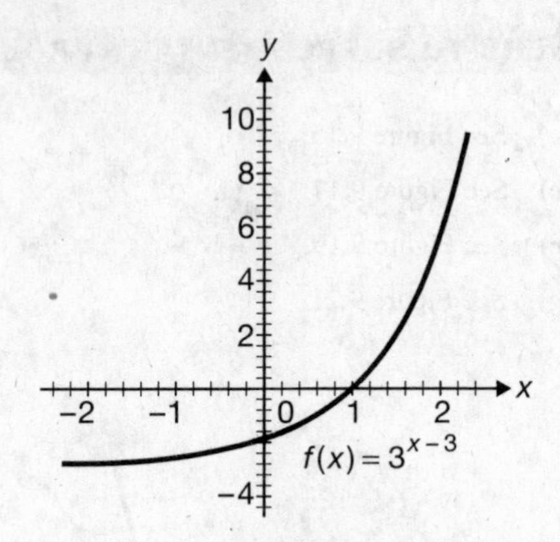

Figure 9.22

9.2 (*a*) $x = 3$ (*b*) $x = \frac{5}{2}$ (*c*) $x = \frac{3}{4}$

(*d*) $x = \frac{-3}{8}$ (*e*) $x = \frac{2}{11}$ (*f*) $x = 4$

9.3 (*a*) $2 = \log_5 25$ (*b*) $x = \log_7 8$ (*c*) $4 = \log 10000$ (*d*) $3 = \ln x$

9.4 (*a*) $8 = 2^3$ (*b*) $9 = x^4$ (*c*) $x = 10^{-2}$ (*d*) $5 = e^{x+1}$

9.5 (*a*) 3 (*b*) 2 (*c*) $\frac{3}{2}$ (*d*) $\frac{2}{3}$ (*e*) $\frac{-1}{2}$ (*f*) −2

9.6 (*a*) $x = 4$ (*b*) $x = 2$ (*c*) $x = \frac{1}{2}$

(*d*) $x = 5^{\frac{3}{2}} = \sqrt{125} = 5\sqrt{5}$ (*e*) $x = 64$ (*f*) $x = \frac{1}{2}$

(*g*) $x = \sqrt[3]{\frac{8}{5}} = \frac{2}{\sqrt[3]{5}} = \frac{2\sqrt[3]{25}}{5}$ (*h*) $x = \frac{2}{3}$ (*i*) $x = -2 + \sqrt[3]{5}$

9.7 (*a*) 3.37383 (*b*) 7.76853 (*c*) −6.93704

(*d*) 1.38629 (*e*) 0.613800 (*f*) 6.68586

9.8 (*a*) See Figure 9.23. (*b*) See Figure 9.24. (*c*) See Figure 9.25. (*d*) See Figure 9.26.

(*e*) See Figure 9.27. (*f*) See Figure 9.28. (*g*) See Figure 9.29. (*h*) See Figure 9.30.

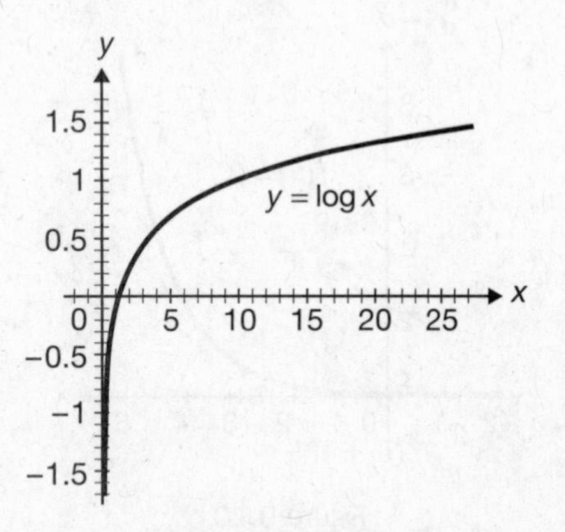

Figure 9.23

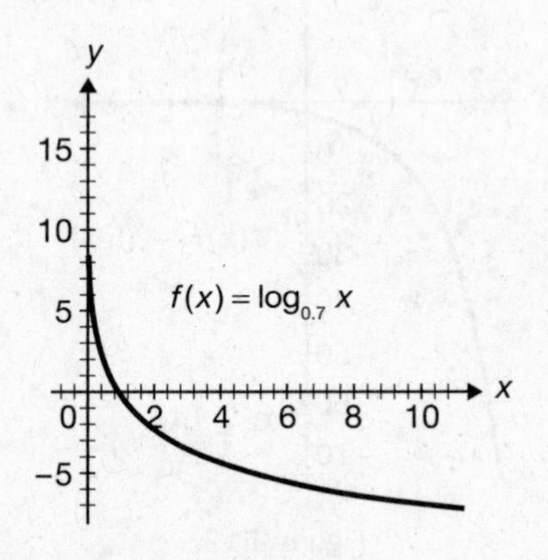

Figure 9.24

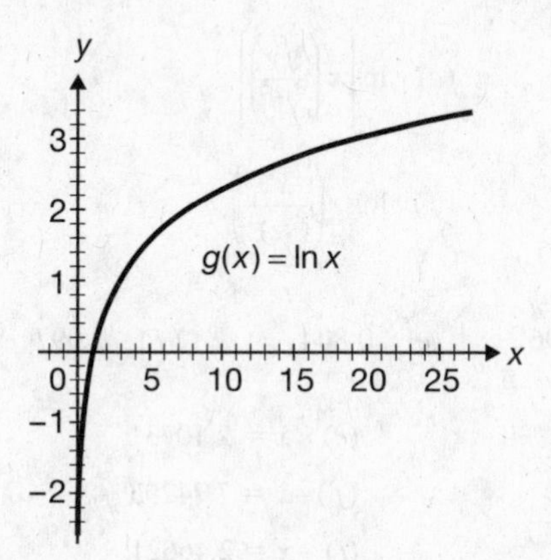

Figure 9.25

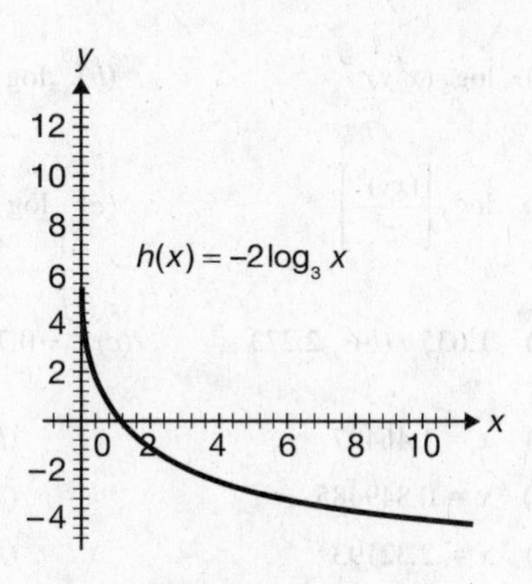

Figure 9.26

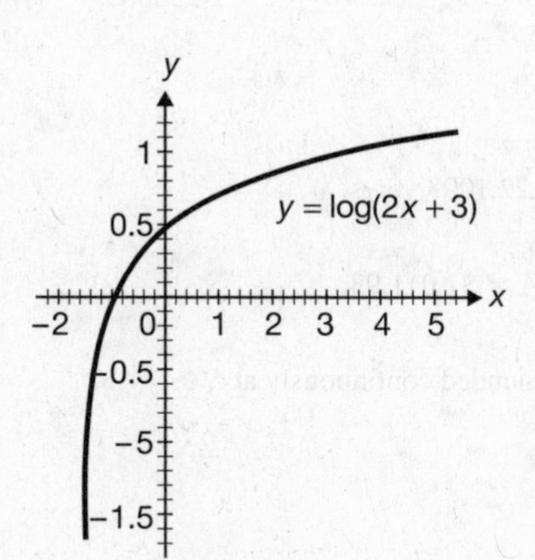

Figure 9.27

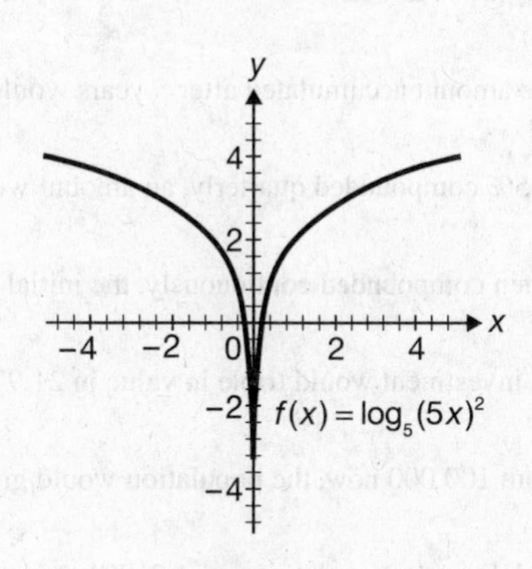

Figure 9.28

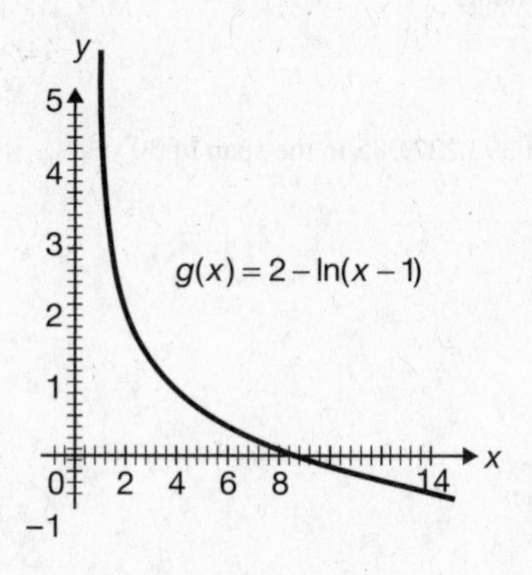

Figure 9.29

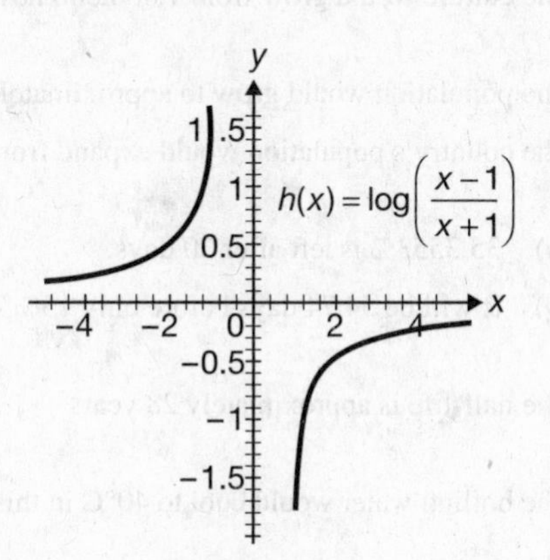

Figure 9.30

9.9 (*a*) 3.54249 (*b*) 1.43068 (*c*) 0.63093
(*d*) 2.58406 (*e*) 1.66934 (*f*) 4.22198

9.10 (*a*) $1 + \log_2 x + \log_2 z$ (*b*) $2 \log_5 x + \log_5 y$ (*c*) $\frac{1}{2}(\log y + \log z)$
(*d*) $\log_7 x + 4 \log_7 y - 6 \log_7 z$ (*e*) $\frac{1}{2} \ln x - 3 \ln y - \ln z$ (*f*) $\frac{1}{5}(3 \log_4 x + \log_4 y - 2 \log_4 z)$

9.11 (*a*) $\log_2 (x^2 y)$ (*b*) $\log_3 \left[x(y+1)^3\right]$ (*c*) $\ln \left[x\left(\frac{\sqrt[3]{y}}{z^2}\right)\right]$

(*d*) $\log_7 \left[\frac{(xy)^8}{z^3}\right]$ (*e*) $\log \left[\left(\sqrt[3]{\left(\frac{x}{y}\right)}\right)^2 z^5\right]$ (*f*) $\log_5 \left(\frac{x^2}{(yz)^3}\right)$

9.12 (*a*) 1.635 (*b*) 2.273 (*c*) −0.721 (*d*) −2.306 (*e*) 0.804 (*f*) 2.097

9.13 (*a*) $x = 1.46497$ (*b*) $x = 2.86135$ (*c*) $x = 2.10731$

(*d*) $x = 0.849485$ (*e*) $x = 1.29248$ (*f*) $x = 7.74293$

(*g*) $x = 2.32193$ (*h*) $x = 3.16993$ (*i*) $x = 2.46621$

(*j*) $x = 0.0539949$ (*k*) $x = 1.25000$ (*l*) $x = 45.0000$

(*m*) $x = 1.21802$ (*n*) $x = 56.0000$ (*o*) $x = 2.25000$

9.14 The amount accumulated after 5 years would be A = \$1,011.46.

9.15 At 5% compounded quarterly, an amount would triple in value in 22.1093 years.

9.16 When compounded continuously, the initial \$800 would grow to A = \$1,011.93.

9.17 An investment would triple in value in 21.9722 years when compounded continuously at 5%.

9.18 From 100,000 now, the population would grow to 117,738.

9.19 Each hour the population was 1.21901 times greater than the hour before. That is equivalent to a 21.901% growth rate.

9.20 The culture would grow from 1,000,000 now to 27,857,617 in 12 hours.

9.21 The population would grow to approximately 77,819,129.
The country's population would expand from 65,000,000 to about 393,227,085 in the span of 30 years.

9.22 (*a*) 35.3553% is left after 30 days.
(*b*) It will be 54.74 days before only 15% is left.

9.23 The half-life is approximately 28 years.

9.24 The boiling water would cool to 40°C in this room in 27.85 minutes.

9.25 The atmospheric pressure at 4000 feet is $P \approx 12.54$ pounds per square inch.

9.26 The pH ≈ 7.5; the solution is alkaline (or basic).

9.27 The 7.6 earthquake was about 12.6 times more powerful than the 6.5 earthquake.

Sequences, Series, and the Binomial Theorem

10.1 Sequences

A *sequence* is an ordered list of numbers. The numbers 3, 6, 9, 12 form one sequence, while 3, 9, 6, 12 form another. The stated sequences are different since their order is different. The *terms* of a sequence consist of the expressions separated by commas. The stated sequences have four terms. A *finite sequence* has a last term. An *infinite sequence* or simply a *sequence* has no last term. The ordered list 3, 6, 9, 12,... is an infinite sequence. The ellipsis symbol, "...", means the terms of the sequence continue in the pattern indicated without end. The list has an infinite number of terms. An ellipsis is sometimes used to represent a finite number of omitted terms also.

The terms of a sequence often follow a particular pattern. In those instances, we can determine the general term that expresses every term of the sequence. For example,

Sequence	General term
$3, 6, 9, 12, \ldots$	$3n$
$1, 3, 5, 7, \ldots$	$2n - 1$
$2, 4, 8, 16, \ldots$	2^n

The variable n represents a positive integer. The first term of the sequence is obtained when $n = 1$, the second term is obtained when $n = 2$, and so on. The general term of a sequence specifies a function that produces the sequence when evaluated at the natural numbers. In other words, a sequence is a function whose domain is the set of natural numbers and range is some subset of real numbers.

It is customary to use a_n to represent the *nth* term or general term of a sequence. Thus, a_1 is the first term, a_2 is the second term, and so on. The entire sequence 3, 6, 9, 12, ... is represented by $a_n = 3n$. We simply replace n by 1,2,3, ... in $3n$ to obtain the successive terms of the sequence.

See solved problem 10.1.

An *arithmetic sequence* or *arithmetic progression* is a sequence such that successive terms differ by the same constant. The constant difference is represented by d and is given by $d = a_{i+1} - a_i$ for all positive integers i. The sequence 2,5,8,11, ... is an arithmetic sequence. Successive terms differ by 3 that is $d = 3$.

There is an explicit formula for the *nth* term of an arithmetic sequence in general.

Arithmetic Sequence

The *nth* term of an arithmetic sequence with common difference d is

$$a_n = a_1 + (n - 1)\,d.$$

See solved problem 10.2.

A *geometric sequence* or *geometric progression* is a sequence such that each successive term is obtained by multiplying a constant times the previous term. Equivalently, the quotient (ratio) of successive terms is the same constant. This constant is called the *common ratio,* and is represented by r. The common ratio is given by $r = a_{i+1}/a_i$ for all positive integers i and $r \neq 0$. The sequence 3, 9, 27, 81, ... is a geometric sequence with common ratio $r = 3$.

The formula for the *nth* term of a geometric sequence is given below.

Geometric Sequence

The *nth* term of a geometric sequence with common ratio r is given by

$$a_n = a_1 r^{n-1}$$

See solved problem 10.3.

10.2 Series

Consider the arithmetic sequence given by 2, 5, 8, ..., $3n - 1$, ... We add successive terms to generate a sequence of partial sums. We employ S_n to represent partial sums. The *nth* partial sum of an arithmetic sequence, S_n, is the sum of the first n terms of the sequence.

$$S_1 = 2$$
$$S_2 = 2 + 5 = 7$$
$$S_3 = 2 + 5 + 8 = 15$$
$$S_4 = 2 + 5 + 8 + 11 = 26$$
$$\vdots$$
$$S_n = 2 + 5 + 8 + \cdots + 3n - 1.$$

There is a formula for the *nth* partial sum of an arithmetic sequence. It is shown below.

Formula for the *nth* Partial Sum of an Arithmetic Sequence

$$S_n = \frac{n}{2}(a_1 + a_n)$$

If a_n is replaced by its equivalent $a_1 + (n - 1)\,d$, an alternate formula is obtained.

Alternate Formula for the *nth* Partial Sum of an Arithmetic Sequence

$$S_n = \frac{n\,[2a_1 + (n - 1)\,d]}{2}$$

See solved problems 10.4–10.5.

The arithmetic sequence 1, 2, 3, ..., n is a sequence with common difference $d = 1$, $a_1 = 1$, and $a_n = n$. The sequence is simply the first n positive integers. A formula for the sum of the first n positive integers or counting numbers can be found using the formula for the *nth* partial sum of an arithmetic sequence. It is $S_n = (n/2)\,(a_1 + a_n) = (n/2)\,(1 + n) = [n(n + 1)]/2$. This useful result is restated below.

The sum of the first n positive integers is

$$S_n = \frac{n\,(n + 1)}{2}.$$

There is a convenient notation that is used for partial sums. It is called the *summation notation*. The sum of the first n terms of a sequence a_n, is represented by

$$S_n = \sum_{i=1}^{n} a_i = a_1 + a_2 + a_3 + \cdots + a_n.$$

In the above notation, the letter i (other letters are employed also) is called the *index of the summation*; 1 is the *lower limit of the summation; n* is the *upper limit of the summation*. The "Σ" symbol is the Greek letter sigma; it tells us to find a sum. The sum of the terms of a finite sequence is called a *finite series*.

The notation above means to find the sum beginning with the term obtained when $i = 1$. We then increase i by one each time to obtain the successive terms in the sum until $i = n$ for the last term. The terms obtained are then added.

See solved problem 10.6.

It is possible to find a general formula for the *nth* partial sum of a geometric sequence also. The proof of the formula is generally left to the study of College Algebra.

Formula for the *nth* Partial Sum of a Geometric Sequence

$$S_n = \frac{a_1(1-r^n)}{1-r} = \frac{a_1(r^n-1)}{r-1}, r \neq 1$$

Use the latter form if $r > 1$ in order to avoid negative numerators and denominators.

See solved problem 10.7.

Consider the geometric sequence $2, \frac{2}{3}, \frac{2}{3^2}, \frac{2}{3^3}, \ldots, \frac{2}{3^{n-1}}, \ldots$. By inspection we observe that $r = \frac{1}{3}$. We wish to find the sum of all of the terms in the sequence. Is it possible to add the infinitely many terms in the sequence? We shall present a convincing (we hope) argument below. We evaluate S_n and r^n for $n = 3, 6, 9$, and 12 and display the results in the table below.

n	S_n	r^n
3	2.888888889	0.037037037
6	2.995884774	0.001371742
9	2.999847584	0.000050805
12	2.999994355	0.000001882

Observe that as n increases, S_n is closer to 3 and r^n is closer to 0. If we calculate more values for S_n and r^n as n gets larger and larger, the pattern continues. That is, S_n is closer and closer to 3 and r^n is closer and closer to 0. It seems reasonable to conclude that the sum of all the terms in the sequence is 3. The result is called a "limit." The concept of a limit is used extensively in higher mathematics courses.

The sum of the terms of an infinite geometric sequence is called an *infinite geometric series*. In general, the sum of all the terms in an infinite sequence, geometric or not, is called an *infinite series*.

The formula for the *nth* partial sum of a geometric sequence was given as $S_n = \frac{a_1(1-r^n)}{1-r}$. It can be shown using calculus that r^n approaches 0 as n gets larger and larger if $|r| < 1$. In that case $S_n = \frac{a_1(1-r^n)}{1-r} \to \frac{a_1(1-0)}{1-r} = \frac{a_1}{1-r}$. We restate this important result below.

Sum of an Infinite Geometric Sequence

If a_n is a geometric sequence with first term a_1 and $|r| < 1$, the sum of all the terms S is given by

$$S = \frac{a_1}{1-r}.$$

We must emphasize the fact that the formula above is <u>not</u> applicable for $|r| \geq 1$. If $|r| \geq 1$, a geometric series has no finite sum.

The sigma notation may be employed to represent an infinite series. The symbolism is $\sum_{i=1}^{\infty} a_i$ .We use the infinity symbol, ∞, for the upper limit on the summation to indicate infinitely many terms are to be added.

See solved problem 10.8.

10.3 The Binomial Theorem

There are circumstances in mathematics in which $(a + b)^n$ is written as the sum of its terms. The process employed is called *expanding the binomial* or *writing the binomial in expanded form*. We now apply the special product forms introduced in Chapter 2 as well as the distributive property to obtain powers of $a + b$ for various n. We are searching for patterns which will be helpful in the future. The following array is obtained after a certain amount of effort.

$n = 0$	$(a+b)^0$	1
$n = 1$	$(a+b)^1$	$a+b$
$n = 2$	$(a+b)^2$	$a^2 + 2ab + b^2$
$n = 3$	$(a+b)^3$	$a^3 + 3a^2b + 3ab^2 + b^3$
$n = 4$	$(a+b)^4$	$a^4 + 4a^3b + 6a^2b^2 + 4ab^3 + b^4$
$n = 5$	$(a+b)^5$	$a^5 + 5a^4b + 10a^3b^2 + 10a^2b^3 + 5ab^4 + b^5$

Observe the variable parts in each expansion for $n = 1, 2, 3, 4,$ and 5.

1. The first term is a^n. The exponent on a decreases by 1 in successive terms.
2. The exponent on b increases by 1 in successive terms. The last term is b^n.
3. The sum of the exponents in each term is n.

Now take note of the numerical coefficients in each expansion. The following array of coefficients is obtained by omitting the variable factors in each term.

$n = 0$	1
$n = 1$	1 1
$n = 2$	1 2 1
$n = 3$	1 3 3 1
$n = 4$	1 4 6 4 1
$n = 5$	1 5 10 10 5 1

The triangular array displayed above is called *Pascal's Triangle*. It is named in honor of Blaise Pascal, a seventeenth-century mathematician and philosopher.

The following patterns in the triangular array can be identified.

1. The first and last coefficient in each row is 1.
2. The coefficients are symmetric with respect to the middle of each row.
3. Each interior coefficient is the sum of the two coefficients above it in the preceding row.

We can now determine the coefficients of the expansion of $(a + b)^n$ for $n = 6$ and $n = 7$. The corresponding rows in Pascal's Triangle are shown below.

$n = 6$		1		6		15		20		15		6		1	
$n = 7$	1		7		21		35		35		21		7		1

See solved problem 10.9.

Pascal's Triangle is useful if n is rather small. Its use is not practical for large n. We shall introduce a more practical method for finding the coefficients regardless of the magnitude of n.

We first need the concept of "factorials." A factorial of a number is simply symbolism that represents a particular extended product.

Definition 1. If n is a positive integer, $n! = n(n - 1)(n - 2) \ldots (3)(2)(1)$.

The $n!$ symbol is read "n factorial." The $n!$ symbol represents the product of all positive integers less than or equal to n.

Definition 2. $0! = 1$.

The $0! = 1$ definition seems arbitrary and illogical, although it will be more apparent subsequently that the definition has merit and is needed for consistency.

See solved problem 10.10.

Definition 3. $\binom{n}{k}$ means $\dfrac{n!}{k!(n-k)!}$ for $n \geq k$.

See solved problem 10.11.

Some calculators have factorial functions on them. Look for a key marked $n!$ or $x!$. Your calculator may also possess the capability to evaluate $\binom{n}{k}$. Refer to your owner's manual for the appropriate technique. Be aware that ${}_nC_k$, ${}^nC_k$, and C^n_k are alternative symbolisms for the same concept. That is, ${}_nC_k = {}^nC_k = C^n_k = \binom{n}{k}$.

We can now use the factorial symbolism to find the numerical coefficients in the expansion of $(a + b)^n$. These coefficients are called the *binomial coefficients*.

Formula for Binomial Coefficients

In the expansion of $(a + b)^n$, the term containing $a^{n-k}b^k$ has coefficient

$$\binom{n}{k} = \frac{n!}{k!(n-k)!}$$

for nonnegative integers n and k and $n \geq k$.

Some useful properties of binomial coefficients follow.

Properties of Binomial Coefficients

For nonnegative integers n and k and $n \geq k$,

1. $\binom{n}{0} = 1$

2. $\binom{n}{n} = 1$

3. $\binom{n}{k} = \binom{n}{n-k}$

Properties 1 and 2 were illustrated in solved problem 10.10 (*c*) and (*d*) above.

We can now state the formula for the expansion of $(a + b)^n$. It is called the *Binomial Theorem.*

Binomial Theorem

If n is a positive integer, $(a+b)^n = \sum_{k=0}^{n} \binom{n}{k} a^{n-k} b^k$

$$= \binom{n}{0}a^n + \binom{n}{1}a^{n-1}b + \binom{n}{2}a^{n-2}b^2 + \binom{n}{3}a^{n-3}b^3 + \cdots + \binom{n}{n-1}ab^{n-1} + \binom{n}{n}b^n$$

See solved problem 10.12.

SOLVED PROBLEMS

10.1 Write the first four terms of the indicated sequences.

(*a*) $a_n = 3n - 1$

$$a_1 = 3 \cdot 1 - 1 = 2$$
$$a_2 = 3 \cdot 2 - 1 = 5$$
$$a_3 = 3 \cdot 3 - 1 = 8$$
$$a_4 = 3 \cdot 4 - 1 = 11$$

(*b*) $a_n = 3^n$

$$a_1 = 3^1 = 3$$
$$a_2 = 3^2 = 9$$
$$a_3 = 3^3 = 27$$
$$a_4 = 3^4 = 81$$

(*c*) $a_n = \dfrac{n}{n+1}$

$$a_1 = \frac{1}{1+1} = \frac{1}{2}$$
$$a_2 = \frac{2}{2+1} = \frac{2}{3}$$
$$a_3 = \frac{3}{3+1} = \frac{3}{4}$$
$$a_4 = \frac{4}{4+1} = \frac{4}{5}$$

(*d*) $a_n = (-1)^{n+1}\sqrt{2n}$

$$a_1 = (-1)^{1+1}\sqrt{2 \cdot 1} = (-1)^2\sqrt{2} = \sqrt{2}$$
$$a_2 = (-1)^{2+1}\sqrt{2 \cdot 2} = (-1)^3\sqrt{4} = -2$$
$$a_3 = (-1)^{3+1}\sqrt{2 \cdot 3} = (-1)^4\sqrt{6} = \sqrt{6}$$
$$a_4 = (-1)^{4+1}\sqrt{2 \cdot 4} = (-1)^5\sqrt{8} = -2\sqrt{2}$$

10.2 Find the *nth* term, a_n, and the 20*th* term, a_{20}, of the following arithmetic sequences.

(*a*) 6,11,16, ...

The common difference $d = a_2 - a_1 = 11 - 6 = 5$. Therefore,

$$a_n = a_1 + (n-1)d = 6 + (n-1)5 = 6 + 5n - 5 = 5n + 1.$$
$$a_n = 5n + 1 \text{ so } a_{20} = 5 \cdot 20 + 1 = 100 + 1 = 101.$$

(*b*) 14, 11, 8, ...

The common difference $d = a_2 - a_1 = 11 - 14 = -3$. Therefore,

$$a_n = a_1 + (n-1)d = 14 + (n-1)(-3) = 14 - 3n + 3 = 17 - 3n.$$
$$a_n = 17 - 3n \text{ so } a_{20} = 17 - 3 \cdot 20 = 17 - 60 = -43.$$

(*c*) $\frac{3}{2}, 1, \frac{1}{2}, \ldots$

The common difference $d = a_2 - a_1 = 1 - \frac{3}{2} = \frac{-1}{2}$ Therefore

$$a_n = a_1 + (n-1)d = \frac{3}{2} + (n-1)\left(\frac{-1}{2}\right) = \frac{3}{2} + \frac{-1}{2}n + \frac{1}{2} = 2 - n/2 = (4-n)/2.$$
$$a_n = (4-n)/2 \text{ so } a_{20} = (4-20)/2 = -16/2 = -8.$$

See supplementary problem 10.1.

10.3 Find the *nth* term, a_n, and the 10*th* term, a_{10}, of the following geometric sequences.

(*a*) 5, 10, 20, ...

The common ratio $r = \frac{a_2}{a_1} = \frac{10}{5} = 2$. Therefore,

$$a_n = a_1 r^{n-1} = 5 \cdot 2^{n-1}. \text{Since } a_n = 5 \cdot 2^{n-1},$$
$$a_{10} = 5 \cdot 2^{10-1} = 5 \cdot 2^9 = 2{,}560.$$

(*b*) $1, \frac{1}{2}, \frac{1}{4}, \ldots$

The common ratio $r = \frac{a_2}{a_1} = \frac{\frac{1}{2}}{1} = \frac{1}{2}$. Therefore,

$$a_n = a_1 r^{n-1} = 1\left(\frac{1}{2}\right)^{n-1} = \left(\frac{1}{2}\right)^{n-1}. \text{Since } a_n = \left(\frac{1}{2}\right)^{n-1},$$
$$a_{10} = \left(\frac{1}{2}\right)^{n-1} = \left(\frac{1}{2}\right)^{9} = \frac{1^9}{2^9} = \frac{1}{2^9} \approx 1.95313 \times 10^{-3}.$$

(*c*) $3, -2, \frac{4}{3}, \ldots$

The common ratio $r = \frac{a_2}{a_1} = \frac{-2}{3}$. Therefore,

$$a_n = a_1 r^{n-1} = 3\left(\frac{-2}{3}\right)^{n-1} = 3 \cdot \frac{(-2)^{n-1}}{3^{n-1}} = \frac{(-2)^{n-1}}{3^{n-2}}, \text{Since } a_n = \frac{(-2)^{n-1}}{3^{n-2}},$$
$$a_{10} = \frac{(-2)^{n-1}}{3^{n-2}} = \frac{(-2)^{10-1}}{3^{10-2}} = \frac{(-2)^9}{3^8} = \frac{(-1 \cdot 2)^9}{3^8} = \frac{(-1)^9 2^9}{3^8} = \frac{-2^9}{3^8} = \frac{-512}{6{,}561} \approx -0.078037$$

(*d*) $1, -x^2, x^4, \ldots$

The common ratio $r = \frac{a_2}{a_1} = \frac{-x^2}{1} = -x^2$. Therefore,

$a_n = a_1 r^{n-1} = 1 \cdot (-x^2)^{n-1} = (-1 \cdot x^2)^{n-1} = (-1)^{n-1}(x^2)^{n-1} = (-1)^{n-1} x^{2n-2}$.
Since $a_n = (-1)^{n-1} x^{2n-2}$, $a_{10} = (-1)^{10-1} x^{2(10)-2} = (-1)^9 x^{18} = -x^{18}$.

See supplementary problem 10.2.

10.4 Verify that S_{10} is the same for $2, 5, 8, \ldots, 3n - 1, \ldots$ when either formula is applied.

We use $S_n = (n/2)(a_1 + a_n)$ first. We are seeking S_{10} so we need a_1 and a_{10} in order to proceed. $a_1 = 2$ and since $a_n = 3n - 1$, we find $a_{10} = 3 \cdot 10 - 1 = 29$. Hence $S_{10} = \frac{10}{2}(2 + 29) = 5\,(31) = 155$.

Now use $S_n = \dfrac{n\,[2a_1 + (n-1)\,d]}{2}$. We need $d = a_2 - a_1 = 5 - 2 = 3$. Therefore, S_{10}

$$= \frac{10\,[2 \cdot 2 + (10-1)\,3]}{2} = \frac{10\,[4 + 27]}{2} = 5[31] = 155.$$

In general, the preferred formula is determined by the known information. That is, if a_1, a_n, and n are known or can readily be found, we use $S_n = (n/2)(a_1 + a_n)$. On the other hand, if a_1, n, and d are known or can readily be found, use $S_n = \dfrac{n\,[2a_1 + (n-1)\,d]}{2}$.

10.5 Find the *nth* partial sum, S_n, of each arithmetic sequence for the given n.

(*a*) $a_n = 6n$; $n = 25$

Solution 1: $a_n = 6n$ so $a_1 = 6 \cdot 1 = 6$ and $a_{25} = 6 \cdot 25 = 150$. Since $S_n = (n/2)(a_1 + a_n)$,

$$S_{25} = \frac{25}{2}(6 + 150) = \frac{25}{2}(156) = 25\,(78) = 1{,}950.$$

Solution 2: $a_n = 6n$ so $a_1 = 6 \cdot 1 = 6$ and $a_2 = 6 \cdot 2 = 12$ so $d = a_2 - a_1 = 12 - 6 = 6$. Since

$$S_n = \frac{n\,[2a_1 + (n-1)\,d]}{2}, S_{25} = \frac{25\,[2 \cdot 6 + (25-1)\,6]}{2} = \frac{25\,[12 + (24)\,6]}{2} = \frac{25\,[12 + 144]}{2}$$
$$= \frac{25\,[156]}{2} = 25\,[78] = 1{,}950.$$

(*b*) $a_n = 3n - 2$; $n = 50$

Since $S_n = (n/2)(a_1 + a_n)$, we need a_1 and a_{50}. It is given that $a_n = 3n - 2$, so $a_1 = 3 \cdot 1 - 2 = 1$ and $a_{50} = 3 \cdot 50 - 2 = 148$. Therefore $S_{50} = (50/2)(1 + 148) = 25\,(149) = 3{,}725$.

(*c*) $a_n = n$; $n = 100$

Proceed as in part (*b*). $a_n = n$ so $a_1 = 1$ and $a_{100} = 100$. $S_n = (n/2)(a_1 + a_n)$ so $S_{100} = (100/2)(1 + 100) = 50\,(101) = 5{,}050$. The result is the sum of the first 100 positive integers.

(*d*) $a_n = \dfrac{1-2n}{n}$; $n = 100$

Proceed as in part (*b*) $a_n = \dfrac{1-2n}{n}$ so $a_1 = \dfrac{1 - 2 \cdot 1}{1} = \dfrac{-1}{1} = -1$ and $a_{100} = \dfrac{1 - 2 \cdot 100}{100}$

$$= \frac{1 - 200}{100} = \frac{-199}{100}. S_n = \frac{n}{2}(a_1 + a_n) \text{ so } S_{100} = \frac{100}{2}\left(-1 + \frac{-199}{100}\right)$$
$$= 50\left(\frac{-100 - 199}{100}\right) = 50\left(\frac{-299}{100}\right) = \frac{-299}{2}.$$

See supplementary problem 10.3.

10.6 Evaluate the following.

(*a*) $\sum_{i=1}^{5} 2i$

$$\sum_{i=1}^{5} 2i = 2\cdot 1 + 2\cdot 2 + 2\cdot 3 + 2\cdot 4 + 2\cdot 5 = 2+4+6+8+10 = 30.$$

(*b*) $\sum_{i=3}^{6} (2i-1)$ Note that the lower limit is 3 and not 1 here.

$$\sum_{i=3}^{6}(2i-1) = (2\cdot 3-1)+(2\cdot 4-1)+(2\cdot 5-1)+(2\cdot 6-1) = 5+7+9+11 = 32.$$

(*c*) $\sum_{k=0}^{4} k^2 x$

$$\sum_{k=0}^{4} k^2 x = 0^2x + 1^2x + 2^2x + 3^2x + 4^2x = 0x + 1x + 4x + 9x + 16x = 30x.$$

(*d*) $\sum_{i=1}^{3} \frac{1}{i(i+1)}$

$$\sum_{i=1}^{3} \frac{1}{i(i+1)} = \frac{1}{1(1+1)} + \frac{1}{2(2+1)} + \frac{1}{3(3+1)} = \frac{1}{2} + \frac{1}{6} + \frac{1}{12} = \frac{6+2+1}{12} = \frac{9}{12} = \frac{3}{4}.$$

Refer to supplementary problem 10.4.

10.7 Find the *nth* partial sum of the following geometric sequences for the given n.

(*a*) $a_n = 5(2)^{n-1}$; $n = 5$

We need a_1, r and n. $a_1 = 5(2)^{1-1} = 5(1) = 5$; $r = \frac{a_2}{a_1} = \frac{5(2)^{2-1}}{5} = 2$; and n is given as 5.

We know that $S_n = \frac{a_1(r^n - 1)}{r-1}$ so $S_5 = \frac{5(2^5-1)}{2-1} = \frac{5(32-1)}{1} = 5(31) = 155.$

(*b*) $\sum_{i=1}^{14} \frac{1}{4}\left(\frac{2}{3}\right)^i = \frac{1}{4}\left(\frac{2}{3}\right)^1 + \frac{1}{4}\left(\frac{2}{3}\right)^2 + \frac{1}{4}\left(\frac{2}{3}\right)^3 + \frac{1}{4}\left(\frac{2}{3}\right)^4 + \cdots$

When $i = 1$, $a_1 = \frac{1}{4}\left(\frac{2}{3}\right)^1 = \frac{1}{6}$ and if $i = 2$, $a_2 = \frac{1}{4}\left(\frac{2}{3}\right)^2 = \frac{1}{9}$. Therefore, $r = a_2/a_1$

$$= \left(\frac{1}{9}\right)/\left(\frac{1}{6}\right) = \frac{1}{9}\cdot\frac{6}{1} = \frac{6}{9} = \frac{2}{3}. \text{Since } S_n = \frac{a_1(1-r^n)}{1-r}, S_{14} = \frac{\frac{1}{6}\left[1-\left(\frac{2}{3}\right)^{14}\right]}{1-\frac{2}{3}}$$

$\approx 0.498287256.$ Approximate values are usually satisfactory when the simplification is cumbersome.

(*c*) $\sum_{k=1}^{10} (-2)^{k-1} = (-2)^0 + (-2)^1 + (-2)^2 + (-2)^3 + \cdots = 1 - 2 + 4 - 8 + \cdots$

$$r = \frac{a_2}{a_1} = \frac{-2}{1} = -2. \text{Since } S_n = \frac{a_1(r^n - 1)}{r-1}, S_{10} = \frac{1[(-2)^{10} - 1]}{-2-1}$$

$$= \frac{(-2)^{10} - 1}{-3} = \frac{1024-1}{-3} = \frac{1023}{-3} = -341.$$

10.8 Find the sum of the following geometric series.

(*a*) $\sum_{n=1}^{\infty} (0.9)^n = (0.9) + (0.9)^2 + (0.9)^3 + \cdots$

The series has first term $a_1 = 0.9$ and $r = 0.9$. Observe that $|r| = |0.9| = 0.9 < 1$. Therefore, the formula $S = \frac{a_1}{1-r}$ is applicable. $S = \frac{a_1}{1-r} = \frac{0.9}{1-0.9} = \frac{0.9}{0.1} = 9$. Hence, $\sum_{n=1}^{\infty}(0.9)^n = 9$.

(*b*) $\sum_{i=1}^{\infty}\left(\frac{2}{5}\right)^{i-1} = 1 + \frac{2}{5} + \left(\frac{2}{5}\right)^2 + \left(\frac{2}{5}\right)^3 + \cdots$

It is apparent that the series has $a_1 = 1$ and $r = \frac{2}{5}$. Also $|r| = \left|\frac{2}{5}\right| = \frac{2}{5} < 1$ so the formula $S = \frac{a_1}{1-r}$ may be employed. Therefore, $S = \frac{a_1}{1-r} = \frac{1}{1-\frac{2}{5}} = \frac{1}{\frac{3}{5}} = \frac{5}{3}$. We write $\sum_{i=1}^{\infty}\left(\frac{2}{5}\right)^{i-1} = \frac{5}{3}$.

(*c*) $\sum_{i=1}^{\infty}\left(\frac{-2}{5}\right)^{i-1} = 1 + \left(\frac{-2}{5}\right)^1 + \left(\frac{-2}{5}\right)^2 + \left(\frac{-2}{5}\right)^3 + \cdots$

The series is similar to the series in part (*b*) above. $a_1 = 1$ and $r = \frac{-2}{5}$. Also $|r| = \left|\frac{-2}{5}\right| = \frac{2}{5} < 1$ so $S = \frac{a_1}{1-r}$ may be employed. Hence, $S = \frac{a_1}{1-r} = \frac{1}{1-\left(\frac{-2}{5}\right)} = \frac{1}{1+\frac{2}{5}} = \frac{1}{\frac{7}{5}} = \frac{5}{7}$.

Therefore, $\sum_{i=1}^{\infty}\left(\frac{-2}{5}\right)^{i-1} = \frac{5}{7}$.

See supplementary problem 10.5 for similar problems.

10.9 Use Pascal's Triangle to determine the coefficients and write the expansion of the following.

(*a*) $(s + t)^6$

$(s + t)^6 = s^6 + 6s^5t + 15s^4t^2 + 20s^3t^3 + 15s^2t^4 + 6st^5 + t^6$

(*b*) $(x - y)^5$

$(x - y)^5$ can be written as $[x + (-y)]^5$ so $(x - y)^5 = x^5 + 5x^4(-y) + 10x^3(-y)^2 + 10x^2(-y)^3 + 5x(-y)^4 + (-y)^5 = x^5 - 5x^4y + 10x^3y^2 - 10x^2y^3 + 5xy^4 - y^5$

(*c*) $(2a + 3b)^4$

$(2a + 3b)^4 = [(2a) + (3b)]^4 = (2a)^4 + 4(2a)^3(3b) + 6(2a)^2(3b)^2 + 4(2a)(3b)^3 + (3b)^4 = 2^4a^4 + 4 \cdot 2^3a^3(3b) + 6 \cdot 2^2a^23^2b^2 + 4 \cdot 2a3^3b^3 + 3^4b^4 = 16a^4 + 96a^3b + 216a^2b^2 + 216ab^3 + 81b^4$

10.10 Evaluate the following.

(*a*) 6!

$6! = 6 \cdot 5 \cdot 4 \cdot 3 \cdot 2 \cdot 1 = 720$

(*b*) 8!

$8! = 8 \cdot 7 \cdot 6 \cdot 5 \cdot 4 \cdot 3 \cdot 2 \cdot 1 = 40{,}320$

(*c*) $6 \cdot 5 \cdot 4!$

$6 \cdot 5 \cdot 4! = 6 \cdot 5 \cdot 4 \cdot 3 \cdot 2 \cdot 1 = 720$

Part (*c*) above illustrates that the factorial process can be terminated with "!" at any point by starting at n and decreasing each factor by 1. That is the reason we normally start at n rather than 1. The simplification of expressions such as those encountered in parts (*d*) and (*f*) below, as well as in solved problem 10.11, is facilitated through the use of this technique.

(*d*) $\dfrac{8!}{5!}$

$$\frac{8!}{5!}=\frac{8\cdot 7\cdot 6\cdot 5!}{5!}=8\cdot 7\cdot 6=336$$

(*e*) $(6-1)!$

$$(6-1)!=5!=5\cdot 4\cdot 3\cdot 2\cdot 1=120$$

(*f*) $\dfrac{7!}{5!2!}$

$$\frac{7!}{5!2!}=\frac{7\cdot 6\cdot 5!}{5!2!}=\frac{7\cdot 6}{2\cdot 1}=7\cdot 3=21$$

10.11 Evaluate.

(*a*) $\dbinom{6}{4}$

$$\binom{6}{4}=\frac{6!}{4!(6-4)!}=\frac{6\cdot 5\cdot 4!}{4!2!}=\frac{6\cdot 5}{2\cdot 1}=15$$

(*b*) $\dbinom{12}{8}$

$$\binom{12}{8}=\frac{12!}{8!(12-8)!}=\frac{12\cdot 11\cdot 10\cdot 9\cdot 8!}{8!4!}=\frac{12\cdot 11\cdot 10\cdot 9}{4\cdot 3\cdot 2\cdot 1}=495$$

(*c*) $\dbinom{7}{7}$

$$\binom{7}{7}=\frac{7!}{7!(7-7)!}=\frac{7!}{7!0!}=\frac{1}{0!}=\frac{1}{1}=1$$

(*d*) $\dbinom{7}{0}$

$$\binom{7}{0}=\frac{7!}{0!(7-0)!}=\frac{7!}{0!7!}=\frac{1}{1}=1$$

See supplementary problem 10.6.

10.12 Use the binomial theorem to expand the following:

(*a*) $(x+y)^7$

$$(x+y)^7=\binom{7}{0}x^7+\binom{7}{1}x^6y+\binom{7}{2}x^5y^2+\binom{7}{3}x^4y^3+\binom{7}{4}x^3y^4+\binom{7}{5}x^2y^5+\binom{7}{6}xy^6$$
$$+\binom{7}{0}y^7=\frac{7!}{0!7!}x^7+\frac{7!}{1!6!}x^6y+\frac{7!}{2!5!}x^5y^2+\frac{7!}{3!4!}x^4y^3+\frac{7!}{4!3!}x^3y^4+\frac{7!}{5!2!}x^2y^5$$
$$+\frac{7!}{6!1!}xy^6+\frac{7!}{7!0!}y^7=x^7+7x^6y+21x^5y^2+35x^4y^3+35x^3y^4+21x^2y^5+7xy^6+y^7$$

(*b*) $(3y+1)^6$

$$(3y+1)^6=\binom{6}{0}(3y)^6+\binom{6}{1}(3y)^5 1+\binom{6}{2}(3y)^4 1^2+\binom{6}{3}(3y)^3 1^3+\binom{6}{4}(3y)^2 1^4$$

$$+\binom{6}{5}(3y)1^5+\binom{6}{6}1^6=\frac{6!}{0!6!}(3y)^6+\frac{6!}{1!5!}(3y)^5\,1+\frac{6!}{2!4!}(3y)^4\,1^2+\frac{6!}{3!3!}(3y)^3\,1^3$$

$$+\frac{6!}{4!2!}(3y)^2\,1^4+\frac{6!}{5!1!}(3y)1^5+\frac{6!}{6!0!}1^6=1\cdot 3^6y^6+6\cdot 3^5y^5 1+15\cdot 3^4y^4 1^2$$

$$+20\cdot 3^3y^3 1^3+15\cdot 3^2y^2 1^4+6\cdot 3y1^5+1\cdot 1^6$$

$$=729y^6+1{,}458y^5+1{,}215y^4+540y^3+135y^2+18y+1$$

(*c*) $(2x-3y)^4$

Think of $(2x-3y)^4$ as $[2x+(-3y)]^4$ and proceed as before.

$$(2x-3y)^4=\binom{4}{0}(2x)^4+\binom{4}{1}(2x)^3(-3y)+\binom{4}{2}(2x)^2(-3y)^2+\binom{4}{3}(2x)(-3y)^3+\binom{4}{4}(-3y)^4$$

$$=1\cdot 2^4x^4+4\cdot 2^3x^3(-3)y+6\cdot 2^2x^2(-3)^2y^2+4\cdot 2x(-3)^3y^3+(-3)^4y^4$$

$$=16x^4-96x^3y+216x^2y^2-216xy^3+81y^4$$

(*d*) $(s+2t^2)^5$

$$(s+2t^2)^5=\binom{5}{0}s^5+\binom{5}{1}s^4(2t^2)+\binom{5}{2}s^3(2t^2)^2+\binom{5}{3}s^2(2t^2)^3+\binom{5}{4}s(2t^2)^4+\binom{5}{5}(2t^2)^5$$

$$=1s^5+5s^4(2t^2)+10s^3 2^2(t^2)^2+10s^2 2^3(t^2)^3+5s2^4(t^2)^4\,1\cdot 2^5(t^2)^5$$

$$=s^5+10s^4t^2+40s^3t^4+80s^2t^6+80st^8+32t^{10}$$

(*e*) $\left(\frac{k}{3}-3\right)^4$

$$\left(\frac{k}{3}-3\right)^4=\left[\frac{k}{3}+(-3)\right]^4=\binom{4}{0}\left(\frac{k}{3}\right)^4+\binom{4}{1}\left(\frac{k}{3}\right)^3(-3)+\binom{4}{2}\left(\frac{k}{3}\right)^2(-3)^2$$

$$+\binom{4}{3}\left(\frac{k}{3}\right)(-3)^3+\binom{4}{4}(-3)^4=1\cdot\frac{k^4}{3^4}+4\cdot\frac{k^3}{3^3}(-3)+6\cdot\frac{k^2}{3^2}(-3)^2$$

$$+4\cdot\frac{k}{3}(-3)^3+1\cdot(-3)^4=\frac{1}{81}k^4-\frac{4}{9}k^3+6k^2-36k+81$$

Work supplementary problem 10.7 to practice expanding binomials.

SUPPLEMENTARY PROBLEMS

10.1 Find the *nth* term, a_n, and the 20*th* term, a_{20}, of the following arithmetic sequences.

(*a*) $2,6,10,\ldots$ (*b*) $11,9,7,\ldots$

(*c*) $\frac{3}{2},\frac{5}{2},\frac{7}{2},\ldots$ (*d*) $\frac{1}{3},\frac{-2}{3},\frac{-5}{3},\ldots$

10.2 Find the *nth* term, a_n, and the 10*th* term, a_{10}, of the following geometric sequences.

(*a*) $3{,}12{,}48,\ldots$ (*b*) $1,\frac{2}{3},\frac{4}{9},\ldots$

(*c*) $2,\frac{-2}{3},\frac{2}{9},\ldots$ (*d*) $-1,\frac{-x}{2},\frac{-x^2}{4},\ldots$

10.3 Find the *nth* partial sum, S_n, of each arithmetic sequence for the given n.

(*a*) $a_n = 5n + 1; \quad n = 25$ (*b*) $a_n = 2n - 4; \quad n = 50$

(*c*) $a_n = n^2; \quad n = 100$ (*d*) $a_n = \dfrac{2-3n}{n+1}; \quad n = 100$

10.4 Evaluate the following.

(*a*) $\sum_{i=1}^{6}(3i-2)$ (*b*) $\sum_{k=2}^{7}k(k+2)$ (*c*) $\sum_{j=1}^{n}S_j$

(*d*) $\sum_{k=2}^{11}(-3)^{k-2}$ (*e*) $\sum_{i=0}^{3}(-1)^i 2^i c$

10.5 Find the sum of the following geometric series.

(*a*) $\sum_{n=1}^{\infty}(0.7)^n$ (*b*) $\sum_{j=1}^{\infty}\left(\frac{3}{7}\right)^j$ (*c*) $\sum_{i=1}^{\infty}(-0.8)^{i-1}$

10.6 Evaluate.

(*a*) $\binom{5}{3}$ (*b*) $\binom{15}{2}$ (*c*) $\binom{10}{5}$

(*d*) $\binom{14}{13}$ (*e*) $\binom{11}{0}$ (*f*) $\binom{100}{100}$

10.7 Use the binomial theorem to expand the following.

(*a*) $(w + 4)^6$ (*b*) $(3x - 2)^5$ (*c*) $(3s + 4t)^4$ (*d*) $(n/2 + 1)^7$

ANSWERS TO SUPPLEMENTARY PROBLEMS

10.1 (*a*) $a_n = 4n - 2;$
$a_{20} = 78$

(*b*) $a_n = 13 - 2n;$
$a_{20} = -27$

(*c*) $a_n = \dfrac{2n+1}{2};$
$a_{20} = \dfrac{41}{2}$

(*d*) $a_n = \dfrac{4-3n}{3};$
$a_{20} = \dfrac{-56}{3}$

10.2 (*a*) $a_n = 3 \cdot 4^{n-1};$
$a_{10} = 3 \cdot 4^9$

(*b*) $a_n = \left(\dfrac{2}{3}\right)^{n-1};$
$a_{10} = \left(\dfrac{2}{3}\right)^9 = \dfrac{2^9}{3^9}$

(*c*) $a_n = 2\left(\dfrac{-1}{3}\right)^{n-1};$
$a_{10} = 2\left(\dfrac{-1}{3}\right)^9 = \dfrac{-2}{3^9}$

(*d*) $a_n = -\left(\dfrac{x}{2}\right)^{n-1};$
$a_{10} = \dfrac{-x^9}{2^9}$

10.3 (*a*) $S_{25} = 1{,}650$ (*b*) $S_{50} = 2{,}350$

(*c*) $S_{100} = 500{,}050$ (*d*) $S_{100} = \dfrac{-17{,}425}{101}$

10.4 (*a*) 51 (*b*) 193 (*c*) $S_1 + S_2 + S_3 + \ldots + S_n$

(*d*) −14,762 (*e*) $-5c$

10.5 (*a*) $\frac{7}{3}$ (*b*) $\frac{3}{4}$ (*c*) $\frac{5}{9}$

10.6 (*a*) 10 (*b*) 105 (*c*) 252

(*d*) 14 (*e*) 1 (*f*) 1

10.7 (*a*) $w^6 + 24w^5 + 240w^4 + 1280w^3 + 3840w^2 + 6144w + 4096$

(*b*) $243x^5 - 810x^4 + 1080x^3 - 720x^2 + 240x - 32$

(*c*) $81s^4 + 432s^3t + 864s^2t^2 + 768st^3 + 256t^4$

(*d*) $\frac{1}{128}n^7 + \frac{7}{64}n^6 + \frac{21}{32}n^5 + \frac{35}{16}n^4 + \frac{35}{8}n^3 + \frac{21}{4}n^2 + \frac{7}{2}n + 1$

Study Tips from the Authors

Your **textbook** provides a wealth of information that can help you succeed in Algebra.

- Become familiar with it. Scan the table of contents. Check out the appendix and the glossary. Make notes in it. Highlight important ideas. Look at the end of chapters for main points summarized by authors.

Did you know?

Most college courses cover more material **in half as much time** as classes you may have taken in high school. Much more is expected in your study time outside of class.

- We recommend 2-4 hours of algebra study each day.

Pre-class Prep

Do a quick reading to **identify the themes of the section** you'll be covering. You'll find the instructor much easier to follow in class.

- Previewing saves you time by helping you to understand algebra concepts the first time around. Identify difficult concepts and question the instructor in class.

After Class

- **Reread the section carefully and slowly,** with paper and pencil in hand.
- **Ask yourself,** How does what the instructor said relate to what I'm reading? What do the examples illustrate?
- **Begin to memorize** definitions of terms, properties, theorems, formulas, and the meaning of symbols. (Statements in the text will become much clearer.)

Consider This:

- If you miss one day of a class that meets three days a week, you've missed one-third or over 33% of the class! If you miss one day of a class that meets four days a week, you've missed one-fourth or 25% of the class. Your success in Algebra begins with the following two essential steps: going to class regularly and participating in class discussion!
- **Don't fall behind!**

Algebra Exercises

- **Read the directions carefully.** Exercises may look alike but instructions may differ.
- **Attempt all assigned exercises and copy the problem correctly.**
- **Think about what you are doing and your rationale for doing it.**
- **Arrange your work systematically,** even on scratch paper. If your work looks haphazard, your thinking probably is too.
- **Do easy steps** mentally and write complex ones down.

Check Your Answers

- After arriving at a solution, **answer all questions.** Is your answer reasonable? Complex problems may require several steps before a question can by answered.

Are You Stuck?

- Take small steps. **Start with the easiest task first.**
- **Analyze the principles involved** and try to apply them to your problem.
- **Reread the text and check your notes** for similar problems.

Still Stumped?

- Take a break and **try again later**.
- **Try a new approach.**
- **Try verbalizing** your approach.
- **Ask classmates** for their ideas.
- **Form a study group.**
- **Ask your instructor for help.** Ask specific questions.

Review, Review, Review!

- Review is **rereading or writing important definitions, properties, theorems, and working exercises from sections and chapters covered in class.**
- Daily review helps cut down on time you need to prepare for exams.

Tuning Up for Tests

- Avoid cramming the night before and **study over several days.** Advance study gives you the opportunity to ask your instructor questions about concepts you may not fully understand.
- **Review your textbook and notes.**
- **Review exams and quizzes.**
- **Drill with a classmate**; make up test questions to ask one another.
- **Do end-of-chapter tests** and time yourself.

Test Time

- **Arrive early** to organize your thoughts, materials, and work space.
- **Listen closely** to verbal instructions and follow written directions carefully.
- **Write down any formulas** you may need.
- **Count the number of questions and divide by the amount of time allotted to determine the average time you can spend on each question.** (Of course some questions will require less time and some more time than the average.)
- **Do the easy problems first** to build confidence.
- **Be neat** and arrange your work in a logical, orderly manner.
- **Avoid writing everything on scratch paper** and transferring work to the exam which can introduce errors and waste time.
- For more difficult problems, think about the concepts in the chapter and how they might relate. **Be systematic and do something different if the method you're using isn't working.** Return to problems you've completed. Do they present a clue? If not, move on to the next problem and return to it later if time permits.
- **Leave yourself time to check your answers.**

Now what?

- **When your test is returned, look it over carefully.** Why were points deducted? Identify your areas of strength and where you need improvement.
- **Clarify problem areas** with your instructor. Try the problem again on a different sheet of paper to test your understanding.
- If possible, **keep your exam for future studying** and preparing for finals.

Work diligently and you will succeed!

INDEX

D

E

F

G

H

I

L

M

N

O

P

Q

R

T

V

Z